The Social Control of Cities?

Studies in Urban and Social Change

Published by Blackwell in association with the *International Journal of Urban and Regional Research*. Series editors: Chris Pickvance, Margit Meyer and John Walton.

Published

Fragmented Societies
Enzo Mingione

Free Markets and Food Riots
John Walton and David Seddon

The Resources of Poverty
Mercedes González de la Rocha

Post-Fordism
Ash Amin (ed.)

The People's Home?
Social Rented Housing in Europe and America
Michael Harloe

Cities After Socialism
Urban and Regional Change and Conflict in Post-Socialist Societies
Gregory Andrusz, Michael Harloe and Ivan Szelenyi (eds)

Urban Poverty and the Underclass: A Reader
Enzo Mingione

Capital Culture
Gender at Work in the City
Linda McDowell

Contemporary Urban Japan
A Sociology of Consumption
John Clammer

Globalizing Cities
A New Spatial Order?
Peter Marcuse and Ronald van Kempen (eds)

The Social Control of Cities?
A Comparative Perspective
Sophie Body-Gendrot

Forthcoming

European Cities in a Global Age
A Comparative Perspective
Alan Harding (ed.)

Urban South Africa
Alan Mabin and Susan Parnell

Urban Social Movements and the State
Margit Mayer

THE SOCIAL CONTROL OF CITIES?

A COMPARATIVE PERSPECTIVE

Sophie Body-Gendrot

First published 2000

2 4 6 8 10 9 7 5 3 1

Blackwell Publishers Ltd
108 Cowley Road
Oxford OX4 1JF
UK

Blackwell Publishers Inc.
350 Main Street
Malden, Massachusetts 02148
USA

British Library Cataloguing in Publication Data

A CIP catalogue record for this book is available from the British Library.

Library of Congress Cataloging-in-Publication Data
Body-Gendrot, Sophie.
The social control of cities? : a comparative perspective / Sophie Body-Gendrot.
p. cm. – (Studies in urban and social change)
Includes bibliographical references and index.
ISBN 0-631-20520-9 (alk. paper) – ISBN 0-631-20521-7 (pbk. : alk. paper)
1. Urban economics. 2. Urban poor. 3. Poverty. 4. Marginality, Social. 5. Social control. 6. Urban policy. I. Title. II. Series.
HT321 .B65 1999
307.76 21 – dc21 99-043568

Typeset in 10½ on 12 pt Baskerville MT
by Best-set Typesetter Ltd, Hong Kong
Printed in Great Britain by MPG Books Ltd, Bodmin, Cornwall

This book is printed on acid-free paper.

Contents

List of Figures

List of Tables

Foreword

The central effort in this book is to examine the possible correlation between the impact of globalization on the city, the growth of inequalities and of power conflicts, and the violence and crime evident in what are often spatially segregated areas. A crucial intermediate variable is whether, and if so how, the new types of welfare and order-maintenance policies of states at the national and local level reflect at least in part the pressures of globalization.

One of the key empirical sites the author uses to explore these issues is cities that have experienced in accentuated forms the impact of economic globalization. These cities have rising numbers of high-income professional households which, through their significant numbers and lifestyles, mark the urban landscape and create strong contrasts with the growing spaces of poverty and segregation. Because they are strategic for the global economy, these cities also develop policies – from development of airports to social order maintenance – that are seen as crucial for the role these cities play as strategic sites for the global economy. An important hypothesis explored by the author is that in this combination of conditions such cities emerge as sites for the production of new norms. Among these norms is the renewed importance of order maintenance as a fundamental responsibility of the state and the novel contents associated with it, given new forms of urban violence.

Against this background, the author examines whether the types of violence and criminality evident today in spatially segregated poor areas in major cities are related to economic globalization and its specific materialization in these cities. And secondly, whether the new welfare and other policies of states at the national level and at the local level that have the effect of raising control over socially marginal populations and areas actually reflect the need for social order maintenance if a city is to fulfill strategic functions for the global economy. Body-Gendrot also examines to what extent these policies may in fact be an enactment of represssion which generates its own forms of violence and criminality as a response to that repression by those who have little if any access to the formal political system.

This is, clearly, a complex argument and one which demands new types of theorization, new types of evidence, and multiple secondary analyses of a range of data sets which have typically been collected with other aims in mind. This book is an enormously important contribution to this new way of studying and interpreting urban violence and state policy. There are today major studies and major findings about this subject, and the author makes a point of discussing these and showing how they contribute to her overall inquiry. This book is, then, centered in the existing scholarship. But she also argues that there is more work to be done. Introducing globalization as part of the explanation adds to this scholarship and provides yet another set of variables to an as yet incomplete explanation of urban violence and the best ways to address it. While using some of the newer types of variable, such as the type of capital present in different social groups and symbolic domination, the author is very focused on developing a set of variables linked to globalization and the intervening role of the state.

The author emphasizes the fact of unequal resources and unequal access in democracies where, in principle, differences of access should not be as strong as they in fact are. She notes how even in large French cities, where poverty and segregation are far less extreme than in US cities, there is nonetheless violence and criminality and the threat of more of it to come. She shows us how global cities are indeed a strategic site for the enactment of new forms of politics by disadvantaged groups with little power, yet with the capacity to exert an "intimidating" presence. Some of the forms of urban violence we are seeing today in large cities are, in the last analysis, one of the ways in which those who lack access to other institutional channels can engage the formal political system and the powerful. Urban violence can, in some situations, be a means to getting a share of the city's resources or respect as subjects, as was illustrated in the Diallo case in NYC in 1999 and in many of the other demonstrations against police brutality in the USA and in several European countries.

Through their actions, these groups introduce alternative normativities in the socio-political and symbolic order of the city. In an earlier book, *Ville et violence*, Body-Gendrot has developed the concept of urban violence as a specific form of violence, one that has elements of a politics and of claims that cannot be easily encompassed by or worked out through the formal political system. In this book, she continues this work through an examination of the specific forms of this new politics by those who are or feel powerless to affect their condition, and she shows us the considerable variance these new politics assume across different cultures.

One of the contributions of this book to our understanding of these issues is the comparison between large French cities, particularly Paris, and large US cities. Such an analysis brings to the fore the specificity of socio-cultural forms through which the general dynamics of urban violence and the state's management of social disorder operate in each of these different contexts. Further, it also shows us that what we associate largely with inner cities in the USA takes place typically in urban peripheral zones in France – often what we would think of as suburban areas in the USA. The author shows us how in France, for instance, it is also the middle class and the farmers who will engage in often violent or violence-prone demonstrations against the state to pressure it into fulfilling its role as regulator and buffer against the negative impacts of unregulated markets, downsizing, and uncertain futures for their children. These are conditions certain sectors of the middle class in France see as connected to economic globalization.

One of the challenges of globalization for social science is that its locational and institutional embeddedness does not correspond to what are standard categories in established social science research. Whether explicitly or implicitly, the boundaries and the scales at which categories are constructed and deployed – the nation-state, the city, the group, the household – do not easily accommodate the cross-border and cross-boundary dynamics of globalization. There is a type of research that confines globalization to the study of macro-level cross-border processes, most especially, cross-border flows of trade and capital; and, one could argue, this type of research can be executed without much concern about these analytic and theoretical issues. But for scholars like myself, who see globalization as a far more complex social phenomenon and one with considerable social thickness, the challenge is centered on how to study its concrete instantiations. That is to say, the challenge is how to study globalization in the specific institutional settings and processes through which it becomes embedded in what we could think of as the non-global. Globalization inhabits specific structurations of the national, the local, the cultural, the subjective. One of the tasks this calls for is the decoding of what has historically been constructed as the national, or the local – in brief, the non-global. The assumption here is that the imbrication of the global and the non-global is complex and variable. The embedding of the global in a specific institutional setting may amplify the features of that setting or it may alter these features, partly or wholly, even as the institutional setting might still be coded/represented as if nothing had changed. In this regard, longstanding dualities such as local–global or national–global are not helpful and indeed do not capture the actual modalities of the locational and institutional embeddedness of globalization.

In this regard, Body-Gendrot's book represents a major contribution to the elaboration of this new type of research agenda. It introduces another sphere for research and theorization: urban violence and the state's order-maintenance function as these materialize in a very specific type of site: large cities that have become strategic places for the global economy. The particular focus on zones of urban violence in major French and US cities further helps us understand the variance in the modes through which certain aspects of globalization get filtered through the specificities of different localities. The subject of urban violence has remained somewhat intractable. There has been a multiplication of explanations which together have contributed enormously to our understanding of the issues involved. Adding the variable of globalization and its impact on conditions of poverty and spatial segregation as well as on the functions of the state introduces an important new dimension to this scholarship.

Saskia Sassen, Professor of Sociology, University of Chicago,
and author of *Globalization and Its Discontents*

Series Preface

In the past three decades there have been dramatic changes in the fortunes of cities and regions, in beliefs about the role of markets and states in society, and in the theories used by social scientists to account for these changes. Many of the cities experiencing crisis in the 1970s have undergone revitalization, while others have continued to decline. In Europe and North America new policies have introduced privatization on a broad scale at the expense of collective consumption, and the viability of the welfare state has been challenged. Eastern Europe has witnessed the collapse of state socialism and the uneven implementation of a globally driven market economy. Meanwhile, the less developed nations have suffered punishing austerity programs that divide a few newly industrializing countries from a great many cases of arrested and negative growth.

Social science theories have struggled to encompass these changes. The earlier social organizational and ecological paradigms were criticized by Marxian and Weberian theories, and these in turn have been disputed as all-embracing narratives. The certainties of the past, such as class theory, are gone and the future of urban and regional studies appears relatively open.

The aim of the series *Studies in Urban and Social Change* is to take forward this agenda of issues and theoretical debates. The series is committed to a number of aims but will not prejudge the development of the field. It encourages theoretical works and research monographs on cities and regions. It explores the spatial dimension of society, including the role of agency and of institutional contexts in shaping urban form. It addresses economic and political change from the household to the state. Cities and regions are understood within an international system, the features of which are revealed in comparative and historical analyses.

The series also serves the interests of university classroom and professional readers. It publishes topical accounts of important policy issues (e.g. global adjustment), reviews of debates (e.g. post-Fordism), and collections that explore various facets of major changes (e.g. cities after socialism or the new urban underclass). The series urges a synthesis of research and theory, teaching and practice. Engaging research monographs (e.g. on women and poverty in Mexico or urban culture in Japan) provide vivid teaching materials just as policy-oriented studies (e.g. of social housing or urban planning) test and redirect theory. The city is analysed from the top down (e.g. through the gendered culture of investment banks) and the bottom up (e.g. in challenging social movements). Taken together, the volumes in the series reflect the latest developments in urban and regional studies.

Subjects which fall within the scope of the series include: explanations for the rise and fall of cities and regions; economic restructuring and its spatial, class, and gender impact; race and identity; convergence and divergence of the "east" and "west" in social and institutional paterns; new divisions of labour and forms of social exclusion; urban and environmental movements; international migration and capital flows; politics of the urban poor in developing countries; cross-national comparisons on housing, planning, and development; debates on post-Fordism, the consumption sector, and the "new" urban poverty.

Studies in Urban and Social Change addresses an international and interdisciplinary audience of researchers, practitioners, students, and urban enthusiasts. Above all, it endeavors to reach the public with compelling accounts of contemporary society.

Editorial Committee
John Walton, Chair
Margit Mayer
Chris Pickvance

Acknowledgements

This book results from investigations carried out in five urban regions: New York, Chicago, and localities around Paris, Lyons, and Marseilles. For each site, whenever sources of data were available, the socio-economic context, the segregative mechanisms, the demographic profile and the presence of "stable" or unstable immigrants and minorities, the political profile of the leadership, the culture and identity of the cities and their neighborhoods, and the interactions of the public, private, and third sector were taken into account to study local governance on the issue of safety.

From January to May 1997, a grant from the French Délégation Interministérielle à la Ville (DIV) and an invitation from the Center of European Studies headed by Martin Schain and from Thomas Bender and Richard Sennett at the head of the Program on Urban Knowledges at the International Center for Advanced Studies of New York University allowed me to conduct interviews with numerous actors involved in social control: policemen, judges, correction officers, mayors, teachers, community organizers, young inmates, ex-convicts, residents of public housing, etc., and to participate in several forums, conferences, beat meetings, and street activities. My field work followed a number of investigations in US depressed neighborhoods, carried out during research trips (concerning the USA) sponsored by various organizations, including the French-American Foundation and the US State Department, or financed by various grants (Fulbright, Hewlett-Packard, Rockefeller Foundation). I would like to acknowledge the very valuable help I received from Ron Schiffmann and his staff at the Pratt Institute and from Lynn Curtis and the Milton Eisenhower Foundation on several of those trips.

The French part of this research was facilitated by an official investigation on urban violence ordered by the French Minister of the Interior, Jean-Pierre Chevènement, which I led with Nicole Le Guennec and Michel Herrou, looking into French and European problematic neighborhoods during the spring of 1998. In addition, my research was aided by numerous exchanges of view which took place over the years at conferences, meetings, research trips, and seminars, including the many insights that came from my CNRS team Urmis (Unité de recherche, migrations et société) at the University of Paris-Sorbonne (Paris VII), the think-tank Profession Banlieue in Seine St-Denis (a suburb close to Paris), the Institut des Hautes Etudes sur la sécurité intérieure, the Ecole Nationale de la Magistrature, the French Union of Public Housing, the French American Foundation, GERN (Groupement européen de recherche sur les normativités) led by Philippe Robert, and from work with community organizations, especially Droit de Cité and Unis-Cité.

I also wish to thank Nicolas Frize who, in the course of my investigations into citizenship, took me on some very unusual visits into prisons and hospitals; Patrick Le Galès, Thierry Paquot, and Patrick Weil for so many stimulating conversations; Edmond Préteceille, Catherine Rhein, and, among others, my graduate students, past and present at Sorbonne-Paris IV (Department of English-American Studies) and the Institut d'Etudes Politiques (Programme d'études avancés – les métiers de la ville). I cannot possibly give all the names, but each in their own way contributed to deepen analysis for which I alone am responsible. Besides these colleagues and friends, others should be thanked for their help: Wayne Barrett, Robert Beauregard, James Bell, Jean-Paul Brodeur, Lynn Curtis, John Devine, Bernardine Dohrn, Joy Dryfoos, Jeff Fagan, Susan and Norman Fainstein, Eric Flamm, Esther Fuchs, Marilyn Gittell, Martin Goldsmith, Diana Gordon, Constance Jewett-Ellis, Nancy Johnstone, John Logan, Peter Lucas, John Mollenkopf, Susan Saegert, Robert Sampson, Saskia Sassen, Stuart Scheingold, Lisbeth Shepherd, Jonathan Simon, Kate Stimpson, Todd Swanstrum, Gretchen Suzy, and Aristide Zolberg.

Special thanks go to my French publisher, Bayard Editions, which gave me permission to reproduce excerpts from a previous book, *Les Villes face à l'insécurité.*

I am especially grateful to Kevin Delaney and Sarah Dancy for their editorial assistance and to Isabelle Bartkowiak for checking the bibliography.

Every effort has been made to trace copyright holders, but if any have been inadvertently overlooked, the publishers will be pleased to make the necessary arrangements at the first opportunity.

Introduction
Tolerance and Suppression

How do societies form their perceptions of urban danger and become alarmed by it? Today it is easy to point to the city-based media which construct the theme and fill their coverage with stories of minor riots, non-violent crimes, and even rumors, to be disseminated throughout the globe. But, even in the past, when the knowledge of a crime would go no further than a horseback ride, people in cities were constantly involved in a dialogue with fear. "Fear everywhere, fear always," L. Fèbvre, a French historian, once remarked.

In recent years, urban violence has been debated in the context of anxieties raised by the slowing of growth in Europe, the marketization of economies all over the Western world, and the prioritization of international economic competition. The discussion in the public debate of violence is euphemistic, displacing more urgent questionings on new forms of inequality and social marginalization that are appearing in cities, as well as a general feeling of precariousness among the middle and working classes. Urban danger is therefore constructed as the discursive displacement of issues already related to family disruptions and institutions' dysfunctions.

My major focus is on the link between the eroding buffer role of the welfare state, the function of the local arena – namely large cities – as the site impacted by the negative consequences of economic restructuring and of rapid social mutations, and that of the "civil society" as cities address the public's insecurity. At stake is a redrawing of what constitutes the legitimate responsibilities of individuals, collectivities, and the state. Partnerships are the sites in which the rearticulation of new socio-political relationships is played out and contested and out of which innovation and change in terms of crime control may emerge (Crawford, 1997). Confronted with new challenges, national elites seem indeed to be confused, particularly in the Old World where the established insiders are advocating a better control of crime and violence frequently associated with "others" – that is, unemployed, unskilled, or redundant workers with different racial or ethnic identities and, more specifically, youth from the inner cities. Excessive flows of information reinforce the idea that societies are becoming more complex, more heterogeneous, and too difficult to manage from traditional centers of power alone.

I look in this book at a period of transition, at a time of crisis, in its etymological meaning of "search for a solution." I wonder, with Jürgen Habermas, if, at least in Europe, we are not observing the exhaustion of a model for state-provided protections against hardship and for the continuity of the social link. I assume that, at least in France, the more the central state devolves its social commitments to the local arena, the more cities become the major spheres in which to understand the treatment of polarization and marginalization associated with the world flux of capital and people. Politics and culture shape civil society and its efficacy. Policies of control are nationally generated, but the decentralization and the "deconcentration" of the French state, added to European Community bulldozing of national idiosyncrasies, give local authorities and communities space to maneuver and innovate, a space sometimes reluctantly seized at a time of fiscal austerity.

The bases of solidarity that have cemented industrial societies' roles as democratic welfare states and rights providers since World War II have indeed been eroded under the twin pressures of global economic and social transformations. In an era of globalization, the processes of disintegration, disempowerment, social invalidation, marginalization – whatever term one wishes to use – fracture post-industrial cities. That rupture may not occur in a dualistic logic, as it is often suggested, but certainly in a myriad of patterns, some of them leading to collective violence and disorder, as shown in Los Angeles in 1992. National societies seem to disarticulate in a strange movement of de-modernization, notes Alain Touraine (1991).

The social control of violence and crime and the defense of some "order" are important stakes for national societies, and for cities in particular, where so many strategic resources, wealth, power, and people are now concentrated and where peaceful collective life is to be maintained in democratic ways.[1] The question-mark in this book's title comes from the fluctuating and diverse interpretations relative to social control. For example, are cities exerting a form of social control when they pursue symbolic politics? Is social control better embodied by a preventative or by a repressive approach, or by a mixture of both? Is social control exercised at city level, at other levels, or by partnerships, however conflictual they might be? As the title of this chapter, Tolerance and Suppression, implies, the social control that local societies exert on people and places varies according to national stakes, ideologies, resources, and actors. Although the goals and the trade-offs are basically the same, some societies put more emphasis on a search for equality, social prevention, welfare, and state intervention for the treatment of marginality and a check of uprisings. Others rely more on individualism and self-help, as well as on deterrence and punishment. In all cases, safety is considered as a fundamental collective goal. Why and how are these choices made? Why and how is the change marked from a more or less recent past?

From the Global to the National Scale

When French society, in accordance with its historical traditions, forcefully and symbolically rejects violence and chooses to fight its root causes through policies of primary prevention, there is an institutional consensus arising from both conservative and liberal elites to support state-sponsored social policies. With the primary prevention approach,[2] welfare, education, city policies and various forms of place-oriented treatment attempt to repair the damage done by economic globalization and reincorporate the marginalized into the mainstream. Pressures on middle and upper classes via both increasing taxes and discursive appeals to citizenship are launched by successive governmental teams and do find some response. (A taxpayer in France is described as "*imposable*" and a policy such as Proposition 13 – which refers to a revolt by taxpayers in California in 1978, leading to approval by referendum and a climbdown by the state, and which was followed by similar action in a dozen other states – has no equivalent, except in the tumultuous, ephemeral social movements in the recent history of the country which the far right tends to revitalize.) The French philosophy that supports such policies is indeed that since society is responsible for its citizens, the state has to emancipate the individual whose nature is fundamentally good and educatable (think, for example, of Rousseau), the troublemaker being not so much to blame as society is. If we wonder whether teachers, prison wardens, and parents are all actually doing the same thing, the answer depends on the kind of responses we are looking for. In this book, I look at organized formal and informal responses to threats of urban violence, crime, and delinquency as they are constructed in the current debate. Some are reactive, others proactive. Some may be directly orchestrated by institutions,[3] others by community mediation; they may also come directly from individual initiatives. Why is the choice of inclusion in the French public discourse preferred to the easier answer of stigmatization and suppression? Is it in the Catholic tradition? Is there also more reliance in France on the efficiency of institutions and of policies, a thesis supported by Elias's (1933) "civilizing process" than on self-help?

The norm of solidarity has been patiently formed throughout centuries of hesitation and confrontation. France has socialized its citizens into this principle dating back to the Revolution, and whether conservative or left-tending regimes are in power does not make much of a difference. At the very start an insurance regime supported by the state gradually expanded to incorporate one vulnerable category after the other. The ordinance of 1945, according to which a youth at risk (as author or victim of a crime) needs education and protection from the state, still prevails. More recently, solidarity has been based on public partnerships; that is, on formal agreements between various sectors of the state and either distressed areas or individuals, under specific conditions.

The primary social prevention policy elaborated at the local level by Bonnemaison in the 1980s was place-oriented: local committees for the prevention of crime and delinquency acted on school programs, financial support to families, cultural development within a national policy of help to cities (*politique de la ville* – see Chapter 3). Yet, in the 1990s, inserted in the processes of globalization and Europeanization, the French central state acting as a regulator met with complex rival priorities, and devoluted the concerns of solidarity to the local sphere, yet without abandoning its regalian control. Local committees on prevention of crime and delinquency have frequently become empty shells. In 1999, it was admitted that only one-third of them actually function efficiently. The general unemployment rate is 11.4 percent in France (at the end of the first semester of 1999), and 22 percent amongst under 25-year-olds. Three-quarters of 18–20-year-olds and 42 percent of 21–23-year-olds have no other choice but to live with their parents (*Le Monde*, 7 July 1999). This situation is very different from the one in America. As a consequence, a dismantling of the welfare state, as in the USA and the UK, would not be tolerated, but what is occurring instead is a devolution from the central state to subnational authorities on social issues, as if a decoupling of economic and social political issues places them on different trajectories. It is not surprising, then, that domestic politics become more and more local and associated with "bread-and-butter" issues.

In that process, despite an official rhetoric based on citizenship and inclusion, solidarity today receives less support from the overtaxed middle classes than during the thirty years following World War II: the poor and "the useless normal" – that is, those who are unfit for or redundant in the labor markets of the post-industrial society, including unskilled immigrants' children – are suspected of taking advantage of dependency. With lower rates of crime in 1997 (1.2 murders per 100,000 people) than in the USA (9 per 100,000), the punitive trend in France is not as dominant as the "inclusive" one, but it is nevertheless growing in public opinion, due to social disturbances in at least one thousand urban areas, amplified by the media, exploited by the national populism of the National Front (which was winning around 15 percent of the national vote in general elections in the 1990s, although 67 percent "empathize" with its opinions).[4] The construction of demands for more surveillance and repression also comes from mayors and professionals in charge of law and order – namely, some bodies of rank-and-file policemen, judges, and prosecutors – exerting corporatist pressures on national decision-makers for more repression. For instance, in twenty years convictions leading to more than five years of detention have grown by 1,020 percent, to more than ten years by 233 percent, and to life by 100 percent (Vital-Durand, 1999: 17). We may therefore suspect that the generous discourse of solicitude and solidarity at the top is modified when confronted with the principle of reality on the ground.

Repression, like solidarity, is the continuation within the multilevel state apparatus of relations and struggles that had their origins and ends elsewhere. That suppression or punishment is a tool of social policy is implicit (Simon, 1993: 12). It is a method of surveillance and control of communities where those defined as offenders live in great numbers. In a federal country, such as the USA, marked at its foundation by a puritanical philosophy, a widely shared idea is that only the work of a community upon itself and by itself can rescue the individual through efforts, deterrence, and constraint. The state's action must remain marginal and residual. It is not supposed to intervene in natural social processes. This implies that, according to an ideology which emphasizes the communities' efforts and their capacity to make moral choices in the foundation of the society they want, the moral insiders, linked by an implicit covenant, do not feel responsible for the fate of delinquent outsiders, except through charity and other private outreach (Elias, 1965). A behavioralist philosophy also supports the idea that setting aside the bodies of dangerous classes (in prisons) will solve the community's problems and that, miraculously, after being released, inmates will be able to rejoin the mainstream. This approach struck Alexis de Tocqueville when visiting American prisons with Beaumont in 1830:

> There are, in America as in Europe, men worthy of esteem whose mind breeds on philosophical dreams and whose extreme sensitivity needs illusions. These men, for whom philanthropy has become a need, find in the penitentiary system, a nutrient for this generous passion: starting from their abstractions which depart more or less from reality, they consider man, however deep he is into crime, as capable of ever being brought back into virtue . . . And pursuing the consequences of this opinion, they foresee a

> time when all criminals being radically reformed, prisons will be entirely emptied and justice will have no more crimes to punish. (quoted in Garland, 1990: 139)

He also added that "while society in the U.S. gives the example of the most extended liberty, the prisons of the same country offer the spectacle of the most complete despotism" (quoted in Garland, 1990: 11). At the time, in the USA as in Europe, rehabilitation through work in prison was thought to take care of marginalization.

Under the conservative regimes and fiscal austerity of the late twentieth century, the "lock-them-up-and-throw-away-the-key" attitude gained popular support and the hope of rehabilitating hard-core delinquents generally vanished, as the representation of the marginalized (most of all, the drug dealer, stereotyped as the child of an urban mother on state benefits) became increasingly demonized. There was a time in the 1960s when therapeutic rehabilitation was in vogue. But a general sense of public frustration ("nothing works") due to rising rates of crime led to a cost-benefit managerialism which reduced welfare and led to higher levels of incarceration (which is a lucrative business). A subtle dialectic of laissez-faire (less state) – deregulation, detaxation, and accumulation – and retaliation (more state) is at work. The counter-discourse of resistance is weak, yet I will show that it exists, especially at the lower levels of the state where policies are implemented. A diversity of voices percolating up from various community-based interests, cities, and states can be heard.

Crime is what society decides it to be; the dangerous classes are defined arbitrarily by institutions in the same recurrent way as in the past. This classification represents stakes within the state apparatus and it may change according to business cycles, as will be shown in Chapter 1. Households with cumulative handicaps are constructed in terms of deprivation and are collectively spatially isolated: the urban poor, minority youth at risk, single-parent families, illegal immigrants, and street people (Sampson and Laub, 1993; Fagan, 1996). No one is eager to live near them, few people identify with them, rarefied groups speak for them; they have little representation and voice, except for their sporadic violence in specific circumstances and the threat they diffuse. Race, ethnicity, age, and class fragmentation increase their isolation.

In the USA, the outcast position of impoverished urban African-Americans and their connection to crime, as victims as well as offenders, creates a singular narrative with no equivalent in other Western countries, not even in those with histories of colonialism. Whether one follows William Julius Wilson's demonstration of the impact on crime deindustrialization has had on African-American impoverished households in specific neighborhoods or Douglas Massey's theory of spatial apartheid, the result is the same: the racialization of criminalization is an American issue with hardly any equivalent (Massey and Denton, 1992; Wilson, 1996).

No doubt, immigrants and their children in France are discursively connected to crime as victims and as offenders. They are also overrepresented in prisons. They mainly serve sentences for less serious, non-violent incidents, often related to their illegal status, or for drugs. But they rarely do so for violent crime, which is scarce. Only 8 percent of French and 4 percent of foreign immigrant entrants to penal institutions are imprisoned for major crimes (Tournier and Robert, 1991: 138–9). There are many reasons that might explain the lower rates of crime in France: weapons are not as free-circulating there as in the USA and violence is not an essential element of French culture, if we do not count revolutions, wars, and political outbursts of social movements at grassroot level. Foreign immigrants, who form 6.4 percent of the national population, represent about 30 percent of the prison population, 30 percent of unemployed in the 15–34 age group (sometimes more, according to nationalities), a figure to be compared with the 50 percent of African-American prison inmates, from a community that forms 12 percent of the national composition and 24 percent of the poor (households with incomes of less than $15,000 a year) – the rate of unemployment is dubious as it does not include the "discouraged" workers who dropped out of the labor market and are estimated at more or less 6 million.

Conceptualizing Urban Violence

In the 1960s, the USA was alarmed by levels of violence which tore cities apart. President L. B. Johnson appointed both a commission on urban disorders (the Kerner Commission) and an 18-month commission to address the causes and prevention of violence under the chairmanship of Milton Eisenhower, the President of Johns Hopkins University. Most of the commissions' recommendations are today ignored. Yet they are worth remembering. In the 1990s a shift occurred in the USA towards concern about violent crime in cities rather than violence per se.[5] But are the two concepts equivalent? What link is there between the Oklahoma bombing in 1995, a mugging on Main Street, the Los Angeles riots of 1992, and ethnic confrontations in Crown Heights? The link is provided by journalists and politicians, who do not distinguish in the words they use – urban violence, urban crime, riots, disorders, disturbances, unrest, rebellion, confrontation, lawlessness, and many more – to characterize such events. As noted by Wittgenstein, language has the power to make all things look alike.

Commentators do not seem to be aware that by simply using blanket words such as "violence," they blend in the same category phenomena which are in fact distinct. Using one term rather than another already gives a subjective and sometimes abusive interpretation of events (Body-Gendrot, 1995). Work on terms such as "crime" or "violence" encompassing different philosophical, sociological, and ethical meanings cannot be avoided.

From birth, everyone grows up knowing what violence is. There is a vital force aspect of violence derived from the Latin term *vis*. Violence becomes a problem when it is not channelled or socialized. It is in fact less a concept than a set of situations all connected one to another, even if their intensity and forms cannot be compared. Most of them refer to behaviors the goal or effect of which is to hurt another's body (affects, mind, goods) or territory physically or symbolically (Héritier, 1999). Violence denies everyone's aspiration to protection, dignity, justice, and inviolability, and this explains why it is such a threat. Moreover, if violence is correlated with urban, there is a collective and philosophical element implied in the sense of failure at social cohesiveness. Yet in France the debate reaches the whole society, a society that is smaller and more integrated than that of the USA. It is this societal debate which, I argue, disappears in the USA after the 1970s with the use of the term "crime." Crime externalizes the issue, it connects it to dangerous classes on specific places which are not part of "us." For the public, however, both "violence" and "crime" refer to disenfranchised youth, perceived as belonging to ethnic or racial minorities, in disempowered urban communities.

In Search of Missing Linkages

The local configuration is an element of differentiation, all the more so in the USA, a federal country where police and welfare functions are within the jurisdiction of fifty states, the counties, and the cities. Punishment and welfare belong by and large to the jurisdiction of local and state governments, a difficulty for international comparison. The importance of the stakes may explain, however, why a federalization of crime control has taken place in the USA. Conversely, in France, welfare policies against social exclusion, as well as national urban policies intended to control violence, reveal reciprocal and feedback influences between the national and the local arenas, as has also been observed in the United Kingdom.

There are theoretical pitfalls, however, when dealing with the local, involving the meaning of spatial scale: if we argue that the city mediates (that is, resists, adapts, acquiesces to) global forces, is it some quality of scales that enables the local to "filter" forces emanating from other spatial scales or "a reference to the frictional interaction of actors with different geographic reaches" (Beauregard, 1995: 238)?

From the National to the Local Scale

Cities are fascinating spheres in which to test the impact of national rhetorics on social control, order, and the treatment of marginalization. If national politics matter and create expectations among anxious voters concerned by economic globalization, what matters most, it seems, are the local arrangements – the governance and the social engineering taking place at the local level (Scheingold, 1991). Local social control may explain why urban societies remain generally peaceful, compared with other arrangements in emerging democratic countries such as Brazil or South Africa.

It is my assumption that it is in the interest of no mayor, police commissioner, judge, community or church leader, or the business community to let antagonistic relations grow between various social components of the city, which make up, one way or another, their clientele. The abstract categories created by the national rhetorics have faces and names in the neighborhoods of cities. Do we not observe mayors attempting to slow down the exodus of the middle and working classes by making the neighborhoods safe, by upgrading education, and playing down criminalization? Even though their stakes may be different, many more actors seem to share the same goals.

I am convinced that the size of cities is a major resource as well as a source of mega-problems, that business cycles and demographic circumstances (cities as mirrors of transnationalization) matter to the social well-being of cities, and that the market accelerates the polarization and segregation of populations and recomposes spaces according to its own stakes, thus creating social chaos and displacement. But at the same time, I also argue that counter-powers oppose or slow down the impact of such market laws and that the historical and cultural specificity of a place matters, in particular the tradition of community activism. In the cities that I observed, the "return of the civil society" refers to new forms of citizenship and participation at the community level. They seem to make a difference in the co-production of security. This concept of co-production is ambiguous, nevertheless, as revealed by individual household strategies and collective actions concerning the threat of social violence. A discussion on the local governance of security will not be avoided. Space is indeed the element that resists, which is the problem as the solution. It is not just the stage of a theater on which a crisis unrolls, but it is a source of problem in itself, making all development processes that exclude it fail (Beauregard, 1995).

The "Community" in Question

Field work reveals the complexity of local governance, its subtle dosage of prevention and repression, the importance of civic cultures and repertoires for problem resolution, and it may offer a theoretical understanding of local social control and of the nature of cities.

If the general erosion of the central welfare state – replaced by a custodial state – is more pronounced in the USA than in France, does it mean that the local civil society can help? The institutional local activity of social control is indeed just one element amid patterns of non-state actions. If we refer to the capacity of a social unit to regulate itself according to collective goals, then we approximate Tilly's definition of collective action (Tilly, 1973) – the application of a community's pooled resources to common ends, in our case, the collective desire to live in a safe environment and the will to take action.

Discourses on civil society and community are ambiguous. The idea of civil society which developed in seventeenth- and eighteenth-century Europe was as a sphere, midway between the state and domestic life, related to the concept of "social contract." According to C. Lloyd, with Marx formulations, the idea of civil society had more conflicting connotations, with a more difficult relationship with state structures (Fine, 1997; Marx, 1968). Gramsci took these ideas further to a terrain of complex networks of social practices and struggling relations, with intellectuals weighing in. An heterogeneous civil society appears, then, as a terrain of struggle and conflict (Lloyd, 1999). According to two opposite trends of thought, its goal, via the strengthening and transformation of micropower relations, can either be to oppose a state seen as evil and impose social transformation or to address the state's dysfunctions and force it to "enforce equality between citizens" (Frost, 1998).

By community, I mean here neighborhoods or local community areas, focussing on disadvantaged low-income urban neighborhoods. I refer to Park's definition of a neighborhood as an ecological sub-unit of a larger community conditioned by cultural and political forces, "a locality with sentiments, traditions and a history of its own" (Park, 1916: 95). And I stress, with Sampson, that it is also a "mosaic of overlapping boundaries," an imbricated structure reflecting the society at large (Sampson, 1997). It is a social construct that plays a role in its own formulation, articulation, and disarticulation. This does not include only processes of incorporation and exclusion, but also processes that adjust relationships among actors even as these actors are being integrated into or marginalized from dominant systems of relationships (Beauregard, 1995: 244).

Case Studies

Case studies allow us to draw comparisons between different contexts within the same country at the same period. For this book, I have chosen to study cities with a "world status" in the USA, that is, key urban nodes concentrating complex transactions and decisions emanating from headquarters of transnational corporations, financial organizations, advanced corporate service producers, and with information-processing centers supporting them. Not only do New York and Chicago comprise the largest metropolitan centers in the USA, but they also have large bodies of political decision-makers, social service institutions, universities, cultural institutions, the mass media, advertisement industries, etc. The selection of these two global cities is not totally unproblematic. Is Chicago a world city?[6] The justification for Chicago over Los Angeles comes from a number of economic indicators related to world city-ness, which puts Chicago second after New York (Abu-Lughod, 1995).

How is social control affected by the characteristics of globalization and by business cycles in Chicago and New York? With notable exceptions (Friedmann and Wolff, 1982; King, 1990; Sassen, 1991; Fainstein et al., 1992), most works on world cities have not been comparative, even less cross-national, and have not correlated the growth of socio-economic, spatial inequalities, social disruptions, and national and local responses. This is what I attempt to do here. In France, I have chosen to study the three major metropolitan areas of Paris, Lyons, and Marseilles.

Three Scenarios

Three potential and overlapping evolutions of the control of crime/violence in cities will be brought under scrutiny in this book to emphasize the importance of the stakes related to these issues. The first one privileges the impact of globalization and leads to a redefinition of crime, not as external to a law-abiding society in the Durkheimian sense, but entirely incorporated in all of its parts, as in the mathematical figure of the fractal. Repression intervenes only to protect the free enterprise of insiders, but outsiders remain free to live in no-go areas as long as they do not threaten the functioning of the global system. Symbolically, marginalized communities may be required to produce peaceful bonding so that productive categories can pursue unencumbered activities.

The second hypothesis is based on extended surveillance in discarded spaces, and strict segregation is enforced. The difference with the previous scenario is that the whole set, unlike networks, is holistic. The insiders and the outsiders share the same references to the law and coexist in a common world. Crime is a necessary construct because, as Durkheim (1933) pointed out, it raises a "*conscience collective*" and a benchmark for the recognition of norms.

The last hypothesis operates on a belief in democratic processes and on the assumption that people tolerate differences and accept people who are unlike themselves ("othered" people). Although evaluative research is lacking, according to that perspective, it is assumed that community mobilizations and internal social control mechanisms based on commonalities alleviate institutional retreat and force institutions to work more closely together and with citizens. Within local institutions, arrangements, mediations, and compromises allow an adaptation to complex and conflicting situations. This scenario is not entirely utopian and proofs of its validity are extracted from research.

In the USA, major foundations, such as the Ford Foundation (Chaskin, 1995) and the MacArthur Foundation (Simmons, 1996), sponsor research on neighborhood and family initiatives in low-income areas and on communities as effective settings for human development. Evaluation of Community Development Corporations (CDCs) and of empowerment zones brings the same perspective in some cities (Briggs, de Souza, and Mueller, 1996; Gittell, 1998a), as does a series of studies of youth in disadvantaged neighborhoods (Body-Gendrot, 1994; Curtis, 1995, 1997). These studies aim at increasing our understanding of how community-level and individual-level factors interact in the development of pro-social and antisocial behavior (Fagan and Wilkinson, 1998) and what amount of social efficacy communities produce to control antisocial behavior (Sampson, 1997; Sampson et al., 1997). Because French elites give less leverage to self-help and grassroots-initiated actions, and as public policies are rarely evaluated, there is a dearth of data on the efficiency of community action versus institutional intervention in France.

The Difficulties of Cross-national Comparisons

Is a cross-national comparison possible on such issues? Tocqueville, it is true, has given legitimacy to comparative work. After his visit to American prisons in 1830, he carried out a pioneering analysis of local democracy. But there is only one Tocqueville. Nevertheless, the tradition of cross-national comparison is strong (see Merryman, 1969; Tilly, 1973; Pickvance and Préteceille, 1991; Le Galès and Mawson, 1994; Hoffmann, 1997;

Sutton, 1997, and many others), allowing us to test a broader range of hypotheses and yielding more forceful generalizations than narrowly focussed studies on one country or one city. However, without even mentioning the discrepancies that occur in data measurement, I am very much aware of the treacherous nature of the comparative exercise. There are a number of problems.

1. Comparing a mid-sized country with a vast territorial, multiracial and multi-ethnic world superpower and attempting to trace a parallel between a long past molded by centuries of centralization and a short history of the forces of law and order of fifty states is bound to lead to oversimplification and misinterpretation.

2. The minefield becomes more dangerous if one compares the nature and rates of urban disorders. It would make no sense to draw cross-national comparisons without bearing in mind the levels of crime experienced in the USA, which are six or seven times higher than in Europe, with variations according to categories of crime and the moral panics which follow from these rates, from the nature of offenses, and the profile of delinquents (Tonry, 1995; Zimring and Hawkins, 1997). Even leaving out the manipulative effects of crime data production (Burnham, 1996: ch. 4), it is the way in which the violence problem is officially constructed and the political positions to which these constructions give rise which are of interest here (Garland, 1990: 20; Scheingold, 1991).

The "land of the brave, home of the free" has one of the highest murder rates in the industrialized world, more inmates in prison than any other Western society, still executes hundreds of offenders each year, and struggles with outbursts of racial violence; its citizens are increasingly polarized, as Marshall aptly remarks:

> the "haves" . . . hide themselves in distant well-kept suburbs and the "have nots" live on the streets, or in poor housing in rural areas or inner cities. There is no doubt: The contrasts in the U.S. are much more extreme than those found in most Western European countries. In many ways, the cultural gap that separates the U.S. and Europe is more formidable than the mere 6,000 miles or so of Atlantic Ocean between the two continents. (1997: x)

That violence should be considered an integral part of the culture, especially in the South, is a main divergence in any comparison with France.

3. Who are the formal agents in charge of social control? In the USA police chiefs are frequently chosen at a local level or appointed via the party caucus. Prisons and correctional facilities are managed by cities, counties, and states, while the police, judges, and sometimes federal penitentiaries

may coexist on the same territories, revealing the dual sovereignty of the system. Conversely, the police, justice, and correctional systems are national in France and appointments are imposed on city officials by a "regalian" state.

Crucial, it seems, is the difference that can be observed in reference to the nature of the law. There are two highly influential legal traditions in the contemporary world: civil law and common law. American laws are based on common law, that is, on tradition and on constitutional rights which are the basis of American identity; civil law in France was established by Church clerks, as early as the thirteenth century, before being submitted to the Code Napoléon which enforced powerful administrative rules – rigid rules that are hard to enforce, as Tocqueville observed, while, conversely, in the USA, flexible rules are enforced rigorously (Tocqueville, 1967: 6, 140). Procedures and/or images produced by any criminal justice system reveal what is "socially implicit" or embedded in the subconscious of a society.

Take criminal law judges. French judges are the heirs to an aristocratic/religious system. Their rites are fundamental: behaving like a cleric whose major function would be to establish a link between the people and the sacred sphere (the law), they are invested into a symbolic action at the demand of the political body in order to repair a social compact that has been disrupted by violence or crime. If law, as conceived in France, is part of a transcendental sphere (Garapon, 1996), nothing of the kind is observed in the USA (except for the Supreme Court), where judges, the police, and politicians are accountable to the people who appointed or elected them. The criminal judge's role is to bring a rationale to conflicts of interests as presented through facts.

In the US accusatorial system, the right of accusation extends to all members of the community. The judge makes sure that the various interests are respected; he does not have inherent investigative powers. The contest takes place between the accuser and the accused. The rules of common law are like the rules of a game which are constantly readjusted and are therefore less metaphysical than the law *à la française*. As a referee, the American judge's function is to uncover a logic from the conflicting interests, while designing a regulation inside "a community of communities" (Normandeau, 1995).

In the French inquisitorial system, an additional step along the path from the system of private vengeance is taken, as Merryman remarks (1969: 127). The private accuser has been replaced by public officials, the judge has investigative powers. Now the contest is between the individual and the state, which conducts the criminal action. If, as in 80 percent of cases, the French examining judge concludes that there is insufficient evidence, the case does not go to trial and victims are frustrated. But legitimate expectations of victims are also frustrated by the American practice of plea-bargaining. Plea-bargaining consists of giving the prosecutor discretion to charge a lesser offense or call for a lighter sentence if the defendant agrees to plead guilty. Plea-bargaining takes care of 90 percent of the criminal law cases.

Which system is the more just?[7] While the judicialization of collective life has always prevailed in the USA, what is new is the taking over of the criminal law system by conservative politicians. Common law judges used to be cult heros and father figures in the USA. But, as in criminal law procedures, they have become operators of a machine designed and built by legislators. Sentencing in the USA is more and more determined and imposed by legislators. Conversely, in France, individualized sentencing left to the judge's discretionary power about mitigating circumstance is the rule, all the more for juvenile court judges. Why has there been such an evolution in France? Is it a question of trust in judges versus legislators?

4. Americans do not seek solidarity from the state. They do not look to the state in search of their identity. Let us remember that, for Jefferson, the best state was the state that governed the less and, for Thoreau, a state that would not govern at all. The right to insurrection was part of the citizens' rights. Americans' membership is found elsewhere, probably in their adhesion to a bill of constitutional rights and their identity in a smaller local community (Sandel, 1996). Citizenship is not a recurring national debate, as it is in France where distinctions can be established between those who call themselves "residents" and others, for instance youth of immigrant origin, who identify themselves as "citizens" (Neveu, 1999). A national intervention in social processes is always suspect in the USA, as it implies a sort of collectivism running contrary to the ideology of laissez-faire. The American dilemma is to balance the preservation of equal rights and the recognition of racial, ethnic, gender, and age differences to be protected by law. At the close of the twentieth century, the politics of identity mark the end of a national imagination.

By contrast, the intervention of a strong and centralized state and of extensive public services in France have been at the foundation of principles of solidarity and collective identity. There is still a shared belief in the state as a motor of progress and the idea that the state could subcontract for certain services to the non-profit sector is not widely accepted. Rather, three out of four French regret that non-profits already play too often a role that the state should play. They also think that they pay enough taxes to have efficient public services and not to have to resort to self-help. Few countries in the Western world have been as much and as long molded by the state apparatus. The message from the centre, diffused according to universalistic principles on a single territory, has been the major agent of the melting of differences, as the state has regulated society according to uniform categorizations and regulations.[8] A community of citizens is thus bonded together by political and judicial principles which transcend them, such as the principle of equal treatment granted to citizens across the whole of French national territory and that of secularization, both of which have become entrenched (Schnapper, 1996). What the French *think* ideologically (on the right/on the left) is therefore more important in the definition of their identity than what they *are*, and consciousness of their rights is not a founding mechanism.

5. That the French state does not actively categorize its citizens according to race or ethnicity also marks a major difference with the USA (Body-Gendrot, 1995). Those referred to in France as *les exclus*, the excluded, form a mix of nationalities, races, ethnicities, ages, and genders, whose common fate tends to be unemployment and loss of benefits (Silver, 1993). Because of housing market differentials, the *exclus* tend to be concentrated in specific neighborhoods and/or at the periphery of cities (Tabard, 1993), whereas, in the USA, the disadvantaged are frequently concentrated and isolated in monoracial inner cities. The French term suburb (*banlieue*) corresponds in effect with the American inner city or ghetto, although it is, by nature and scope, different.

In France and in the USA, research shows that links exist between racial minorities, poverty, single-parent families, school drop-outs, disinvested urban zones, and potential collective disorders.[9] But two consequences of diverging approaches make cross-national comparisons complex. First, the absence of reliable specific statistics on race and ethnicity in France prevents researchers from drawing conclusions about poor immigrant youth's rates of arrest and mobilizations. Once they get French citizenship, they are no longer counted in the statistics. Parallels cannot therefore be accurately drawn with minorities' arrests or rates of incarceration in the USA. France does not recognize the existence of minority groups in a "one and indivisible" nation and they have not been statistically tabulated, except in one single longitudinal study which did not look at crime and violence (Tribalat, 1995). The binary distinction of nationals and foreigners, as limited as it is, prevails. Statistics in France concerning *immigrés* and their children are therefore necessarily biased.[10] The marginality faced by some of them, especially unskilled and illiterate populations from the Maghreb (North African countries), sub-Sahara, and Turkey, is a very serious problem in French post-industrial society: "an urban management nightmare with several facets: a high level of unemployment among the young, hostility engendered by the impression that many are in France illegally, the failure of these groups in school and their juvenile delinquency" (Haut Conseil à l'Intégration, 1991: 17–21; Irving Jackson, 1997: 133). Following the negative perceptions shared by a majority of the French concerning "Arabs," the comparison with US poor racial urban minorities (black on black crime) seems in general more appropriate than with US immigrants.

The second consequence has to do with space dispersal. Where it is possible to analyze how the mayors of New York and Chicago and their teams address crime problems, establishing a comparison with urban regions in Paris and Lyons, we would have to – but we will not – enumerate the problems and solutions found by the mayors of the dozens of fragmented localities of such metropolitan regions (except for Marseilles, where the problems are in the city), often from diverse political backgrounds. We should also analyze the difficult relationship of some of those mayors of poor localities with the mayors of wealthy Paris and Lyons.

Other major differences also make a cross-national comparison of France and of the USA difficult. Cities are creatures of the states in the USA, themselves in competition on various issues with the federal state. This tension, the sharing of functions in a federal system, sometimes evoked under the image of a marble cake, are very different from the Jacobin centralization that the administration in France exerts on its 36,000 localities, despite the decentralization trends already mentioned. The strong intervention of the state, however, does not mean that French mayors, police chiefs, or judges are less efficient in terms of social control than their American counterparts; on the contrary – and this point will be analyzed later. But in the USA, the size of the country, the heterogeneity of the population, and its history of rebellion against central authority imply that each local arena has an autonomy and capacities of its own and that diversity abounds. Modelization is possible concerning French national and local policies, because of the extensive discursive production of public civil servants in the explanation of their action, whereas repertoires of savoir faire – i.e. tools from a kit chosen for each appropriate case – and catalogues used by multiple actors are more appropriate to the USA.

Keeping these obstacles in mind, I still defend the cross-fertilization, the awareness of diversity, the clarification of concepts, and the theoretical challenges brought by international comparative analysis. We know so much more than we did fifty years ago about the global developments that shape crime and violence at the local level, and about national and local social controls, differences of rationales, the ethnicization of social relations, and community actions in both Europe and the USA. The analyses of changes, trends, and processes observed all over the postmodern world of the late 1990s in Western countries have made cross-national differentiations somewhat less formidable for researchers.

Methodology

Questions relating to urban crime and violence control are complex. Modern penology has for a long time occupied the field alone, narrowed our perceptions, and obscured the ramifications of these issues (Garland, 1990: 1). But issues revolving around the exercise of power, control, and citizenship, their social meaning, their sources, and the basis of their support are too important to belong to a single discipline. They hail society at large. This book is thus an argument for a multidimensional social science approach to urban social control. Increasingly, scholars from across numerous disciplines, including anthropology, political science, sociology, economics, and philosophy, as well as authors of novels, films, and photographic montages have begun probing themes related to cities and social justice.

Both macropolitical tools to study the politics of depacification generated by globalization and microethnographic ones to observe the social engineering elaborated by cities have been used here. So far, the association of studies on social disintegration, the criminal justice system (police, justice, prisons), and of economic, sociological, and political data have been weak in the study of cities, all the more so in cross-national comparisons. Silver's (1993) cross-national analysis of social exclusion, Mayer's (1999) examination of urban social movements in Germany and the USA, and Green's (1998) studies of immigrants in the garment industry in Paris and New York have been exceptions. There is a variety of comparative studies in urban politics, especially of the English-speaking countries (Wolman and Goldsmith, 1992; Fainstein et al., 1992; Judge et al., 1995), but if they include the USA, they usually omit "difficult" countries such as France or Germany, and if they focus on Europe (Bagnasco and Le Galès, 1997), they omit the USA.

It is beyond the scope of this book to go back to the grand narratives on social control as found in the works of Emile Durkheim (1933), Norbert Elias (1933), Karl Marx (1968), David Harvey (1973), Michel Foucault (1977), David Rothman (1980), and others. My focus is more modest, relating to cities and their numerous actors, to the knowledge that they have accumulated about social threats, and to the "repertoires of deterrent" that they use to deal formally and informally with them.

Outline of the Book

In Part I, Chapter 1 sets the economic parameters of the late twentieth century in the USA and in France and their impact on cities. In a context of destabilization of the state and denationalization of space (more so in the USA), the global/local interactions and their consequences on violence and crime are examined, as well as the growth of inequalities, of polarization, and of segregation according to national contexts. A connection is established between the flux of capital, the new international division of labor, and increasing levels of migrations, and with long-term unemployment, and the incarceration of "mismatched" males. A first scenario linking crime and the network society is envisioned.

Chapters 2 and 3 move on to discuss national discursive practices and laws and to examine their possible influence on cities. The punitive discourse and the policies of national conservative elites, making the USA a case apart in the Western world, are analyzed in Chapter 2. The discourse's forms, its evolution, its timing, the media's role, and opinion polls are examined, as well as the federal policies relative to violence, crime, and welfare cuts. The politicization of the criminology of exclusion seems directly related to the legitimated "exit" of privileged categories from the public space. Official help to low-income categories and distressed places exists, but it is not a dominant trend. A second scenario of increased surveillance is elaborated at the end of the chapter. Chapter 3 analyzes how, in the last twenty years, successive administrations in France have emphasized the struggle against social exclusion and economic marginalization in a context of high unemployment. Implicitly, it was suggested that if social prevention was abandoned, chaos and disorder would follow, starting in low-income areas and reverberating through the whole country, as in May 1968. There is also an instrumental logic shared by both left- and right-wing parties at preserving many welfare service provider jobs, with the state as a major employer in France of five and a half million people. The forms, the evolution, and the timing of this top-down rhetoric supported by intellectuals, the media, and opinion polls are examined here, as well as national welfare and city policies. These policies of the third type are high in rhetoric and low in budget. They mask a repressive trend – still minor, but growing rapidly – due to the construction of collective disturbances in urban contested areas which are less and less tolerated and to the propaganda of the far right. Security is a high stake when a crisis of authority at all levels of the state is glaringly apparent. The far right party produces a discourse of exclusion focussing on moral panics and on threats from immigrants' children in the inner cities. It is reproduced by other political parties.

Where does all this lead us? Staying at the national and global levels of the law and order strategies would make us miss most of the subtleties of changes taking place in the "real" world of cities. In Part II, "The Politics of Reconciliation," Chapters 4 and 5 take the socio-economic, demographic, political, cultural, and crime/violence data into account and attempt to evaluate the social control exercised in New York, Chicago, metropolitan Paris, Lyons, and Marseilles. Not only do the criminal justice (police, justice, corrections) and the welfare systems reveal different practices, but the interpretation of rules are stakes, within institutions subject to the ebb and flow of resources. The role of non-governmental actors appears as central in American cities, counterbalancing the repressive practices of law-enforcement actors and reconciling the city with itself.

Chapter 6 attempts to clarify theoretical and comparative assumptions on cities' social control. It examines interactions, governance, the mediations operated by low-income communities and individuals from those communities. It looks at what is expected here and what is not there, what is essential and trivial. Some mobilizations may lead to a subsumption of the general interest by special interests (the NIMBY – not in my back yard – syndrome), while others force national and local political elites to develop social justice and actions in favor of social citizenship and a local general interest. From contrasting national backgrounds, local convergences, however, are observed at the neighborhood level.

My previous comparative research (1993a) dealt with acts and authors of crime and violence in US, British, and French cities. This present book does not take the same approach, but attempts to answer the question: "What is being done in cities confronted with the growing threat of delinquency and urban unrest in an era of globalization?" Having now set out the argument of the book, we will go on to examine the connection of global forces and of urban disorders, then we will move to the national spheres, comparing the USA and France's different postures.

Notes

1 By "social control" (or "social regulation," as the French translation would be), I refer to a set of means specifically used by structured communities and institutions to prevent or limit collective threats of violence and disruptions.

2 By "primary prevention," I refer to the work done by institutions and to their interventions in the social and physical environment. Primary prevention is opposed to "secondary prevention," i.e. work with "at risk" groups of potential offenders and to "situational prevention" as developed in the works of Jane Jacobs (1961), Oscar Newman and defensible spaces (1972), Wilson and Kelling's broken window thesis (1982), and Ron Clarke's (1995) crime opportunity reduction, which we will discuss later.

3 Following J. Sutton (1997), I argue that institutions bring order to experience by organizing phenomena – persons, actions, objects – into collectively shared

and symbolically meaningful categories that individuals recognize as natural and take for granted in everyday interaction.

4 *Le Monde*, 6 July 1998.

5 For instance, Professor Alfred Blumstein in an article for the National Institute of Justice called "Violence by Young People" (1996) writes about the rise in juvenile crime. A lot of criminologists choose one term or the other.

6 One might object that Los Angeles is the second largest center in the USA. But is it a city in the way that New York, Chicago, or even San Francisco and Miami are – that is, with an accountability related to clearly identifiable social policies linked by a mayor, a city council, and other teams of actors? "Los Angeles is seventy-two suburbs in search of a city," Dorothy Parker remarked once, a quote to which Jean-Paul Sartre added: "It offers no organization. If one travels through this huge urban area, one will come across 20 cities . . ." Sartre was short-sighted: more than 80 enclaves within Los Angeles County have declared their independence and set up their own municipalities and administration. But although this institutional fuzziness and the historical reformist tradition of Los Angeles compared with machine-oriented cities are indeed problems for a cross-national comparison, a host of talented authors do include Los Angeles in their comparisons.

7 A statement made by an eminent comparative scholar after a long and careful study is instructive: he said that were he innocent, he would prefer to be judged by a civil law court, but if guilty, by a common law court (the guilty and the innocent being more accurately distinguished by criminal proceedings in the civil law world) (Merryman, 1969: 132).

8 The French state has also co-opted civil society via generous public allowances. One-third of the French people's incomes comes from public benefits; over a quarter of the working class received unemployment benefits in 1995 (Beuve-Méry, 1997). The French will not give up "Womb-to-Tomb" perks quietly. Unlike Americans, who consider government to be a nuisance, they have been willing to endure the highest taxes in the Western world to support a most generous welfare state. Today, the French state employs more than a quarter of the workforce – a level twice as high as in Germany or the USA. There are 5.3 million active and 4.2 retired civil servants, and half a million public jobs have been created for young people since 1998. Public spending consumes more than half of the country's GNP, a proportion greater than in any other industrial country, including the Scandinavian states and former socialist societies of Eastern Europe (Drozdiak, 1996).

9 The term *banlieue* goes back to the thirteenth century. It referred then to the perimeter of one league (*lieue*) around the city. Its jurisdiction was a *ban* which is what the residents in the Middle Ages called the geographical perimeter of one league where they lived. Since the middle of the nineteenth century the term *banlieue* has defined a portion of space sprawling between the city and more distant and loosely urbanized zones and bound into a compact whole (Soulignac, 1993).

10 Immigrants (*immigrés* or guest workers) are people born abroad and living in France. Their children born in France are French. Not all *immigrés* are

necessarily foreigners: some of them become French citizens (around 30,000 a year). Conversely, not all foreigners are *immigrés*. As diplomats, businessmen, students, etc., they are not perceived negatively as *immigrés*. In 1990 31% of *immigrés* are not foreigners and 21% of foreigners are not *immigrés*. To give just one example illustrating this statistical difficulty: in the department of Seine St-Denis bordering Paris, the share of foreign children between 0 and 14 years old is 19% but the share of children in an immigrant household is 38%.

Part I
The Politics of Depacification

1
Economic Globalization and Urban Unrest

The purpose of this chapter is to test the possible correlation between the impact of the globalization of the economy on the city, the growth of inequalities and of power conflicts, and the violence and crime which may ensue in specific segregated urban areas. Welfare and repressive policies as national tools used to manage marginal categories nationally will be the objects of the next two chapters.

The combination of the spatial dispersal of numerous economic activities and global integration contributes to a strategic role for major cities in the current phase of the world economy. A set of complex hypotheses have emphasized that these processes take place in a number of cities, "world cities" (Friedmann and Wolff, 1982; Fainstein et al., 1992; Mollenkopf and Castells, 1991; Sassen, 1991). After a time, once the growth of the nation-state had relegated cities to the back seat, it is my argument that the subsequent hollowing-out of the state led to the development of global cities. Cities are indeed influenced by macroeconomic mutations: the flows of capital and labor; the growing importance of finance; information and computerization, marking the advent of a post-industrial society; the exodus of firms to developing countries; and the expansion of categories of technicians and executives in producer services at the expense of manufacturing skilled and unskilled jobs.

At first glance, global cities seem to take advantage of the accelerated internationalization of economies. They have been defined by Sassen as command points, as key locations for the leading industries, finance, and specialized services for businesses, and as sites for innovations in those industries (1996).[1] But, as correctly mentioned by Sutton (1997), we lack a systematic explanation relative to the interaction of macroeconomic processes, inequalities, urban violence/crime, and the structuring of a moral order. Is there a connection between globalization, unemployment, the growth of the GNP per capita, crime, and incarceration? Do welfare policies mediate crime? How are the norms and authority negotiated to manage marginal populations? Do inequalities have an impact on crime? Are crime and rates of imprisonment correlated?[2] In the first section of this chapter, my analysis relates to inequalities and spatial segregation in the "globalized cities" and to their possible correlation with violence, crime, and repression. In the second section, the novelty of the phenomenon of globalization and its impact on crime is questioned by snapshots from the past. And, in the final section, a scenario for the future linking of globalization, cities, and crime is examined.

Inequalities and Crime: Ecological or Behavioral Effects?

My contextual analysis takes into account the production of new normativities enacted in places, namely cities. Two pictures synthesize what this chapter is about. In the first, the JFK, the O'Hare, or the Charles de Gaulle airports, with their complex and sophisticated machineries in terms of security. An airport is one of the most flagrant examples of a necessary and successful global place, where security is taken care of, whatever its cost and the means required. Few passengers are scared to walk into an airport, yet the flux and diversity of people and the world exchanges in this confined zone are extreme. An airport is a global actor, with a global agenda, in a local space. In the second picture, we look at nearby areas which could be East New York or Brownsville, NY, or the South Side of Chicago, or Aulnay-sous-Bois or St-Denis, north of Paris, plagued by delinquency, drugs, and daily insecurity. Those pictures are elements of global cities.

The role of places reveals the construction of spaces for power within cities and their strategic geography (Sassen, 1999). Claims made by certain actors on highly valorized spaces are encouraged in global cities by the denationalization and deregulation or privatization of such spaces which are "highjacked" by liberal projects. State and local authorities in the USA, influenced by the conservative ideology of "rolling back the state", participate in legitimating the growth and strength of the global economy. They are agents in the implementation of global processes and emerge as altered by their participation in this implementation. The impact of a global economy is felt on the particular form of the articulation between sovereignty and territoriality which has marked the history of the modern state and its apparatus. Now a major institutional discontinuity has occurred. Sovereignty is decentered and free-trade zones operate inside the sovereign territory of the public space, as Sassen (1999) observes. The globalization of the economy has existed for a long time, but what has changed since the mid-1980s is the deregulating practice, the globalization of finances, and the growth of direct transnational investment in global cities. We may then wonder if cities' participation in the global economy produces better conditions and collective benefits for them. In some cities, as robust as New York and Paris, it probably does. This point will be examined in Part II. The correlation between immigration and unemployment does not prove to be a drag on the economy of these regions, whereas it may in many other cities, (Lyons, Marseilles, and, possibly, Chicago) and in parts of the neighborhoods of the aforementioned cities. The same can be said for foreign investment which, on the whole, benefits suburban markets more than central cities, except in the two mentioned cities, New York and Paris, which prosper.

Yet there is a cost to prosperity. Despite the recent decline in crime rates, global cities in the USA remain problematic places for a number of people, such as suburbanites and people holding recurrent "anti-city" views (Beauregard, 1995).

It has been argued that crime and violence were indeed direct responses to the erosion of significant economic boundaries around the nation-state. In so far as the globalization of the economy has swung economic power away from nationally defined economies and has, at the same time, resulted in a local decentering, institutions have found themselves destabilized in a variety of arenas that had previously been contained.

Whether places or residents suffering unemployment or underemployment and spatial segregation are the motors of crime has divided social scientists from Engels to the School of Chicago. On the one hand, for some analysts – neo-Weberians such as Saunders (1986), Coleman on social capital (1988), Bourdieu expanding on the notion of different capitals held by individuals and on symbolic domination (1984), or Castel on the redundant worker (1996), and those debating the underclass – the economy has a reduced impact on social stratification and its consequences. On the other hand, for "global economists" (Sassen, 1991) and the analysts of regulation (Amin, 1994), the city and its social and spatial relations are molded by the economy.

Potential dangers in urban areas send us back to unequal resources, to processes supposedly open to all in a democracy but which are, in practice, extremely unequal. If, as will be shown, pauperization and segregation are less extreme in large French cities, violence remains a threat. The global city is indeed a strategic site for the enactment of politics of disadvantaged groups from below which have little power but to exert an "intimidating" presence. Under favorable circumstances, violence can become their potential mode of expression when they want to gain something, such as their share of the city or their respect as subjects (cf. the Diallo case in New York City in 1999 – see Chapter 4 – but also various protest demonstrations against police brutality in a number of countries). They introduce alternative normativity to the city. Two possible sources may disrupt the collective space; on the one hand, as in the USA, the poor intimidate the poor, or in France, urban youth, with no prospects in life, attack institutions, civil servants, and "those who are not them" as symbols; they have nothing to lose; they act visibly, in the public space, in the same way, with the same words, and with the same rationalization used by angry farmers or public employees exasperated by the lack of response to their claims, all the more so as the global economy has passed them by; on the other hand, as in France, the middle classes, threatened by precariousness, an uncertain future for their children, and possible downsizings, also take to the streets to force the state to keep its role as regulator and as buffer against market destructions.

Since European cities are smaller in size than most of those in the USA, have experienced fewer architectural shifts, and retained many historical and cultural traditions, they seem to display more philosophical and social cohesiveness. The car seems less prominent physically and mentally, and public transport allows the interaction of diverse populations in and between peripheral areas and the centers. It should be re-emphasized here that if we are to compare European and American cities on the theme of globalization, social order, and the management of social disintegration, we need to compare American inner cities with French peripheral urban zones (except for a few cities like Marseilles) where urban disorders are concentrated.

Hot spots versus violent youths: the theoretical debate

It seems necessary at this point to bring forward the theoretical debate related to the correlation between the structural causes of inequality versus individual responsibilities which would lead to violence and crime in large cities (Hagan and Peterson, 1995: 55). Linkages between social inequality and crime have been subjected to speculation since the early days of criminology. In the USA, DuBois wrote about it as early as 1899, Merton in 1938, not to mention Shaw and McKay and the Chicago School of sociology (Merton, 1938; Shaw and McKay, 1942; DuBois, 1961).

Criminogenic places? The early and influential theory of Shaw and McKay (1942) on social disorganization integrated Park et al.'s ecological theory of cities (1925) by focussing on neighborhood characteristics associated with high rates of delinquency. Clifford Shaw was a former probation officer who became fascinated by the delinquents he dealt with. In a story published in 1929, he wrote about "Stanley the Ripper" as if he were himself Stanley. Life stories become a new approach in the study of delinquency. Frequently, he said, there is a big difference between the situation as seen by others and by the individual. If people define their situation as real, it becomes real in its consequences. What is striking for current observers is that whether Shaw talks of "Stanley the Ripper" or "Jack the Roller" (whose story was published later), these delinquents are not murderers but small-time robbers in Chicago.

In *Juvenile Delinquency and Urban Areas* (1942), Shaw, now associated with McKay, states that some places are more criminogenic than others. The two authors quote, as an example, Corsica in 1825, where there was, on average, one homicide to every 2,000 residents, as opposed to one in every 37,000 in Creuse, a *département* at the center of France. They cite contagion effects, the seduction of crime, and the attraction of vice. Working on maps of delinquency in Chicago, which they analyze over time (between 1900 and 1930), they found that about 100,000 boys under the age of 17 are taken each year to juvenile courts, one-third of them for burglaries, 20 percent for larceny, and 10 percent for cart thefts.

Three structural factors – low economic status correlated with poverty, racial or ethnic heterogeneity, and residential instability – are seen as consistent predictors of delinquency. Shaw and McKay anticipated that such places could not be easily lifted out of their condition. They lack community-based social controls, and this absence contributes to crime. The same phenomena, argued Shaw and McKay, are intergenerationally transmitted in criminogenic neighborhoods, so that the spatial clustering of social problems persists in the same areas over time. This is an important finding: high rates of delinquency persist in hot spots over the years regardless of changes in immigrant and minority populations. It would, then, be less individual behaviors that explain delinquency than processes of transmission of delinquent socialization in certain areas. Instead of racially stereotyping dangerous classes, Shaw and McKay have demonstrated that black neighborhoods (there were 4 percent of blacks in Chicago in 1920) do not form a homogeneous category, nor do black young males. Variations in crime rates correspond to heterogeneous black neighborhoods. "The important fact about rates of delinquents for Negro boys," they write, "is that they vary by type of area (as whites do)" (quoted in Sampson and Lauritsen, 1997: 334). For instance, in low-income areas with low social organization, the residents exert less control over unsupervised teenagers. The social agencies are inefficient; they are outsiders and do not understand the residents. Newcomers result in delinquency: first the Germans, then the Irish, followed by the Poles and the Italians. The transition from a rural life to the complexity of urban life adds to the teenagers' destructuration. Delinquency can be interpreted as a mode of adaptation to confront problems of rapid mutation in a city like Chicago. But after families move to a second place of residence, the rates of delinquency generally diminish. Mobility tends to solve such problems.

Black neighborhoods with stabilized families experience lower rates of crime, Shaw and McKay conclude.

One of the best syntheses on the topic of criminogenic places has been written by criminologist Robert Sampson and sociologist W. J. Wilson (1995). Starting with the Chicago School research, they emphasize the importance of space or neighborhood effects. Crime remains attached to certain places regardless of what populations experience them. Hot spots, unsafe housing projects and streets, and interstitial zones are well known to the police; location does matter. The place stigmatizes the residents, who become ashamed to give their address as they know it will penalize them in the eyes of the police and in the search for a job, for relationships, and for any entry outside the area. According to Jargowsky, who studied thousands of ghettos and barrios, the more a poor neighborhood is surrounded by other spatial areas of poverty, the harder it will be for this neighborhood to lift itself out of poverty and associated problems (1996), a fact that is all the more true because spatial segregation is produced by global trends.

As for youth subcultures, they are competing with modes of socialization which may be less appealing or in crisis. A language, codes, tags, rites, internal hierarchies, and scripts establish a belonging, an identity, a protection against the outside environment and an enclosure. For Doreen Massey, spatiality is always and everywhere full of power, because it is constituted out of social relations (Massey, 1997: 114). Identities and spatialities are established in and through relations of dominance and subordination. She analyzes the case of English lads from a public housing estate who appropriate the public space as their own after 10 o'clock at night, in order to establish a strong identity for themselves and for the women and children whom they intimidate. Their move can be interpreted as a resistance to spatial entrapment, as the expression of young males' dissenting voices – what I call the return of the warriors fighting over honor and respect – within a homogeneous community. Such elements interact with all the variables already being taken into account. The neighborhood must be seen as a unit, submitted, reacting, resisting, or yielding to both internal and external forces.

A further line of approach is offered by Jeffrey Fagan and Deanna Wilkinson, who are currently engaged in research to reconstruct the stages and the transactions surrounding firearms incidents among inner-city adolescent males (16–24-year-olds) in East New York, Brooklyn, and the South Bronx, neighborhoods that were submitted to an epidemic of gun fights between 1985 and 1992.[3] They listen to the youths' narratives at length, including those just out of jail or in hospital emergency rooms, and they try to understand their "scripted" behaviors[4] sustaining violence and the context of "situated transactions" in which disputes are settled with or without guns, with shots or not (Fagan and Wilkinson, 1998). Among other factors, they analyze the importance of bystanders, of drinking, and of drugs. As in the life stories told by Clifford Shaw, and as confirmed by the French field work, patterns and functions of adolescent violence emerge: achieving and maintaining status, "respect," and identity; acquiring material goods as a source of status; exerting coercion, domination, and power; experiencing pleasure; managing conflict; expressing an oppositional culture, etc. Fagan and Wilkinson point out that their approach also requires an analysis of events, of "the person-event," and "person-place," and "person-context" interactions that shape the outcome of events.

Delinquents with criminal careers? Other experts choose to emphasize the impact of social marginalization on urban disorders. Marginalization or social exclusion are elusive concepts, as shown in Silver's analysis (1993). In the 1970s, Carol Stack's ethnographic research revealed that people in ghettoized areas were bifocal subjects. They knew exactly how the other half lived, but to survive daily in their environment, they had to stretch their values and develop a schizoid approach. In this way, "marginal" populations are highly integrated, as Stack pointed out (1975). Their disintegration is seen as problematic only from outsiders' norms and their diversity and plurality are ignored. The design of their integration is always formulated by political leaders who require from them change and who design utopian schemes with this aim. But when they fail to adapt accordingly, the dominated are harshly judged. In the words of one French mayor, "you could put 'these' people on the Champs Elysées, and they would manage to make a mess out of it."[5]

In the USA, crime is perceived as a problem of young, disadvantaged, jobless, minority males. "While inequality promotes violence, racial inequalities are especially productive of violence because of feelings of resentment," observes criminologist John Hagan (Hagan and Peterson, 1995: 22), an observation also recurrent in France. The disastrous consequences of isolation, racial discrimination, and the concentration of jobless individuals and of gangs of teenagers in specific neighborhoods can foster a subculture of violence leading in some cases to criminal careers (Fagan, 1997). While public concern with crime and racial outbursts is primarily focussed on African-Americans and slum areas, other groups seem to catch up. Hispanics – another fuzzy concept – figure prominently in the youth gang literature and their subculture of violence is among the more popular explanations of gang membership. While this is not necessarily correlated, they constituted the fastest growing minority group in prison from 1980 to 1993. Because of extremely high unemployment rates, depressing poverty, and disheartening living and social conditions of the people living in

Chinatowns, Asian youth gangs also cumulate high crime rates and acts of destruction (Marshall, 1997: 15; Mann, 1993: 97).

According to the theory of relative deprivation, cities and neighborhoods in which high- and low-income communities live in close proximity experience high crime rates. This is precisely the configuration of global cities. "The concentration of wealth and poverty in the same geographical area is more exacerbated in the American city," Sullivan observes, "and it constitutes the precondition of street crime in the city" (1991). Seeing the wealth and possessions of the neighboring communities, the delinquents could be motivated to commit crimes because of emotional frustration, latent animosities, and lack of opportunities (Sampson, 1985: 8; Brantingham and Brantingham, 1980). According to one sociologist, who studies Chicago neighborhoods, the proximity of poor and wealthy families may account for 56 percent of the variance in homicide rates and almost 40 percent of the variance in robbery and assault rates (Block, 1997: 52–5). But if low status youths living in high status areas are likely to commit more crimes than those living segregated in poor areas, does not this seem to justify practices of segregation and a rationalization for gated communities?

Outside forces?

Writing about France, Irving Jackson makes a connection between places with a high number of foreign immigrants, their high rates of unemployment, and crime:

> For the *étrangers* population nationally, the rate of unemployment in 1990 was over 19.5%. Those who took French citizenship – 3% of the population – had a slightly lower rate of 14.3%. France's overall unemployment rate of 11% underscored the unemployment problem of the North African groups, especially where their population was the greatest . . . Overall, it is apparent, then, that official crime rates in France are the highest in those *départements* with the largest official counts of *étrangers* . . . it does help to explain French perceptions of trouble in those locations with the largest minority populations. (Irving Jackson, 1997: 137, 139)

She then draws a parallel with US transitional areas in which poor, unemployed minorities and immigrants have been found to have the highest crime rates, "not only because their social disorganization destabilizes conventional normative structures, allowing deviant norms to prevail, but also because those outside of these areas recognize that these are places where residents are less likely to initiate contact with the police to report drug sales or other criminal behavior" (1997: 140). Is there a correlation between minorities, localities, unemployment, and crime?

According to conventional representations, at the individual level unemployment opens a potential pathway to a criminal career; at the collective level, it allows institutions to create meaningful, status-loaded categories which resonate with police officers, welfare workers, judges, etc. sorting out deviants and shaping the moral order. Criminals and victims are often the same people, exchanging roles from situation to situation.

Yet theories that link polarized cities and crime in relation to unemployment are inconclusive at this point and far too deterministic. "The same problems that plague time series analyses of wages, interest rates, and unemployment plague time series analyses of crime," Freeman observes. "Differences in years covered or in the model chosen or in the particular measures used affect results substantively. The safest conclusion is that the time series are not a robust way to determine the job market-crime link" (1995: ch. 8, p. 10). The 1960s were affluent, yet crime was rising in US cities. The events of May 1968 erupted at a time of high consumption in French cities. At the end of the twentieth century, juvenile delinquency rates remain high in prosperous Denmark, whereas Spain, with a 20 percent rate of unemployment, has a low crime rate. Other factors, such as the proliferation of weapons, social integration/stigmatization of socially excluded categories, societal responses to households' hardships, institutional priorities in terms of prevention and repression, and the acquiescence of their clienteles, have to be included in the analysis.

The connection between business cycles, polarization, and urban threats (violence and crime) thus remains a riddle for social scientists. Another question related to economic inequalities is whether cities were more threatening in the past for those who lived there and for the authorities in charge of them?

Lessons from the Past

Descriptions of social and economic inequalities in New York, by Edgar Allan Poe in 1844, by Jacob Riis in *How the Other Half Lives* (1885), and by Charles Dickens in *American Notes* (1842), could be used here. All three opposed the wealth and control of the dominant "half" to the pauperization and deprivation of the rest. The poor experienced probably much worse living conditions than today, as shown by their life span and mortality rates. They had fewer rights than now and risk was part of their daily life. Yet we also intuitively grasp that the worlds of the rich and the poor and of the intermediate categories, in pre-Haussmannian Paris, for instance (i.e. before the mid-nineteenth century), were more intertwined than we might suppose, even in terms of economic relationships. Comparative analysis of the issue of inequality and crime is made difficult by the scarcity of historical studies on ethnic and racial crime patterns and by the lack of consistent official data or even coherent newspaper reporting.

Were cities more dangerous in the past for their residents? According to Ted Gurr, who surveyed a large body of research, current high rates of homicide and of other criminal violence are a relatively recent phenomenon, especially if compared with an idealized view of the past. However, he remarks that our "medieval ancestors had few inhibitions against clubbing and knifing their neighbors during angry brawls" (Gurr, 1989: 11). In the fourteenth century, they killed one another at rates at least ten times higher than those in Britain today and twice that of the USA today. The decline in violence is due to the civilizing influence of humanitarian values, according to the well-known thesis of Norbert Elias (1965), and to the stabilization of the frontier in the USA – American urban violence decreased at the end of the nineteenth century and the first half of the twentieth century, according to historian Roger Lane (1997).

For his part, historian Eric Monnoken (1995: 102), examining New York City homicides between 1800 and 1874, claims that in those days guns were rare. As now in emerging countries, death was also accepted as an ordinary part of physical violence amongst men. Coroners' juries would play a major role in determining whether a death was natural or not. Consequently, the cause of death of a man hit in a bar could as easily be referred to as a fall to the floor or a weak heart rather than homicide.[6]

The serious historical study of crime is less than a generation old, according to Lane (1997). To what factors can we attribute a seeming lack of interest in the study of crime in New York City, an atypical, but the major city of the USA (Hawkins, 1995: 36)? The paucity of studies on crime in white national and ethnic groups and a tendency to downplay differences may have been affected by an ideological resistance to eugenicists and anti-immigrant groups at the turn of the century. This tendency was counterbalanced in subsequent decades by preconceived assumptions of the relationship between crime and non-white races, also linked with the production of official data on the white/black dichotomy, rather than on differences between all kinds of groups, including gender, age, and locality.

A snapshot of historical New York

Luc Sante has attempted to recreate the conditions of life of the dispossessed in the city at the end of the nineteenth century, with its eternal repetition of poverty, low life, and carnival traffic (Sante, 1992: xiv). The city's "unconscious," he says, is the repository of the repressed history of bribery and corruption, misery, crime and despair, chaos and vice. For his part, he emphasizes the continuities. The city today is like it was a century ago; what has changed it is more technology and higher costs. I will come back to this point.

Among Sante's findings is the extreme importance of the social diversity within neighborhoods which led to a complex economy of games. Janet Abu-Lughod, in her comparison of three cities, has pointed to the special distributions of racial and ethnic groups in New York as being chequered, and to the existence of "intervening" ethnic and racial groupings, making the city less subject to neat polarization. Wheeling and dealing, pay-offs, and Tammany Hall politics may all have played some role in keeping New York from massive civil strife, and at least damping down the fire. The major riots which mark the city's past were not, indeed, revolts of the underclass. The riots of the anti-abolitionists in 1834, of the Astor Place Theater in 1849, of the draft in 1863, or of Tompkins Square in 1874 were all uprisings due to tensions running high between ethnic groups, Catholics and Protestants, the Irish and the blacks (Sante, 1992: 18, 201). Race riots began in the 1870s and culminated in 1901, all the way down to the Harlem riots of 1936 and 1943 (Abu-Lughod, 1995: 188).

Frequent clashes were initiated by poorer gangs fighting for days on end over turf boundaries with every weapon then available; they were sometimes repressed by the National Guard. What comes to mind here is, first, the prevalence of collective violence over individual crime, at least as told by historians, although "contract jobs" were also a business for specialized gangsters; and, second, the connection between poor populations and Tammany Hall and between the political machines and the police, as shown by this description after a battle involving a hundred gangsters at Five Pointers in 1903:

> Tammany politicians intervened and forced the gangs to accept a truce . . . under the eye of the prominent fixer Tom Foley at a meeting held . . . at the Palm Café. The strip of turf between the Pelham Café on Pell Street and the Bowery sidewalk was deemed neutral. A grand racket was held and the gangsters danced with each other's girls. (Sante, 1992: 223)

It sounds like an act in *West Side Story*. Could it happen today?

As for the police, there was a time when the entire criminal justice system in New York City was embodied in one person, the "schout fiscal" who served as sheriff to the 270 Dutch settlers in 1625, performing a range of functions, including maintaining order, keeping the peace, enforcing the laws passed for the common good of all, arresting, and prosecuting.

The District Attorney's Office was created in 1818, jails in 1830, the police force in 1844, and specialized courts in 1838. At the turn of the twentieth century, the police were often squeezed between the various power structures and the protection of one part of the population was carried out at the expense of another. As today, the meaning of laws shifted then according to whoever was in a position to interpret them, and also according to moral panics. The police were go-betweens, halfway between the gangsters and the politicians, frequently serving as interpreters between the two. They were expected to be pure in a system that was corrupt, and were certainly underpaid. They would also often choose corruption, according to historical accounts (Chevigny, 1995; Muir, 1977: 271–2). On foot or on horseback, the duty of those on patrol was principally to keep a mental file of the population on their beat (as the maps of today's police); to be aware of whorehouses and gambling houses, of unlocked homes, and to watch out for potential mob riots. They were instructed to use their nightsticks, not on the head, but over the arms and legs of the "enemies of society," that is, thieves, crooks, street peddlers, hoboes, "non-criminals who did not speak English," and blacks (Sante 1992: 243, 250).

Issues of ethnic differences in crime and punishment emerged with the immigration of Europeans to America during the nineteenth and early twentieth centuries. Monnoken (1995) compares crime rates and punishment of African-Americans with those of specific nationality groups and finds that New York City experienced a relatively high rate of fatal intergroup aggression between 1831 and 1874 (1,560 murders). African-Americans, who represented only 2 percent of the population, were not significantly more likely than recently arrived white immigrants to be involved in such violence, although, because of their colour, they experienced more bias and discrimination. They were responsible for, respectively, 23 and 27 murderers of white and black victims, out of a total of 1,482 murders in the city. In a smaller sample of 300 killers, they were outnumbered by the 95 Irish who were responsible for 83 percent of murders amongst their own, and the 39 Germans who were responsible for 56 percent of the murders amongst their own. These statistics reveal ethnic clustering. Given our current perceptions of black criminality, this study is particularly interesting. Monnoken also adds that immigrants who had arrived in the USA in childhood were more violent than those who arrived as adults (1995: 18). For historians, native-born Americans were more delinquent, once age and locality were taken into account (Sutherland, 1934: 113; Sellin, 1938: 74–107).

In brief, New York was a violent city, even if not as violent as today, because of firearms. African-Americans lived in a dangerous city, where the main danger came from other groups. It has been suggested that the police were used by dominant ethnic groups to target minorities with utmost discretion in arrests, in order to reduce the "threat" of inter-ethnic economic competition. This point brings us back to the distribution of jobs, to market inequalities, and to the patterns of crime, urban unrest, and social control that I mentioned earlier, and to which I will come back later. I will not develop here the themes of Rothman's pioneering history of the deviancy control system (1980), but I will simply reaffirm the increasing involvement of the state at the time in the business of categorizing, setting apart deviant, recalcitrant, and dependent groups, punishing and sending them into overcrowded, corrupt, and certainly not rehabilitative, custodial institutions. When Cohen writes that "not just the prison but the crime system as a whole is part of the larger rationalization of social relations in nascent capitalism" and that "a new technology of repression emerges to legitimate and strengthen ruling-class control of the work-force and to deal with various redundant, superfluous, and marginal populations" (Cohen, 1985: 23), we may wonder what is new.

One of the important differences that can be noticed between the past and now comes from the linkage – or its absence – between the institutional system and the populations at risk. We have seen that at the turn of the twentieth century, immigrant populations and the police force were connected to political machines. Even gangsters worked for them. There was a time during the Progressive Era, between 1880 and 1920, as brilliantly shown by David Rothman (1980), when new and powerful portrayals and representations of crime and responses to crime emerged. Reforms provided an articulation of a cultural imagery based on popular and plausible portrayals of crime and the techniques of social control and rehabilitation which expressed the conscience of elites and their sense of accommodation. Unlike what is occurring now, there was no divorce between representations of crime and the techniques of social control.

Three types of representation prevailed during the Progressive Era, as Simon and Feeley point out (1994). The first, formulated by Park et al. (1925) and by Shaw and McKay (1942), the early sociologists of the School of Chicago in the 1920s, associated criminogenic neighborhoods or delinquency areas with immigrants from Southern and Eastern Europe. In the slums, the environment was seen as shaped by pathological forces imported in part from these areas. But this problem of delinquency was seen as only temporary and could be solved with adequate programs, according to the Chicago School sociologists. The bad news all related to the immigrant slums. The good news was that the rest of American society was strong, prosperous, and stable enough to meet the challenges (Rothman, 1980: 52).

The second form of representation, also developed by Shaw and McKay (1942), related to the offender with a specific developmental history leading to deviance and law-breaking. Popular acceptance of this notion facilitated the development of new programs and helped gain popular support for them. This offender, after appropriate treatment, would be rehabilitated and would join the mainstream.

The last representation had to do with "born criminals." The eugenic concerns of the time about the racial make-up of the USA as a result of immigration justified life imprisonment as a way of preventing the reproduction of a possible criminal class (Simon and Feeley, 1994).

All these diagnoses offered solutions: the individual was seen as educatable, treatable, or at times incurable. The suggested reforms were based on case-analysis and ethnographic research. This basis of information was critical in winning support from the influential class of college-educated journalists and professionals who shaped public opinion. These methods and analyses had great cultural resonance for the new professional managers who were beginning to emerge as power-holders in American institutions. The success of Progressive penology was that it shaped public discourse about crime, and shaped crime policy, as well. The problems of the working-class neighborhoods were heard, people had a sense of entitlement to solutions to their problems from the government, a modus vivendi between liberal thinkers, decision-makers, and the various classes could be accomplished via an "imagined community." This linkage is what seems to have disappeared. There is a chasm today between professionals and academics and their knowledge, on the one hand, and politicians, the media, and the public on the other.

What about Paris? Was it a more dangerous city in the past?

Nineteenth-century Paris was by no means just a quiet and beautiful city. Like New York, it was immersed in globalization, trade, sophisticated services, imports and exports, and the flux of migration. The main issues that were discussed at the time were prosperity and extreme poverty, violence and insecurity. Those topics were nothing but the continuation of a century of popular uneasiness and political and social unrest, the consequences of what would later be called "Paris's urban pathology."

Throughout the nineteenth century, Paris absorbed a continuous supply of migrants from the provinces and from adjacent countries, and the city was unfortunately unable to adapt to this massive influx of people, who were in search of a better way of life (that is to say, mainly housing and work). The living conditions were disastrous for the poor and kept deteriorating as people arrived, especially after Haussmann's big renovation works in the 1860s which precipitated the revolt of the lower classes during the Paris Commune of 1871. Paris's poor neighborhoods rapidly became overcrowded, dark, unhealthy, filled with unemployed and homeless people, and characterized by a growing number of clashes within the working class, mainly due to social inequalities and racial differences.

Many writers have documented the misery and despair that brought crime to the city in the nineteenth century. Literary quotations confirm the necessity for the poor to steal, beg, prostitute, or commit murder to survive. Those quotations, from both the century's beginning and its end, show that the situation did not change in a hundred years. George Sand, in 1827, was among the first French authors to expose the link between poverty and crime: "There are more poor people on the street. You have forbidden them to beg on the outside, and the resourceless man begs at night, a knife in the hand" (quoted in Chevalier, 1984: 122). Proudhon, in 1851, even connected crime with the desperate economic condition of the working class:

> When the hand worker has been abated by work, engines, ignoring instructions, bitty division of work, when he has been discouraged by his salary, depressed by unemployment, starved by monopole, when he has no more bread, nor pie, no more money, no more stitch, nor fire, nor place, then he begs, he thieves, he cheats, he steals, he kills. (quoted in Chevalier, 1984: 445)

Crime, then, represented the problems facing a worker during his career, bringing them to their most simple and dramatic expression.

It is since those days that the working class has often been considered as the "dangerous class." This opposition of the dominant class and the "dangerous" pauperized one (which represented one-third of Paris's population) grew throughout the nineteenth century. For the intellectuals and the elite of the time, who were mostly locked up in their ivory towers, blind to the difficult urban situation of the other classes, the lower class was a distinct part of society. Trying to overcome their miserable conditions, those "marginal people" were perceived as fomenting revolts. The elite clung to the traditional view that crime was intimately connected to the working class (Marchand, 1993: 66). But this would be much too simple an explanation for Paris's social unrest.

There was also racial antagonism within the working class itself. The "good Parisian workers" would reject the nasty *provinciaux*, who came to Paris to steal job opportunities from them. The question of separating and categorizing the established and the newcomers within the working class grew and developed throughout the century. In each case, animosity went both ways.

These are the main characteristics of the frequent outbursts of violence in nineteenth-century Paris. The working classes (and especially the "strangers") were considered by the bourgeoisie to be wild, barbarians, and nomads – all words that expressed the truly racial aspect of social unrest in the city.[7] On the one hand, there were the rich and the poor, and on the other, there were Parisians and the "strangers" (those who were not born there, didn't live or get married in the same way, who didn't have the same jobs, or live on the same street, or whose lodgings were not equally oriented towards the sun or the clean air of the Seine – those that police records, diverse descriptions, and demographic studies depicted as a "racially different" population).

The opposition of classes in the nineteenth century was not only a social matter. It was also expressed by a geographical cleavage in the city: roughly a right bank/left bank separation. Throughout the repeated urban renovation and reconstruction of the city, the poor concentrated in the center and on the left side of the river. The dominant class moved in a north/west direction, in isolated wealthy areas, among them the famous "New Athens." The gathering of poor populations around City Hall (Hôtel de Ville) and the open market (Les Halles) that was noticed in the 1820s continued after the 1840s. At the end of the century, the poorest and the most "problematic" part of the population concentrated in very specific places (mostly in the southern and eastern parts of the city): the Hôtel de Ville, the Cité, the Île St-Louis, Porte St-Denis, and the twelfth arrondissement. In the 1840s there was such a geographical dissymmetry in Paris that people said there was a rich western half and a poor eastern one. This separation lasted throughout the second half of the century.

Haussmann's famous reorganization of the city did much to isolate the left bank once more. But Paris's main heritage from the Haussmann renovations was the creation of a real *banlieue,* peripheric sites where, from the Second Empire (1850–70) onwards, the workers were sent in successive waves. These were the places where poor citizens, used to urban life, were "exiled."

It is not surprising, then, that the main "hot spots" of Paris coincided with the places where the poorest people were gathered. These places were mainly the ones mentioned above. Studying the particularities of those hot spots, we have to consider their past. Let us take Porte St-Denis as an example of spatial continuation. In the first part of the nineteenth century, crimes were more numerous there than in most parts of the other neighborhoods. Delinquents and violent criminals lived there. They did not become criminal offenders by choice, but were driven to commit crimes by "their laziness, stupidity or misfortunes" (Chevalier, 1984: 500). The Place de Grève was known for its executions and its guillotine. It was very often the place where troubles (fights or the settling of scores) broke out. Those troubles were mainly the urban expression of conflicts between the guilds which were becoming more and more numerous at the time.

The police in those days were not very concerned with working-class clashes. "We were following each other in groups. The police was scarcely after us," troublemakers would say (Chevalier, 1984: 698). Pity and mercy were actually the principal reactions during disturbances involving poor people: "We have been told today, that many national guards, shopkeepers or workmen were reluctant in taking arms during the troubles because, they said, they could do nothing but feel sorry for those unhappy workers driven to despair by their great poverty", Chevalier quoted the people of Paris as saying (1984: 438). The police dealt more with the hard-core criminals who supposedly caused harm for no apparent reason (or for their personal enrichment or interest) and the outsiders known for their sinister reputation as killers and murderers. The police had their names, their addresses, and conscientiously followed them, hoping to catch them red-handed. Other types of person sometimes prosecuted by the police were petty delinquents, known as anarchists or ruffians (the last ones known currently as *appaches*), who used to behave as if they "owned" a particular street or neighborhood.

For a long time during the nineteenth century, crime was a question of lonely, personal, brutal intent. From the 1860s onward, murderers and thieves enrolled in pseudo-gangs, obeying disciplinary rules, and creating codes of conduct. They acted in groups, in a tactical manner. The most famous gangs in those days were known as the Stranglers, the Beguilers, the Gang Charpentier, Courvoisier, and Gauthier-Perez. During the Commune of 1871, the fights between classes were frequently due to a fear of Barbarians coming to Paris to take jobs and bring their violence with them (bitterness and xenophobia towards the non-Parisians were strong even before the Dreyfus Affair). But, in the following years, violence decreased significantly and the barricade fights were replaced by noisy, but peaceful demonstrations. There was a sudden peak of violence in the 1890s, when social struggles increased again. But these were different from the political revolts of 1827, when poor people ran through the streets carrying stones in their aprons, breaking shop windows, hurting citizens in order to get a reaction to their miserable conditions. It was also less threatening for the residents and authorities than the 1830s, when people slaughtered unfortunate pale-faced passers-by, whom they suspected of disseminating disease or of poisoning fountains (killing most of them or throwing them into the Seine).

Crime at the end of the nineteenth century was similar to what Greater Paris experiences today, with organized gangs and delinquents, and hot spots where the police dare not venture for fear of being hurt or igniting further disorder. Such areas are roughly the same: St-Denis is still a problematic place, and the north-east *banlieue* is famous for its unrest.

Such accounts denounce the tendency to evaluate the current situation of polarization in large cities with respect to a mythical past, whether in the Progessive Era in the USA or in what the French call "the thirty glorious years," after World War II, a time of growth, social homogeneity, and welfare redistribution which took place roughly between the end of World War II and the first global oil crisis in 1973. In cities as we remember them, Mollenkopf points out, poverty was not so intensely concentrated in certain places and racial divisions were a less defining characteristic of urban ecology (1997). The quality of shared public spaces and public services was higher. Increasingly international capital cities strengthened the national base (Smith, 1997: 124). In France, an almost fully employed society provided the working class with protections it had never experienced hitherto and the whole nation was more closely knit by universal social benefits acting as a cement in the name of solidarity (Castel, 1996). Employment and welfare transfers for those on the margins of society, it was argued, mitigated the possibility of violence. Yet we may wonder if this account is not an a posteriori reconstruction?

It is after the second global oil crisis (1979) that the term (social) "exclusion" appears in France (Silver, 1993). At the beginning of the 1980s, the link between the social disorganization of peripheral zones, their economic handicaps at a time of restructuring and job losses, and their riots is established. In the USA, it is also after the second oil crisis, in a Republican-dominated era that was cutting social policies, that the term "underclass" became widely used. The urban minorities, more disempowered and disenfranchised than ever, to which it referred, had not enough clout to mobilize and protest against such a stigmatizing stereotype. Besides, who ever interviewed the underclass, an artefact constructed for political purposes? Poverty increased and became the common fate for one out of seven census tracts of the one hundred largest central cities, while racial segregation did not abate (Wilson, 1996: 14).

Global political responses

As Sutton (1997) points out, welfare and incarceration policies are used as alternatives to deal with the unemployment rates produced by erratic business cycles. In the USA, as will be seen in the next chapter, a connection can be established between the growth of unskilled male unemployment and rates of imprisonment, but not with crime. In contrast, French governmental elites multiply welfare transfers to mitigate the social disintegration of inner cities and of peripheral zones and to limit the threat of conflictual violence from such depressed neighborhoods. Yet unemployment grew in the 1980s year after year in France, as did the electoral scores of the far right. In a situation of crisis and confusion, simple answers were given to complex questions. Although serious crime rates remained limited and stable, urban violence linked to a loss of meaning and self-worth was revealed in many forms. It is difficult at this point to establish what the independent variables among these elements are and if the theoretical basis is sound. An examination of concrete cases may clarify these enigmas in Chapters 4 and 5.

Society, Globalization, and Crime

A tentative theory has been elaborated by a French magistrate, Jean de Maillard, relative to globalization, society, and crime (Maillard, 1997). Let us apply this theory to our topic. In this scenario, an analogy can indeed be drawn between processes going on all over the planet between core and peripheric countries on the one hand and, on the other, between the new centralities and the new marginalities within and among cities.

In our post-industrial era, the theory goes, societies can no longer be symbolized by a pyramid, an egg, a sandglass, not even by a simple geometrical figure or by Russian dolls. The design of a multidimensional polycentric and indefinitely fragmented set, and of complex nodal points where networks intersect is more to the point. Maillard borrows the concept of "fractalization" from MIT mathematician B. Mandelbrojt. If applied to our complex society, the fractal figure is more than just a useful metaphor, it defines a model of social organization for the future which could lead us to revise our collective representations of crime. According to B. Mandelbrojt, the characteristics of a fractal figure go as follows: its parts have the same shape or structure as the whole, but on a different scale. The shape can be slightly distorted, it can be either irregular or fragmented. The structure of snowflakes or of tree branches gives the general idea: each part reproduces the whole of the snowflake or of the tree, the whole and its parts influence each other. It seems that our current societies are undergoing the same process of differentiation leading to continuous fractalization and to the same unbalance, "chaotic order," and constant readjustment.

In a global city or large metropolitan region, differentiated financial, economic, cultural, and intellectual networks proliferate in discontinuous spaces. The question relates to their links. For some analysts, it is less and less obvious that the wealth of the rich is generated by the deprivation of the poor. The rich speculators of Wall Street do not need the poor to be rich, they are more autonomous than ever. Does Paris need French Corsica or Milan the Italian Mezzogiorno to expand? The unemployed, the elderly poor, the single-parent families are not currently incorporated in an economy based on exploitation as in the nineteenth century, neither are they part of the architecture of global information and exchange systems. Yet other observers – and it is also my point here – contend that the power-holders need the have-nots, even more so than before. They do not need them to exploit their labor force, but to make use of their crimes which serve markets outreaching to the rich, as the cases of arson and the subsequent transformation of poor areas for the benefit of real estate interests or the drug market network demonstrate (Castells, 1998a: vol. 2, ch. 3). As remarked by Friedmann and Wolff, global cities "are luxurious, splendid cities whose very splendor obscures the poverty on which their wealth is based. The juxtaposition is not merely spatial; it is a functional relation: rich and poor define each other" (1982: 32). A set of extremely complicated relationships within poor neighborhoods functioning as retail drug, weapons, touristic, cultural, or other goods and service markets and between them and other metropolitan areas in terms of demand and supply has been observed by Sullivan (1991) and Bourgois (1992) who conducted ethnographic studies of specific distressed zones. In some areas, the traffic remains internal to the neighborhood. Much street crime simply circulates resources within poor neighborhoods. In his work on Brooklyn neighborhoods, Sullivan reports that "the implications of the redistributive aspects of crime for local social control patterns" are apparent when, despite generalized disapproval of crime by many citizens in these neighborhoods, their actual responses to crime depend on their perceptions of whether particular crimes or crime patterns are endangering them or, conversely, bringing cash and cheap merchandise into their households and neighborhoods. The community is not anomic, it produces its own rules, other rules, and other modes of organization (Sullivan, 1991: 239–41; Stack, 1975). I will come back to this point later. The same phenomenon is observed in some French housing projects where a law of silence prevails and where a few young druglords exert their law.

Yet in other neighborhoods, the enormous profits of the trade, especially the drug trade, could not be sustained without a broadly based clientele, including many working people, both white collar and blue collar. In this sense, the drug economy spans society and links these differentiated sub-neighborhoods in the frame of the global city. "Diversity weaves itself into the economy by reentering the space of the dominant economic sector as merchandise and marketing . . . the new *glocal* culture is absorptive, a continuously changing terrain that incorporates the cultural elements whenever it can" (Sassen, 1995). Gangs' transnational involvement in heroin-trafficking, money-laundering, high-tech thievery, and other racketeering activities makes them wealthy and no longer subordinated to domestic police control. "Crime theories need to incorporate the influence of global development on the nature and distribution of crime in the U.S., including in the inner cities" (Marshall, 1997: 29). France is included in this landscape. Marginalized elements in terms of income and status (women, immigrants, and minority youth) are not then residual, but tied up to the capital accumulation process; they perform specific services, create new spatial dynamics in the urban social structure and are absorbed by the machinery of the global system.

At the core of our historical memory, we had been convinced that affluence, the major condition needed to guarantee security, relied on the ability to control people and scarce goods. We had seen immigrants and minorities linked to each other via the political machines, some of them part of organized crime in the USA, then eventually incorporated in the mainstream via labor markets and conformism to the norms of social order. Discipline and surveillance systems and the social support brought by the welfare state intervened more or less into poor neighborhoods, linking all elements of society. Currently, power and control rely on the capacity to organize communication between specialized experts and, except in the case of the custodial state, the power of state intervention into the social sphere hollows out, bringing the privatization of space and other modes of organization (Zukin, 1997). According to this scenario, the opposition is no longer between law-abiding citizens and those who ignore the rules, but about the use that is made of rules. Suppose, for instance, that a South Bronx gang decides to wear jackets manufactured by a California firm inspired by inmates' uniforms. The gang appropriates these jackets and gives them a specific and opaque meaning. For its part, the firm markets the jackets all over the planet, making them as visible as possible. A competition takes place between the firm and the gang about the management of symbols and norms, their production and their use. "Competitive *angst* is built into the world city politics," Friedmann writes, "as it is constantly engaged in an equilibrating act to adjust its economy to the processes of creative destruction and the relentless competition comprised in its very essence" (1995: 23).

Similarly, a single meta-narrative cannot give an account of the processes going on in the metropolis. In the exchange networks, communities with different norms, codes, and rules bring their own separate narratives and it is the dynamics between them via crime and violence which now has to be analyzed.

Maillard's elaboration of the crime landscape of tomorrow marks a complete departure from Durkheim, who had a holistic conception of society. I will come back to this conception in my next chapter. Society can no longer be reduced to individual behaviors and decisions. What is the social meaning of the importance given to crime and violence in our fractalized societies?

In the decentered and interdependent set of urban areas, which are either integrated or gray or no-go areas with myriads of possible configurations, crime can no longer be seen as a marginal and peripheric phenomenon but as intrinsically internal to the process of the global city. Not only is the "two-cities" metaphor deceptively simplistic, not only can a simple dichotomy between prospering white Manhattan corporate service professionals and a "lumpentrash" in the surrounding boroughs be deceptive, but all these parts of the city are the deeply intertwined products of the same and underlying processes. Crime is not to be analyzed as the isolated actions of a so-called "underclass" but as a relationship which is not immediately visible. It is no longer crime and violence at the margins that are frightening for the working and middle classes, but crime at the center in stable/unstable spaces, in the flux of order/disorder, and inside financial and communications networks. Still, traditional forms of crime and violence remain visible – a murder, an assault, a riot – but other forms, because of their complexity, do not immediately reveal that they are part of a whole process. They are not the negative reflection of a normal side, as a binary rhetoric wants the public to believe, but the complementary sides of the economic and social functioning of the global city. The whole is in each of its parts, the golden boys and the druglords and a whole variety of individuals in between, the police high-tech maps and the criminal maps, all influencing each other in a constant readjustment.

This proposal leads to redefinitions. Deviance is what penal regimes consider it to be. It is part of an ongoing negotiation of normative boundaries and political authority. In the same manner, as the public space, the common ground, even the nation-states are being eroded under the blows of privatization and deregulation, common law is also yielding to an interstitial law allowing the delinquent to be judged by people of his/her own community, under the pressure of communitarianism. The O. J. Simpson affair, Crown Heights, and the subsequent Goetz trials reveal that groups have acquired the legitimacy to refer to their own norms and that what they have in common is less the rule of the law than a corpus of common rules processing diverging norms. The rapid development of hung juries goes in that direction as does the proliferation of mediators, that is, of private community representatives intervening as informal judges in ad hoc conflicts. Currents laws reflect norms or rules which emphasize the fractal character of the global set of networks. Other norms are internally produced, such as in walled communities or in gangs or among large corporations. If individuals or groups disagree with the norms of their community and become the minority within a minority, they can withdraw to differentiate themselves. The existence of a common world based on a universally shared covenant is negated.

In brief, according to this scenario, crime is no longer just a visible transgression of social norms, the delinquent is no longer just a marginal individual, crime is no longer anomic, a marker between the established and the outsiders. What is being repressed is border crossing, the impact that crime may have on the status of individuals and groups in their own arenas, and the interference of the offender with those exclusive groups and spaces. The 1993 urban crime on the Long Island commuter train (when a mentally unstable young West Indian, claiming he had received no recognition in America because of his race, shot commuters at random on a Manhattan–Long Island train) is a good example of this kind of transgression, the intrusion of someone who threatened a supposedly safe space to which he did not belong. The bombing of the World Trade Center in 1993, resulting in six deaths and a thousand wounded, is another example.

> The internationalization and globalization of criminality implies that U.S. citizens can no longer deal with their fears by moving, buying a gun or installing sophisticated security equipment in their homes. Keeping the borders sealed will not work. Stiffening prison sentences will not work . . . The economic deprivation/inequality theory applies not only to the U.S., but to all members of the global village. (Marshall, 1997: 30)

It is the interference with the dynamics of the fractal society (i.e. order/disorder; criminal/non-criminal networks) and destructive criminal capacities towards the superordinated networks at the core that are under high surveillance and strenuously repressed, and less the fact that people live and work in ghettoized drug zones. This is why the images of international airports next to impoverished urban zones at the beginning of this chapter and of the apparatus of security that the former deploy are relevant to our discussion. Besides, the police no longer have the power to maintain order within the marginal categories, just as the border patrols cannot seal the frontiers. They attempt to contain the criminalized poor in their no-go areas. But, as police officers themselves admit, all they do is give them a hard time. Their function is not to deal with root causes.

This scenario for the future may appear excessive: it is one out of three hypotheses that will be tested in this study. They are not mutually exclusive, but overlap according to time and space. Marx had not anticipated that poverty would be more profitable than ever, thanks to crime outside any mode of production. Crime in poor areas is a way of making a profit out of capital via consumption.[8] It would seem that, via crime, capitalism, less in need of an industrial labor force, is reinventing a manner to create profit through the consumption of the poor. Crime is not the only example. Gangsta rap is another (Berman, 1997), as are the visits by tourists to Harlem, who change the city's landscape by their very presence.

In conclusion, attitudes toward inequalities have an impact on crime and violence. Localities offer spaces which are contested by various sets of actors. Disinvested places tend to breed social disruptions, but actors from below may also exert a disquieting presence. Inequalities, which have been worsened by economic globalization, play an important part in shaping policies. In French large cities, as will be seen in Chapter 5, social polarization and segregation are revealed by the growth in wealth at the top, but not by the impoverishment of the bottom. Not only are the laws of the market less brutally exerted than in American cities, but governmental welfare transfers improve the income of the least well off. As a tool of social prevention, such policies are meant to mitigate the violence which could result from social and spatial divisions increased by economic globalization. Segregation also appears less severe than in American cities, where the intensity of ethno-racial specialization is more pronounced. In European larger cities, political and cultural resources, the valorization of historical and territorial assets, and the presence of middle classes, a large proportion of whom work for the public sector, allow the rules of globalization to be kept at bay, at least temporarily (Body-Gendrot and Beauregard, 1999).

In societies where inequalities loom large, stress and crime can eventually hinder social success. This concern was largely expressed at the Davos Summit in 1997. The NATO leadership is also worried by the disenfranchisement of inner cities, which appears to threaten domestic security, "an enemy from within," and by "immigrants at our door," in San Diego or El Paso. A militarization of the police and a policization of the army are new processes to be analyzed. Will the current sophisticated and high-tech repressive tools in use always be able to bar access of the have-nots to the "theaters of accumulation," as more actors are pushed into acts of destructive violence, and at what price? And can New York and Chicago, Paris or Marseilles, after all, never experience the recent fate of LA or of Third World megalopoles? This is a central question in terms of social control.

National logics and laws and their impact on urban conditions in relation to social control will be addressed in the next chapters: are national elites confident enough that myths of success, individual rights, and free enterprise can mitigate the effects of unacceptable inequalities? Is the choice of welfare transfers a rational option to counter ineluctable globalizing processes and their social consequences, as is the case in France? Is the choice of incarceration the counterpart of the mismatch of unskilled categories in a post-industrial economy or a Durkheimian way for the haves to constitute a collective indentity in an increasingly fragmented society?

Notes

1 My purpose at this point is not to discuss whether or not such and such a city is a "global city" and whether the thesis of rising inequalities and socio-economic polarization holds or not. There is abundant literature on this point (Mollenkopf and Castells, 1991; Hamnett, 1996).

2 Imprisonment rates are defined as the proportion of inmates per 100,000 in the general population. John Sutton (1997) remarks that in cross-national comparisons, this operational definition, while comprehensive and consistent across countries, is nonetheless flawed because it aggregates demographic processes that are probably causally distinct. Imprisonment rates are the product of rates of admission determined by criminal statutes and the processing capacity of the courts on the one hand and, on the other, the average time served by inmates according to the severity of sentences and available prison capacity.

3 Interviews with Jeffrey Fagan, spring and fall 1998.

4 Cornish's "script" focusses in detail on the step-by-step procedures of committing crimes that are learned, stored in memory, and enacted when situational clues are present (Cornish, 1994: 8, quoted in Fagan and Wilkinson, 1998).

5 Interview by the author, spring 1998.

6 Few remember that, prior to his election to the presidency, Andrew Jackson murdered one man in a gun fight and that he had tried to murder others in violent attacks. These were understood as exemplary of manly virtues.

7 The proletariat was seen as a race rather than as a class and the word connoted a savage way of living and dying rather than an occupational distribution or economic characteristic.

8 Opium bought at between $25 and $125 a pound in the Golden Triangle was being sold for $100,000 in 1997 on the American market after its transformation into heroin.

2
Law-Enforcement in the USA

In this chapter, I will show that by their nature and their breadth, as well as by the support they get from public opinion, repressive policies confer an undeniable stamp of exceptionalism on the United States. America has gradually separated into two societies with a highly stratified workforce, which, despite its heterogeneity of privileges, shares in the general increase in American wealth, and an underprivileged category locked out of the economic expansion of the postwar era altogether. The stratification of society has translated into an increasing spatial separation. The recurrent fear of crime and lack of trust in urban institutions responsible for law and order has indeed been used by the upper and middle classes to rationalize privatization and a withdrawal into enclosed spaces. There are now more private than public police officers, the sign of a distrust for public institutions. The "disqualification of disinvested areas" has found a place in political rhetoric alongside the old idea that moving is a way to solve individual social and racial problems. My argument at this point is that mutations in penality – i.e., astronomical rates of incarceration, especially for young blacks – do not arise from internal requirements of punishing offenders but from transformations in social and political structures. The politicization of the issue of law and order becomes then highly profitable.

The Social Context: A Segmented Society

The history of American cities is deeply marked by violence that is as much the responsibility of majorities and their institutions as that of the dominated categories, due to the strong legacy of prejudice against racial minorities (Body-Gendrot, 1993a). With juries systematically discriminating against non-whites, with a soft apartheid supported by institutional complacency segregating them in specific areas, and with a system of representation that shut out non-whites until the 1960s, the American rule of law has weighed on poor minorities' fates to a degree that is difficult to exaggerate. Skolnick reminds us that at one time it was a felony in thirty-nine states for blacks and other non-whites to marry whites and that these anti-miscegenation laws remained in effect in thirteen states until 1968 (1998: 90). In addition, "nothing has more eroded confidence in the criminal justice system than the long history of willful refusals to punish white anti-black vigilantes", Kennedy, a professor at Harvard Law School, writes (1997: 25).

Yet through the years more minorities moved to the cities of the northeast and of the west, while affluent categories were leaving the cities, following jobs to the suburbs or the southern and western parts of the country. The diffuse clashes between ethnic and racial groups living on circumscribed territories have been gradually replaced by forms of criminality sanctioned by the legal system today, such as drug offenses. And, while the French are looking for the roots of urban violence and of juvenile delinquency in a societal context (cf. Chapter 3), in the USA, the issue of crime has replaced the debate on collective racial violence which marked the 1960s.[1] Nowadays, the term "violence" is used for episodic riots, such as those that took place in Miami and Los Angeles in the 1980s and 1990s, and for slaughter perpetrated in schools. Segregation links ghettos and crime in the popular imagination, and therefore externalizes the problem from the broader society. So, in the end, unlike France today, risk management and its association with what Simon and Feeley describe as "actuarial justice" (1994) displaces the societal debate over the structural causes of violence.[2]

At the same time, trust in technologies of control becomes an increasingly important, yet fragile element of the risk society. We move from a paradigm based on deviance-control-order to that of knowledge-risk-security (Crawford, 1997: 87; Ericson, 1994: 167–8; Beck, 1992).

Another factor has to be taken into account: since 1993, fear of stereotypical crime (that is, crime that lands its victims in the emergency room or the funeral parlor) ranks first among Americans' concerns; fear of juvenile offenders strongly stirs imaginations. This may change, but in recent decades the media worked continuously to create the perception of the city as a dangerous place, which provoked lower- and middle-class urban "flight" and avoidance of specific areas.

Mutual avoidances

A new geography of inequalities characterizes urban America today.[3] Not only did the affluent categories pull apart socially and mentally from the poor, but spatially as well (Massey, 1995: 27). Segregation is not a new phenomenon in the USA and geographical mobility is among the major trends revealing the society's dynamism, in contrast to France. Throughout this century, Americans have followed jobs, and jobs, in turn, have moved to where the land was cheaper and where unions were not threatening. But the causes of this new urban geography, which concentrates affluence and poverty, are more complex and the consequences profound.

Over the past thirty years, a remarkable urban transformation has occurred. In 1960, residents were equally divided between central cities, suburbs, and non-metropolitan areas (see table 2.1). By 1990, more than half of all white Americans (50.3 percent) lived outside of central cities but within metropolitan areas (see table 2.2). With three white Americans out of four no longer living in central cities, political representation leaned toward suburban and rural interests. In contrast, for blacks and Hispanics, life was essentially urban: 57 percent of blacks and 52 percent of Latinos currently live in central cities. Whites and minorities inhabit worlds that do not meet either socially or spatially (see table 2.2).

Table 2.1 Metropolitan location of Americans, 1960–1990 (%)

	1960	*1970*	*1980*	*1990*
Metro area	66.7	66.6	74.8	77.5
central city	33.4	31.4	30.0	31.3
suburb	33.3	37.2	44.8	46.2
Non-metro area	33.3	29.4	25.2	22.5
Total numbers	179,323	203,323	226,323	248,710

Source: 1990 US Census of population

Table 2.2 Metropolitan location of Americans by racial and Hispanic origin, 1990 (%)

	Whites	*Blacks*	*Hispanics*	*Asians*
Metro area	74.7	83.5	90.4	93.8
central city	24.4	56.9	51.5	46.4
suburb	50.3	26.6	38.9	47.4
Non-metro	25.3	16.5	9.6	6.2
Total numbers	188,128	29,216	22,354	7,274

Source: 1990 US Census of population

Mobility as a major freedom

The residential environment of suburban whites is overwhelmingly white (82 percent), native born (92 percent) and non-poor (94 percent). In contrast, the living environment of most minorities is non-white, foreign, and disadvantaged. City-dwellers are twice as likely as suburbanites to live in female-headed families, 56 percent more likely to be unemployed, and their incomes are about 26 percent lower than those in the suburbs (Massey, 1995: 28). As one American out of five moves each year – frequently within the same state (Lamm, 1996: 29, n. 59) – one of the most common practices consists of voting with one's feet. Moving is the American way of responding to social problems. Unwilling to strongly enforce anti-segregation policies, President Richard Nixon made a motto out of it: an open society, he said, is based on mobility, that is, on the right and opportunity for all to be able to live where and how they want to live, whether it is in an ethnic enclave or in an open society. Thus, the importance of threshold effects under which only "captive" populations will remain in bleak neighborhoods. The economic *pull* may explain the fact that people follow jobs, but the *push* comes from urban crime. According to the New York quip: a liberal is a person who has not yet been mugged (Silberman, 1978: 6).

Inside over-protected, "gated" communities – a new form of urban tribalism – the affluent categories redefine the concept of community. The broadening of the "me generation," privatization, and the consensus according to which American adults have to protect themselves and their kin: these are phenomena that are also apparent in France. In the USA, however, institutions support mobility by means of voucher policies or by implementing section 8 measures,[4] according to which people move out of the ghettos' public housing units and settle in better areas. Moreover, the multiple laws and judicial decisions aimed at fighting segregation in the 1970s were not rigorously enforced during the 1980s. On that score, the federal government has not really challenged local traditions, institutions, or sentiment (King, 1995: 196). The weight of local self-government, which, to some extent, is also visible in France, minimizes its chances of success in developing racially mixed neighborhoods. Through zoning restrictions, bans on multi-family dwellings, vetoes on public housing and mobile homes, and stringent building codes, local authorities prevent "undesirable categories" from moving into their jurisdictions and thus faithfully represent the majority of their constituents' NIMBY ("not in my back yard") sentiments.

The preference for living out in the suburbs and for the levy of separate taxes can be regarded as a new American phenomenon. Around

42 million residents, mostly white, live in such autonomous communities and about 8 million live in electronically or physically gated communities (McKenzie, 1993). Each year, Americans spend about $65 billion for their private security (Donzinger, 1996). Inside cities such as Los Angeles, New York, Chicago, Bridgeport, Dayton, and elsewhere, there is no public access to the "defensible" spaces that are protected by new technologies (cameras, digicodes, gates, etc.). According to Jacobs (1961), based on her observations of Greenwich Village at the end of the 1950s, crime could be reduced if buildings were oriented to encourage surveillance by "the eyes of the streets," that is, residents, and if public and private spaces were clearly differentiated. Newman (1972) elaborated the concept of defensible spaces a few years later. He advocated the use of real or symbolic barriers to make neighborhoods manageable, to erect walls and gates to fight drug-dealing and crime, to film the precincts with video cameras in order to reassure the residents, and to have private police on duty 24 hours a day.

As in big Brazilian cities, such urban areas claim to have private police on duty 24 hours a day. The property value of such spaces is rising and more demands from diversified social categories by age, ethnicity, and income are registered every year. The privatization of safety – defensive communalism – and the freedom to carry weapons in the public space (taken advantage of by one-third of US citizens) distinguish the American landscape.

The role of the media and the "mean world" syndrome

The media's influence on the American public's fear of street crime cannot be underestimated. The huge difference between public perception and the reality of crime is likely the result of media reporting. In 1994, 73 percent of Americans perceived crime to be rising when in reality it had declined by 3 percent (*NYT*, 15 January 1994). According to surveys of victimization, 95 percent of Americans have never been seriously victimized, a proportion equal to that of the French. When they were victimized, the crime was, in general, committed by someone they knew. There is little doubt that America's information about crime comes from the media.

Stories are biased toward scapegoat offenders, as happened with the Bernard Goetz vigilante story in New York in 1981. Some narratives have been deliberately made up, such as the crime waves against the elderly in New York in 1976. They did not actually happen, but it was an election year and law-enforcement officials, who wanted more funding, helped concoct them. There was also false coverage of missing children and serial killers in the mid-1980s.

Table 2.3 Number of crimes per 100,000 people, 1984

Country	*Homicide*	*Rape*	*Robbery*	*Burglary*	*Auto theft*
Canada	2.7	–	92.8	1,420	304
France	4.8*	5.2	105.6	809	483
Germany	1.5	9.7	45.8	1,554	118
Japan	0.8	1.6	1.8	231	29
UK	1.1	2.7	44.6	1,708	460
USA	7.9	35.7	205.0	1,263	437

* includes attempted homicides. The rate of homicides is around 1.1
Source: Fairchild, 1993: 8

Crime dominates the network news in the USA, its coverage tripling since 1991 and surpassing the time devoted to economic and presidential news. This is not yet the case in Europe. An analysis of ten American network and cable channels in 1992 reported 1,846 incidents of violence in one day. The Annenberg School of Communication found that seven out of ten prime-time television shows in the last ten years depicted violence. Crime is entertainment, as Sissela Bok (1998) forcefully demonstrates. Some major programs present real-life crime cases and invite the public to participate. What is the impact of this approach on public perceptions at a time when a majority of Americans no longer lives in city centers and is cut off from everyday interaction with those outside of their own community?

No one denies that the fear of crime is legitimate – the US urban experience is in fact much more lethal than that of European countries (see table 2.3). And the discourse about dangerous classes has always pervaded the lives of American cities. To use journalists' rhetoric, at the peak in 1991, one murder occurred every 22 minutes, one rape every 5 minutes, and one armed assault every 30 seconds in the USA.[5]

Although crime rates have decreased in recent years, Americans continue to fear their cities. Few journalists take the time to explain the complexity of cases. Take, for instance, the victims of violence in US schools (314 young people murdered in 1995). While this figure seems very high, it should be correlated with the number of secondary public schools (20,000) and with the number of the 10–17 years old in public schools (38 million) and also, as a comparison, with the number of people over the age of 70 who were murdered (630).

What is the impact of violent images on perceptions? Experts are divided. George Gerbner, an expert on television violence, suggests that a simplistic coverage of crime leads to a "mean world" syndrome. As a result, heavy TV viewers feel that their lives are under siege, support the perception that no one can be trusted, and arm themselves (1994: 385–97). But French researcher Philippe Robert argued that, in France at least, the media heighten fears of crime only when the viewer or the reader easily identifies with the message delivered by the medium (Robert and Pottier, 1997: 623).[6] Much of crime-reporting is focussed on youth homicides, especially those committed by blacks. For Jewelle Taylor Gibbs, the representation of black males as dangerous categories does not depart from historical representations of the dangerous classes:

> Black males are portrayed by the mass media in a limited number of roles, most of them deviant, dangerous, and dysfunctional. This constant barrage of predominantly disturbing images inevitably contributes to the public's negative stereotypes of black men, particularly of those who are perceived as young, hostile and impulsive. (1988: 2–3)

The externalization of a category as if it were not part of the American society contributes to the aggravation of the problem related to the roots of crime.

Disinvested areas

According to a logic of disqualification of disinvested areas, once an area is crime-infested, it should be abandoned. Just as the Greeks burned their ships upon arriving in a new land, those who flee to other neighborhoods (suburban areas, for the most part, in the USA) sever their ties with the disinvested areas that they have left.

Racial poverty in urban America has been the subject of considerable recent controversy. Some experts have described the new formation of the poor as an "underclass," a population for whom the primary means of social organization in mainstream society are inaccessible (Auletta, 1982; W. J. Wilson, 1987, 1996; Gibbs, 1988). For Wilson, who no longer uses the term, two profound changes in American social and economic life have created this "underclass": first, the deindustrialization of American cities which placed the stability of the traditional working class out of reach for inner-city residents; second, the opening up of housing opportunities for working- and middle-class minorities outside of the inner cities, which led to a bifurcation by economic status of minority communities and the social isolation of the poor minority (W. J. Wilson, 1987; Duneier, 1992). The urban poor are thus defined by both a shortage of direct employment opportunity and an absence of normative structures.

This category is spatially isolated because of the mental fear it inspires. Zones of social distress[7] shelter only 2 percent of white Americans and 3 percent of Latinos (Massey, 1995). For American blacks, the situation has worsened. In 1970, 28 percent of blacks under the poverty line lived in such distressed zones; by 1990 the majority of them lived there. Without being disorganized, they had to arrange themselves to fit their circumstances, and to stretch their support networks. For instance, welfare mothers would have to share their checks with their boyfriends. Young males and their girlfriends would double up in public housing, living in their mothers' apartments. Although this behavior made sense in the social "no man's land" where it evolved, their survival strategies appear repulsive to the majority of Americans who approve of the "one strike and you're out" enforcement of regulations in public housing.[8]

Without resources, the growing inability of inner-city communities to sustain their most marginal members (young men) is manifest in the growth of the now all-too-familiar phenomenon of homelessness and of the extremely poor (those with incomes at or below one-half the poverty level) who still benefit from temporary support systems. The capacity of these support systems is not endless. One crucial parameter is the flow of income into the community, primarily through welfare. Cutbacks in welfare since the early 1970s were a major reason for increased homelessness in the 1980s. Decline in welfare income directly stunts the capacity of the poorest to absorb and sustain those without resources, such as those just released from jail.

Conservative rhetoric has ignored the link between disinvestment, deindustrialization, racial segregation in the inner city, and their potential criminal consequences. Rather, the right wing has concentrated on behavioral features – family disintegration, high-school drop-outs, teenage pregnancies, and crime – which set the urban underclass apart as a category. It has linked "the tangle of urban pathologies" to liberal government policies which, according to Murray (1984), have subsidized the lifestyles of those who are not attached to the labor market. Stemming from this rhetoric, the 1994 bill suppressing "welfare as we know it" had its ideological basis in the perceived link between idleness and crime and the desire to restore personal responsibility.

The anxious middle classes

The difficulties experienced by the poor in large American cities are indirectly related to the challenges of economic globalization and deregulation, to the ferocious competition to which the middle class is subjected, and to the middle class's withdrawal from public space. The rules

Table 2.4 Efficiency of revenue transfers in different countries

	% of poor households before transfers	*% of poor households after transfers*
Canada	24.9	12.5
France	36.4	7.9
Germany	31.0	6.8
Sweden	36.5	5.6
UK	30.0	8.2
USA	27.1	17.0

Source: Mitchell, 1991: 47

of the new financial game strongly reward those who are well-positioned to meet the global challenges. Federal and state governments acknowledge an obligation to ensure the security and well-being of those citizens with economic and social capital. But, in contrast, the disadvantaged have been ignored by policies of redistribution, except under the New Deal and during the 1960s. Rather, the underprivileged have generally been perceived by law-makers as a distinct and separate group requiring careful control.

More than any other Western country, the USA has seen the greatest increase in inequalities since the beginning of the 1990s (see table 2.4). This phenomenon is characterized at the extreme ends of the economic scale: a growth of wealth and a growth of poverty. Among the evidence, during that period:

- 40 percent of the profits made on the stock markets were captured by 9 percent of the players.
- The wealth of the richest 1 percent doubled and their income increased by 91 percent, while the income of the poorest 1 percent diminished by 21 percent.
- A quarter of American workers have an income inferior to those of their German, Swedish, Belgian, and Swiss counterparts (see table 2.4).
- Out of 119 million jobs, 22 million are held by working poor, with an over-representation of minorities (Bok, 1996: 90).

Because of the brutality of the American capitalist system, "anxious" middle classes do not believe that the state should redistribute their hard-earned taxes to ghetto residents who are stereotyped as not sharing the same work ethics. Two opposed types of analysis are usually applied to these issues. On the one hand, a French approach held in the 1990s both by government officials and neo-marxist researchers sees poor communities and their residents as "the structural victims" of deindustrialization, of public and private disinvestment, and of institutional racism. The other approach puts the emphasis instead on individual opportunities, on the incapacity of some households to lift themselves out of poverty at a time of economic prosperity, and on the threats that those people represent to the dynamics of society. Aggressive incarceration policies are a response to such threats.

US domestic policy in the 1980s and 1990s has been based on this second approach. While Robert Reich, the former US Labor Secretary, demonstrated that this short-sighted view was disastrous for the civic values of the nation, he was ignored and the structural interpretations of crime presented as "liberal" are constantly marginalized by current political discourse. Street crime has become a prominent political issue in presidential, congressional, and gubernatorial politics.

The Politicization of Street Crime

The interference of the federal state in the issue of crime

Unlike many European countries, there is no cabinet official in charge of national crime policy in the USA. Each of the fifty state legislatures determines its own crime policy separately. Some states refuse to send a person to jail for possessing a small quantity of drugs, while others impose a strict sentence. Some states choose to have the death penalty. Some have mandatory minimum sentences for drug crimes, while others allow judges discretion to impose sentences. Some judge juveniles as adults, others have no age limit, and others consider them as juveniles until they are 18. "Get-tough" politics on crime and drugs have animated American social policy for at least a generation (Gordon, 1994: ix). The national level has an impact on cities, when it launches a major war on drugs supported by a very broad consensus and when it withholds money from the states unless they accept certain "get-tough" policies favored by the federal Congress. This is the "top-down" aspect of the system. "Truth in sentencing" – that is, set sentences for particular categories of offense – can be seen as a significant federalization of a traditionally local issue.

The early politicization of law and order

Criminologists have traced the contemporary politicization of crime back to Barry Goldwater, a Republican presidential candidate in 1964. Language is an exercise in policy-making power, legitimizing the state's definitions of morality and the proper measures to be taken to defend it

(Gordon, 1994: 11). Goldwater is the one who fused together negative perceptions about the civil rights movement and the liberal welfare state and the real or perceived loss of personal security. He also gave Republicans the key to a strategy that made "crime in the streets" a central issue in 1964:[9] "Many of our citizens – citizens of all races – accept as normal the use of riots, crime, demonstrations, boycotts, violence, pressures, civil disorder, and disobedience as an approach to serious national problems", he said (quoted in Flamm, 1996: 5). Governmental paternalism runs counter to individual responsibility. Linking all kinds of lawful and unlawful disorders, actual crime, and irrational fear to progressive politics, in a famous discourse, he decried the Democrats' laxness and the pathetic situation of cities:

> Tonight there is violence in our streets, corruption in our highest offices, aimlessness among our youth, anxiety among our elderly . . . the growing menace in our country tonight, to personal safety, to life, to limb and property, in homes, in churches, on the playgrounds and places of business, particularly, in our great cities, is the mounting concern of every thoughtful citizen in the United States. (quoted in Feeley and Sarat, 1980: 35)[10]

In 1961, John F. Kennedy formed a Presidential Commission on Youth Crime and declared "a war on delinquency" in response to looming urban and racial crises (Cloward and Ohlin, 1960).[11] With the threat of racial riots increasing, it was surprising, Flamm (1996) remarks, that President L. B. Johnson waited until late October 1964 to link the issues of poverty reduction and crime control. "The war on poverty . . . is a war against crime and a war against disorder," Johnson declared (quoted in Flamm, 1996: 29). In 1965, he said: "We must arrest and reverse the trend towards lawlessness," "crime has become a malignant enemy in America's midst," "crime scratches the face of America. It appears as a threat in our streets . . . it corrupts our youth . . . we must pledge not only to reduce it but to banish it." At the time, a fight against the root causes of crime was on the Democrats' agenda. Johnson pledged that his anti-poverty legislation would take one million young people off the streets and prepare them for productive careers.

But the next four years would see the administration consumed by the fight against street crime and racial disorder (not to mention the cost of the Vietnam War). By 1968, Richard Nixon would be able to launch a populist call: "Our judges have gone too far in weakening the peace forces as against criminal forces. Our opinion-makers have gone too far in promoting the doctrine that when a law is broken, society – not the criminal – is to blame" (quoted in Scheingold, 1984: 78). He would be followed

Table 2.5 Respondents reporting fear of walking alone at night, within a mile of home, 1965–1982

Year	*% reporting fear*
1968	35
1976	45
1982	48

Source: Scheingold, 1984: 39

along these lines by President Ford: "How much longer shall we abdicate law and order . . . in favor of a soft social theory, according to which the guy who throws a brick across your windshield is simply the misunderstood and unprivileged product of a broken home?" (quoted in Sunquist, 1968: 286). Thus, in the debate over law and order, conservatives had inverted the arguments and appropriated the language of liberals, claiming to free "the forgotten American" from urban violence and developing federal programs and dispensing money for that purpose. In the process, they had reshaped national politics (Edsall and Edsall, 1991).

There was, of course, a foundation to these symbolic politics. In 1969, the National Commission on the Causes and Prevention of Violence launched by President Johnson was to conclude its investigation on a somber note: "Among all the democratic stable nations, the U.S. is visibly ahead per capita in terms of homicides, assaults, rapes and burglaries" (1969: xv). Another commission on victimization, the Kazenbach Commission, was to reach the same conclusions. And the fact is Americans felt more and more insecure. People were generally afraid to walk in their neighborhoods at night (see table 2.5) – women more so than men.[12]

The evidence suggests that levels of fear of street crime were high and rose significantly between 1965 and 1982, a time when affluent households fled the cities for the suburbs in great numbers. But, as Scheingold aptly remarks, "substantial fear of crime may be something that Americans have lived with, off and on since Frontier days. Without pre-1965 data, we do not know much" (1984: 40). In the meanwhile, according to the Uniform Crime Reports of the FBI between 1960 and 1970, violent crime increased by 126 percent. Of course, crime statistics are notoriously unreliable (Burnham, 1996: ch. 4), since they rely on voluntary reporting by the police departments, many of whom under- or over-report crime. But these statistics, commented heavily upon by the media, contribute to the fear of street crime (see table 2.6). Figure 2.1 shows how the 1960s

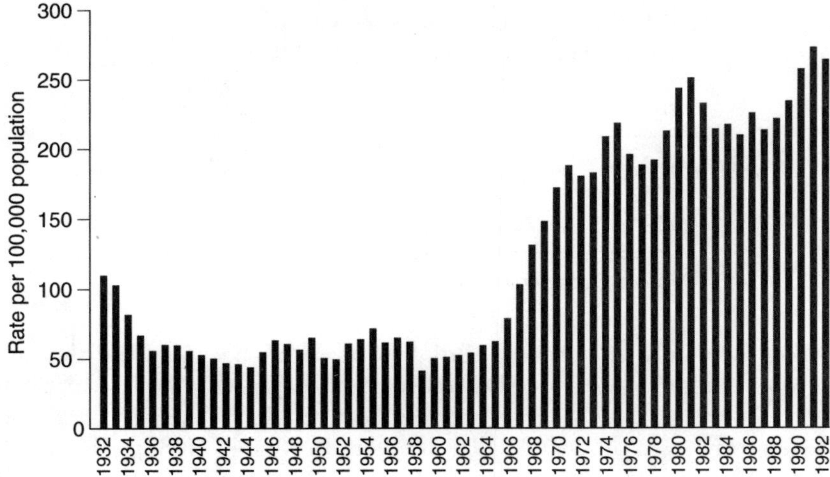

Figure 2.1 FBI's Uniform Crime Reports. Robbery complaints to police, 1932–1992

Source: Burnham, 1996: 115

Table 2.6 Crime rates per 100,000 US residents, 1969–1981

Year	*Crime rate*
1969	328.7
1975	481.5
1981	576.9

Source: FBI, Uniform Crime Reports

marked a broad increase in registered robbery as reported by the FBI over a stretch of years.

The legislative response to the FBI's reported rise in crime in the 1960s was swift. As early as 1965, spurred by President Johnson, Congress passed an anti-crime law. The Law Enforcement and Assistance Act (LEAA) was meant to offer states and localities financial and technical support. Billions of dollars were spent on such programs: from 1965 to 1992, federal expenditures for justice programs rose from $535 million to about $12 billion.

The 1968 Crime Control and Safe Streets Act was the most expensive anti-crime legislation in history to date. Richard Nixon gave voice to the anxieties of "the forgotten American" and fashioned those anxieties into a get-tough program. The law provided for emergency wire-tapping, tightened controls over interstate firearms transfers, and allocated hundreds of millions of dollars to localities to upgrade their law-enforcement capability (Gordon, 1991). Out-of-favor liberals had little to offer as a counterpart. Nixon's first budget in 1970 appropriated $300 million, but the following year it tripled to $900 milllion when the President made the fight on crime a national priority and reinforced police and justice resources. Thousands of drug rehabilitation programs were also launched then. We see how decisive that decade was in terms of perceptions and federal policies. Before examining the evolution of national representations and federal policies, let us briefly retrace the three historical stages characterizing the American way of punishment and prevention.

The historical evolution of punishment

Despite major transformations, continuities appear to feature more than discontinuities in the history of punishment in the United States. Conceptual analogies of the past and the present are perceptible. The ideology of the past contains a hidden agenda for the present, as ideas take successive forms and institutions adapt in the light of changing sensibilities and socio-economic circumstances. Here, I follow Simon's (1993) major contribution.

The disciplinary model Simon argues that between 1850 and 1950 industrialization used the discipline of the labor market as an anchor for control. The rise in the prison population is closely tied to apprehensions among elite groups in societies experiencing the stirring of capitalist economic relations in the late eighteenth and early nineteenth centuries. The geographic mobility of uprooted rural inhabitants and the flux of immigrants attracted by urban industrial jobs created tremendous fears of crime and disorder, while undermining traditional modes of punishment. I have already alluded in Chapter 1 to such changes taking place in New York City and Paris. The prison, with its emphasis on confinement, separation, and discipline, offered an appealing lever in that period of deep economic change. New programs were implemented to bring discipline and order to bear on dangerous classes (Simon, 1993: 4–5).

The emergence of the modern prison in the eighteenth century can be interpreted – in the aftermath of Elias's thesis (1965) and the work of Spirenburg (1991) – as an attempt to discourage communities from carrying out self-policing and their own forms of justice, and as heralding the emergence of state institutions as guardians of morals in society. The modern prison also expresses both a more autonomous view of the individual prisoner, who may be "normalized" after his time of confinement (the Philadelphia approach, focussing on isolation), and as in Auburn, NY, an emphasis on coercion and the discipline of the industrial state. It is the latter which prevailed.

Since Max Weber first formulated the idea in 1904–5, it has been demonstrated that the Protestant ethic is connected to capitalism and that industriousness is linked with the election of the happy few. Conversely, idleness and crime are seen as related evils which could be checked by forced labor.

The penitentiary appealed to white Anglo-Saxon Protestant elites as a technology which could solve many of their anxieties about the collapse of colonial social institutions and the emerging urban industrial society (Simon, 1993: 34). It would reinforce social control at a time of change. At the beginning, private contractors imposed labor on penitentiary inmates, then the states took over that function. When prisons were overcrowded, parole became a technique with which to reconnect offenders with the discipline of industrial labor that was on offer in the outside community.

The Great Depression of the 1930s undermined the foundations of a regime tied to employment. It became increasingly difficult to obtain jobs for inmates at the penitentiaries and for parolees in the communities. The latter would not be released because there was no opportunity "to go straight" via jobs. World War II started the job machine rolling again.

The therapeutic model A new model was created in the 1950s thanks to an influential book written by Maxwell Jones (1953) which explained how the norms shared by the mainstream could be supported by the inmates' individualized treatment. Research programs developed in the correction system at a time when drug-addiction had become a public concern. For instance, studying California, Simon points out that "the fury over narcotics in the 1950s provided a bridge between politicians anxious to cultivate and service moral panics and the California correctional management's drive to intensify its professionalism and reputation as a leading innovator in rehabilitative methods" (1993: 85). Instead of seeing delinquents as idlers meant to be incapacitated through extended sentences, the clinical approach would emphasize medical treatment and control programs. It would focus less on disciplining the offenders than on adjusting them to working life, to getting along, and feeling positive. Group meetings would be organized to discuss problems of parolees in a community context. This system did not replace the other. New agencies were simply added on top of old ones. Prison surveillance was expanded via new techniques.

The therapeutic approach was soon discredited for many reasons. It left a lot to the discretion of the people working with the inmates, it required an enormous amount of paperwork, and it met the resistance of an inmate culture which refused to be seen as needing psychological help. The tools that had been efficient in the 1920s, as analyzed by Park et al. (1925), to reintegrate offenders in the community – the settlement houses and social work – were no longer relevant, as problems worsened in the inner city. The massive migration of blacks from the south disrupted the neighborhoods where they concentrated for lack of choice, while remaining at the same time largely "invisible" to the rest of the city. According to 1950 census data, the unemployment rate of African-American men was 190 percent greater than that of whites; it peaked at 225 percent in 1980. For Simon,

> it is likely that correctional officers were some of the first social managers to become aware of the degree to which America was separating into two societies – the mainstream, the highly stratified work force which, despite its heterogeneity of privileges, would share in the general increase in American wealth through collective bargaining agreements, social insurance, credit, pensions, and an underclass locked out of the economic expansion of the postwar era altogether. (1993: 100)

Another explanation given to the lack of legitimacy of the clinical approach came from the crime rates. As they kept increasing in inner cities (the same story will be told in the next chapter, but in France), public opinion suffered from compassion fatigue. An influential article – titled "What works?" – questioned the whole approach and concluded that in fact nothing worked, a point of view which had a colossal impact on penology as well as on the criminal justice policies (Martison, 1974). The experts' discourse had criticized the efficiency of the system in combatting the fear of crime so much that people had come to regard it as dysfunctional. Very soon, funds for rehabilitation programs were cut and the managerial model which prevails today came into force and added one more stratum to the already hypertrophied system of crime control (Austin and Krisberg, 1981; Cohen, 1985).

It is interesting to observe that, instead of being personified by a blindfolded woman, holding a scale in her hand, Justice came more and more to be symbolized by a funnel. In 1976 another criminologist, von Hirsh, claimed that punishment depended on the type and seriousness of the crime rather than on the delinquent's capacity to be reincorporated in the mainstream. The "just deserts" approach meant that the inmate had to do all his or her time, and it is in that context that the LEAA was dismantled. From the 1970s onward, the collapse of the inner-city economy and increased legal and political demands for accountability produced a new construct, the managerial model with performance parameters displacing the goal of normalization (Simon, 1993: 9). The means of control in this narrative are as revealing as their ends and intentions.

The managerial model of surveillance Professionalization and accountability are the imperatives of the current punitive system, "based on scientific technical knowledge and instrumental rationality in decision-making". It involves highly systematized and codified forms of knowledge ("science") and their systematic application in terms of technology, social engineering, information-processing, decision-making, and work procedures. "Pressures for efficiency, productivity, and rational control emanate not only from managers, experts and technocrats, but also from cost-conscious and vote-conscious politicians as well as from counter-cultural, ecology-conscious or populist critics of bigness, waste, formalism and irrationality" (Heydebrand, 1979: 32, 52). With the management of cases, the assessment and reassessment of risks and needs, the development and updating of action plans, the securing of controls, and the provision of services, the individual offender becomes somewhat lost, pushed from view (Simon, 1993: 131). New technologies are instituted, such as classification based on statistics, offender-based computerized data systems, and drug-testing as a method of surveillance. The search for new grounds for credibility is generated by public anxiety about criminal violence and the lack of confidence in professionals. But as one older employee of the penal system sees it: "we don't supervise people any more, we are supervised by the computer . . . It tells us when to have reports done and it sets the priorities" (Simon, 1993: 174, 180).

Because the theme of crime repression is politically rewarding, all kinds of political actors attempt to benefit from it. But while attempts have been made to "federalize" the crime issue, conversely, states and cities have been eager to prove that they were already "tough" and did not need an incentive from Washington to institute repressive measures. The process of contagion has been multilateral. The recent policy of "three strikes and you're out"[13] was first tried by the state of Washington in December 1993 after a popular referendum showed 60 percent approval for it. The governor of California proposed the same law in his state, and it was approved by referendum in March 1994. Since then, twenty other states have adopted the three strikes laws. The same could be said of the cancellation of automatic release, the extension of technological surveillance, and the changes in parole policy: a linear approach (as a penal rhizome) prevails.

Sometimes, measures have been passed in a few states only. In Texas, for example, as a result of National Rifle Association lobbying, concealed weapons are legal despite police unions' disapproval. Proposition 187 in California and restrictive measures against undocumented immigrants have been limited to that state. The death penalty for juveniles is still restricted to a few states. In other words, multilateral experimentation characterizes the USA, in contrast with France.

The current politicization of law and order

If "politics" means "who gets what, when, how", as Lasswell (1950) argued, then the symbolic elements of policy-making on crime are "political." A particularly notorious attempt at the politicization of crime relates to George Bush's presidential campaign and its use of the so-called Willie Horton television advertisement. Many analysts are convinced that the strategy was sucessful in destroying Democratic candidate Michael Dukakis, then governor of Massachusetts, who was accused of being "soft" on crime (Ellsworth and Gross, 1994: 43).

In May 1988, during the presidential campaign, respondents to a national Gallup poll were asked which candidate "comes closer to your way of thinking" on the death penalty; 18 percent chose Dukakis, 21 percent chose Bush, and 61 percent said that there was no difference or they had no opinion. By October, however, 63 percent of another sample agreed that Dukakis "is too soft on law enforcement because he is against the death penalty and because he left a dangerous murderer [Willie Horton] go on furlough in Massachusetts, who then committed rape" (Harris poll, October 1988). Besides the fact that the very formulation of the question prompted support, what happened between May and October is that the Bush campaign strategists used the Willie Horton episode to "drum home the theme" that Dukakis was soft on crime. According to Ellsworth and Gross, the lesson of this campaign is that candidates for the US presidency must support the death penalty.

This lesson was not lost on Governor Clinton, who presided over two executions in Arkansas during his 1992 presidential campaign. There is also a racial element in this issue. Is it a coincidence that Willie Horton is black and that the Bush campaign "did everything humanly possible to make sure that every American voter got to see his picture"?

In their analysis of the 1988 campaign, Ellsworth and Gross (1994) also imply that incapacitation and retribution may not be the sole reasons behind support for "tough" law-enforcement. "What would people say if asked, 'Would you rather have Willie Horton sentenced for life to a pleasant but utterly secure prison in Tahiti, with no possibility of ever returning to society, or would you rather he was sentenced to a vicious overcrowded maximum security prison for 30 years?'" By asking this question, the researchers were trying to demonstrate that fear is not the public's only driving emotion. Retaliation and a desire to see the offender suffer may also be present. What politicians do when they raise the issue of crime is to arouse a satisfying sense of outrage and power in the voter (Ellsworth and Gross, 1994: 44). The current politicization of the crime issue is based on complex processes influenced by political and media intervention. According to Bartkowiak (1998), social disorganization and the rise in crime rates:

(1) attract the concern of politicians and the media;
(2) affect public opinion;
(3) affect perceptions of potential victimization;
(4) lead to an increase in the fear of crime;
(5) give legitimacy to repressive policies;
(6) encourage initiatives on repressive policies;
(7) and result in bills on repressive policies.

It is on point 5 that national policies differ between the USA and France, where there is a reluctance by politicians, except on the far right, to stigmatize offenders from inner cities and to advocate repressive policies about which their constituencies would be uneasy.

"The debate over crime and drugs has become less important for the possibilities of concrete public action than for its breadth as a contested terrain on which battles over power and principle rage," remarks Gordon (1994: 5). The search for federal control on the issue of crime has been demonstrated by the transformation of 3,000 crimes into "federal" offenses since 1994, revealing the fight that takes place between Republican and Democratic lobbies in Congress to get hold of the crime issue.

Connecting crime and welfare The discourse associating ghettos with crime involves the construction of a set of narratives which allow the kept, the keepers, and the public to believe in a capacity to control (Simon, 1993: 9). Words do not come from the skies, as Mao once asserted, and they cannot be taken as explanations of what happens. Yet, as ideological constructions, they are full of contradictions and paradoxes which may reveal a shadow agenda. James Allen Fox, a close advisor to President Clinton was among the first to designate street youths "superpredators", and this term has been used by Bob Dole and other politicians to conjure up images of an unholy and monstrous generation created by over-liberal policies.[14]

The American welfare state is segmented by race, gender, and class. Programs in support of the disadvantaged have been repeatedly attacked by conservative spokespeople. Hostile stereotypes diffused by the media damage welfare policies.

According to conservatives, social policies of redistribution cost the nation an astronomical sum, while the number of poor recipients (13 percent) remained the same. What is surprising is that this discourse has met with such success – again, another difference from France – and that Democrats were unable to produce a counter-discourse, or that when they did, they did not receive the attention that they deserved (Ellwood and Bane, 1985; Edsall and Edsall, 1991; Jencks, 1992). Between 1980 and 1993, it is remarkable to observe that while expenditure for prisons benefitted from a 4.6 percent increase ($31.9 billion), in the meantime the AFDC's (Aid to Families with Dependent Children) budget just doubled ($20.3 billion in 1993). The food stamps program followed a similar trend to the AFDC ($26.3 billion in 1993) (Committee of Ways and Means, 1997: 459, 861; Maguire and Pastore, 1997: 2). These budget allocations show clearly the reordering of priorities and prompted some of the Clinton assistant secretaries, such as Peter Edelman from the Department of Health and Human Services, to resign in protest at the 1996 welfare law. The Personal Responsibility and Work Opportunity Reconciliation Act of 1996 linked dependency to a negative street culture associated with crime. According to this logic, welfare must be terminated and replaced by workfare. But four years later, at the time of writing, the debate over the success of the welfare reform is still running. For Edelman (1999), persistent poverty is endemic in cities and rural areas and even in the suburbs. Welfare rolls are down 40 percent but the 6 million beneficiaries who left the rolls are, on balance, worse off, with less earnings than before. Around 30–50 percent of those who leave welfare do not find jobs, Edelman adds, and cannot get health coverage or food stamps (which have also been cut by a third). And all these problems hit minorities the hardest. The number of very poor people (those whose earnings are half the poverty line, that is, $6.750 for a family of three) has actually increased (to 14.6 million in 1997 from 13.9 million in 1995, with the poorest 10 percent of single mothers losing 15.2 percent of their income in those two years). The problem comes from poorly paid jobs, held by former welfare recipients, which cannot lift them out of poverty.

In the European view, the failure of the post-industrial economy to reach the inner city, where so many clients of the penal system are based, should have been compensated for by extended safety nets. On the contrary, under the Clinton Administration and with a Republican majority in Congress, affirmative action programs intended to accelerate the mobility of minorities were threatened and cuts in federal support to cities jeopardized public housing. Clinton defended a "get-tough" attitude for public housing tenants: "one strike and you're out." Welfare cutbacks made the environment less stable.

The politics of punishment With the rise of conservative forces in Congress and in the White House, calls to punish those perceived as destroying the American dream of prosperity and security took on new vigor. Congress, through the power of the purse, can put pressure on states to follow the policies it favors. The crime issue has thus been significantly federalized. Take "truth in sentencing" policies, under which inmates would be required to serve at least 85 percent of their time. Because federal prisoners – who already served at least 85 percent of their sentences – represented only 8 percent of the overall prison population, Congress wanted to extend such policy to states (dangerous convicts imprisoned by the average state served only 48 percent of their sentences). Incentive grants were the solution (Lin, 1998: 340, 342).

The most recent legislation, the 1994 Anti-Crime Bill, was designed to send a message to the public that crime is being taken care of both by Congress and the President. Endorsed by the Clinton Administration, the law has:

- extended the death penalty to cover 50 federal offenses;
- given cities money to hire 100,000 new police officers and develop community policing (COPS program);
- allocated $23 billion for law-enforcement, including $9.7 billion for prisons;
- directed another $6.1 billion for crime-prevention programs.[15]

It was in the interest of localities to get community policing grants from the federal government, since they lacked the tax base to fund the hiring of additional police officers. Over six years, $8.8 billion from the COPS program was to go directly to the local police departments without passing through the states' political apparatus. A similar phenomenon has been observed in France, as urban policies initiated by the state are frequently hastily approved at local level in order to get funding, which may then be diverted to other purposes. The national level's involvement with the local arena on the issue of violence prevention was accelerated by the Reagan years. But which level influences which?

At first sight, the balance weighs in favor of the federal level. Under the Clinton Administration, the rights of death-row inmates to appeal against their convictions in a federal court were limited. Like three-quarters of Americans, the President supported the death penalty. He has praised community curfews. In the name of fighting terrorism, he sponsored and signed into law a bill that limits the centuries-old writ of habeas corpus, which allows federal courts to review the constitutionality of criminal trials in state courts. The terrorism law has also expanded the power of the government to deport aliens and to extend surveillance systems. On some occasions, the Clinton Administration went against the current. For example, it confronted the National Rifle Association and its powerful allies in Congress to advocate gun control, supporting a waiting period before people would be allowed to a handgun – legislation known as the Brady Bill – as well as endorsing a ban on the sale of certain types of semi-automatic assault weapons. Of those surveyed in an opinion poll, 54 percent said that "the President made a real effort to reduce crime" (*NYT*, 1 August 1996). The President made a point of showing that, thanks to the Brady Bill, "60,000 felons, fugitives and stalkers did not get weapons."

But limits have been imposed on such interventions, coming mostly from the courts, the Republicans in Congress, and public opinion. Ruling on a legal challenge to the Brady Bill, the US Supreme Court, in *Printz* v. *United States* in 1997, prohibited the federal government from requiring local police and sheriffs to check the backgrounds of gun purchasers, demonstrating the power of states to counter the federal influence (*NYT*, 1 April 1996).[16] After President Clinton convinced Congress of the need to prohibit firearms near schools, the majority opinion of the Supreme Court stated, in the case *United States* v. *Lopez* in 1995, that the prohibition of a firearm in a school zone by a federal law was a trespass upon state authority (*NYT*, 28 March 1996; Lin, 1998). Even after the tragedy of Littleton, Colorado, where two students killed fourteen students and a teacher in spring 1999, Democrats as well as Republicans in Congress were reluctant to follow the President and pass effective gun-control legislation.

President Clinton also attempted to shift money to drug-treatment programs. Congress resisted Vice-President Al Gore's efforts to this end, despite the fact that one-half to two-thirds of those caught in the criminal justice system have drug problems. The President also supported preventative measures, after-school activities, and midnight basketball programs, but again the Republicans attacked those proposals, fearing that they would disproportionately benefit cities and Democratic constituencies. Congress also slashed the legal aid budget. For almost thirty years, class-action lawsuits funded by this budget had allowed lawyers to defend the indigent and minorities and their rights of redress.

In May 1997, Clinton supported a bill offering $1.5 billion to the states if they met new standards of harshness in the way they treated young offenders. The states would have to ensure that any juvenile charged with a violent crime was tried as an adult, unless the state attorney-general found that public safety would be "best served" by trial in a juvenile court. Those convicted would in some instances have to be confined with hardened adult inmates (Lewis, *NYT*, 19 May 1997). House Republicans turned down elements of the bill, including prevention measures and a proposal for safety locks on handguns. But most states have subsequently passed laws to the effect that violent juvenile offenders should be tried as adults. Whether they will reduce crime is an open question. A study by Jeff Fagan (1997) shows that young defendants tried in adult courts are more likely to be arrested again, and sooner rather than later.

Did the President make these decisions in response to a demand? According to Anthony Lewis, "Congress and President Clinton are driven by the same thing: politics. They are intruding the Federal government into an area that has been and should be left to local decision . . . For the sake of looking 'tough', they are putting an iron national grip on a diverse problem that needs diverse local solutions" (*NYT*, 19 May 1997). This federalization of a local issue goes hand in hand with politicization and racialization of crime.

Justifications for the extension of prisons

As in the nineteenth century,[17] penitentiaries have become machines to manage the human debris of economic restructuring, that is, the redundant or unqualified workers perceived as groups at risk (Jean, 1995). The justification presented for incarceration is the threat of lethal violence and the need to check it vigorously. In the early 1990s there were six times more burglaries in London – a city of 6.6 million – than in New York (7.3 million), but America remains exceptional for its homicide rates. In 1989, there were 11 times as many murders in New York as there were in London (Zimring and Hawkins, 1997), and in 1997, one American out of 10 carried a gun outside of the home. On any given day, one adult in 50 will carry a handgun (*The Economist*, 26 September 1998). There are approximately 9 homicides for every 100,000 people in the USA each year, and 1.2 in France. American elites justify their policies by comparing the risks. But more interesting is the comparative evolution of risk-management (as manifested in incarceration rates): a 148 percent increase in the USA, unmatched by any other developed country (see table 2.7).

If we look at the correlation between violent crime and incarceration, we find a number of interesting points. First, in 1992, the rate of serious violent crime was 16 percent below its peak in the mid-1970s.[18] The term "serious" has to be clarified. Few people are aware that only 1 percent of the victims of violent crime nationally require a hospital stay of one day or more (Donzinger, 1996: 3, 11). The media focus overtly on that 1 percent. But the mystery that makes the USA unique is the amount of

Table 2.7 Rates of incarceration in the G7 countries (1960–1990)

	1960	*1990*	*% Change*
USA	118	292	+148
UK	58	94	+62
France	59	82	+39
Germany	85	78	–8
Italy	79	57	–27
Japan	66	32	–51

Source: Zimring and Hawkins, 1997

Table 2.8 A profile of prisons in the USA, 1993–1994

Jurisdiction	*Cities, counties*	*States*	*Federal government*
	jails	prisons	prisons
Number of inmates	490,422	958,704	100,438
Characteristics	people awaiting trial or sentences under 1 year	sentenced 1 year or more	federal offenses
Non-violent inmates	66%	53%	89%
Total:	10 million admissions per year		

Source: Donzinger, 1996: 33

policy that also stems from that 1 percent. The policy-makers behave as if the country is under siege and do not actually work on the roots of violence.

Second, the USA builds more prisons to lock up more people than any other country in the world, except maybe for China. Between 1980 and 1999, the prison population more than tripled from 500,000 to 1.8 million (see table 2.8).

Third, correctional facilities are filled with non-violent offenders. Of the increase in federal prison admissions since 1980, 89 percent were accounted for by non-violent offenders. To fill the new jails, policy-makers have broadened the definition of crime to include conduct which is either merely undesirable or breaks administrative rules, such as panhandling.

The number of people under some form of correctional supervision surpassed 5 million at the end of 1994, or 2.7 percent of the adult population. Prison admissions data show the dynamic nature of the system, ever growing even though the crime rate is declining.

Finally, African-American men are arrested at six times the rate of white men (1,947 per 100,000) and incarcerated seven times more often. Are they six times more guilty or do they live in more criminogenic spots or are judicial districts six times more punitive? In 1930, 75 percent of those admitted to prisons were white and 22 percent were African-American. At the end of the century, though African-American males make up 7 percent of the population, they constitute half the prison and jail population.

Black victimization, on the other hand, is ignored by the system. From 1985 to 1991, the rate of crimes committed against African-Americans increased by 154 percent. One black out of twenty-two can expect to be murdered, according to Kennedy (1997). That is twice the fatality rate of American soldiers during World War II.

One of the main explanations for the over-incarceration of African-Americans, according to Tonry, is that "urban African-Americans have borne the brunt of the War on Drugs. They have been arrested, prosecuted, convicted, and imprisoned at increasing rates since the early 1980s and grossly out of proportion to their numbers in the general population or among drug users" (1995: 105).[19] At the height of the drug war in 1989, five times more African-Americans were arrested than whites (see figure 2.2), despite the fact that both populations were using drugs at the same rate (though not in the same manner – speaking very broadly, African-Americans would use crack in the streets and whites cocaine at home or in their office).

Some judges in New York and Maryland have refused to enforce the harsher penalties.[20] Other judges have resigned in protest. And a few politicians have attempted to eliminate the disparity of punishment again and again. But this law prevails at the federal level and in many states. No president or governor will risk looking lenient on such an issue. The racialization of the crime issue in the politicization of drugs cannot be minimized. The consequences are easy to anticipate: when half the young minority men in a city are "criminalized," their disenfranchisement toward institutions and mainstream society is not hard to imagine. Research shows that children whose parents have been imprisoned are more likely to be incarcerated themselves than children whose parents did not go to jail (Johnston, 1993). Between one and two million children in America have at least one parent in prison at any given time (Breen, 1994).

It is our task to understand why this harsh policy, which costs $100 billion a year in taxes (Donzinger, 1996: 85), is preferred to policies of

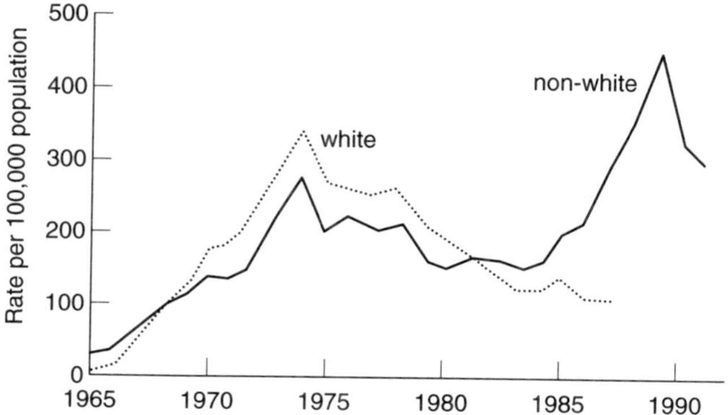

Figure 2.2 Arrest rate for drug offenses, juveniles, by race, 1965–1991
Source: Burnham, 1995: 199

prevention. But we can only present hypotheses here. A cynical approach, is to see in these high imprisonment figures the management of polarization. According to the *Wall Street Journal*, if male prisoners were counted as part of the workforce, the jobless rate would rise by 2 percent (29 August 1995). One cannot avoid associating the low percentage of unemployment (4.8 percent at the end of 1997) with the 2.7 percent of the active population (more than 6 million) either incarcerated (1.8 million) or under judicial control (5 million). But this policy of massive incarceration becomes an extremely costly burden for the states and the localities. "Corrections has become the Pentagon of state budgets, pushing other service priorities to the side and sending ostensibly conservative governments into a massive buildup of debt," as sociologist Jonathan Simon observed (1993: 2).

Electorally, crime pays and is an easy answer to social anxieties. Not only is it more difficult to appeal to solidarity when the majority of the social body feels unsafe, but the search for scapegoats becomes a temptation which few elected officials at the higher levels resist in such contexts. Paradoxically, the times when repression is the most narrowly associated with social control are those when it is the most difficult to logically account for continuities and changes. The current policies of repression targeting visible categories do not take into account the advice of many criminology experts, who argue that incarceration without rehabilitation does not solve the root problems of crime. Instead, new penological theories recommend risk control and situational prevention, carrying whatever measures seem most likely to restore social trust. Today, emotions seem to lead decisions.[21]

Institutions in charge of repression benefit from huge budgets (Gordon, 1994: 38–9). Not only did the annual amount spent on policing increase from $5 to $27 billion between 1980 and 1992, but the justice budget quadrupled, while the number of prosecutors grew by 230 percent. In order to get more prisons in states and cities, the Justice Department has been caught resorting to gross exaggerations (Zimring and Hawkins, 1988: 425–36).[22] The government and private security companies now spend almost as much on crime-control each year as the Pentagon spends on national defense.

Repression is a lucrative business for the private sector, as well as for interest groups such as the National Rifle Association, an organization of 2.8 million members with an annual budget of $140 million. Because American politics are decentralized and segmented, a national interest group of this kind can intervene powerfully in the electoral campaigns of dispersed candidates. Other private industries benefit from the construction of public and private prisons, the hiring of guards, the revenues in correctional management, cafeteria and geriatric services, and a host of high-tech devices, bullet-proof vests, computer technology, etc. sold to law-enforcement authorities. As a result, many small towns compete to convince state or federal officials that their location is the best site for a prison. In return, police and correctional associations are large contributors to "their" political candidates' campaigns. An iron triangle made up of the police and justice administrations, private industry, and political candidates capitalizes on the public's fears and frustration, campaigns on "get tough" platforms, and has frequently won in the recent past.

The American Public in Question

Is American public opinion more punitive than that of Europe? On what grounds should we, as Europeans, morally blame Americans and their way of justice?

The notion of public opinion must, of course, be deconstructed. Bourdieu remarked that public opinion does not exist (1984: 222). What we find are blocks of opinion, diversely intense, mobilized around individual interests and affiliations that are more or less punitive, based on factors such as age, gender, race, urban or non-urban, heavy TV-watcher or not, political attitudes, etc.[23] When it comes to the importance granted to polls, relativism should prevail. Polls reveal, indeed, how plural so-called public opinion can be. We find at least four paradoxes.

First, between 1982 and 1991, 83 percent of Americans believed that courts were not tough enough, revealing a large consensus in favor of repression (53 percent would send a young recidivist burglar to jail, compared to 37 percent of the British and 13 percent of the French (*The Economist*, 14 February 1998)). Yet 86 percent of Americans recognize that harshness does not work to reduce delinquency and that prison hardens offenders.

The second paradox is that 75 percent of Americans support the death penalty (compared with 48 percent of the French (*Le Monde*, 16 August 1999), yet the film *Dead Man Walking* (1996), a vibrant plea against capital punishment, was a huge success with American movie-goers (and readers of the book on which it was based). On the other hand, the question of the death penalty exposes the hardening of attitudes toward young offenders. Two million juveniles are arrested each year, 3,000 of them for homicides. Support for the execution of young people (teenagers) has risen dramatically since 1936 when the question was first asked.[24] How can the change be explained? Perhaps, thanks to films and media reports, the popular image of a violent killer is that of an adolescent, which diminishes the reluctance to execute young people (Ellsworth and Gross, 1994: 39). Yet, when a Gallup poll asked what could be done to reduce recidivism rates, only 31 percent of the respondents said "enforce capital punishment."

The third paradox is that while 36 percent defended stronger rehabilitation, 89 percent agreed that "increasing employment opportunities for youths could prevent a lot of serious crimes." Furthermore, 89 percent of those who want more incarceration agree that jobs for juveniles are the best policy against crime (Gaubatz, 1995: 6). They name drugs and single-parent families as the root causes of crime (in France, unemployment is most often cited). "To say that there is a public consensus about harshness does not mean that Americans across the board hold thoroughly harsh views. A majority holds views that contain both harsh and mild elements" (Gaubatz, 1995: 7).

And, finally, a fourth paradox is that three out of four Americans polled require longer sentences and additional penitentiaries. But at the same time they support education and training programs and, for them, the treatment of economic and social problems should be a priority.

Most polls ask simplistic questions that produce anticipated answers, which are then used for political platforms. But other polling reveals more complex public concerns about crime. When the issues are fairly presented, most Americans are willing to consider alternative approaches relying less on prisons and more on alternative solutions. In a national sample, 79 percent chose the statement: "in most cases, society would be better served if non-violent criminals were not jailed but were put to work and made to repay their victims" over the assumption "non-violent criminals must be kept in jail because allowing them out represents too great a risk for society" (Gaubatz, 1995: 7). It seems here that what a majority advocates is not formulated by law-makers. The same sort of non-decision-making is observed in France.

The negative consequences of economic liberalism, of assigned racial situations, of segregation, and of society as a whole's responsibility for the creation of so many drifters are almost never publicly debated.

Several hypotheses – none of them really convincing – attempt to explain the paradoxes. First, repression could be a symbolic response to larger individual insecurities linked to social change, family frustrations, and job precariousness. Second, as the conservative "moral majority" adjusts to liberal changes (abortion, homosexuality, affirmative action, freedom of speech, etc.), it may concentrate its resentment on "undeserving categories" for whom there is little empathy. Third, prison is prized for its functions of incapacitation (by 33–40 percent of those polled), of retribution to the victims and to society (19–29 percent), of education (15–26 percent), and of deterrence (8–20 percent). Fourth, there is also the feeling that nothing else has worked and that with time this policy could work. Fifth, there is a punitive trend that goes way back into history (Friedman, 1993).[25] Finally, studies reveal that public opinion is misinformed: while most people believe that street crime is increasing and is more violent, experts tend to think that the contrary is true.

To conclude this point, liberals think that opinion can be educated about the inefficiency and the harm caused by incarceration and the death penalty's ineffectiveness as a deterrent.[26] Enlightened minds probably underestimate the need and desire of law-abiding citizens to punish deviants. Punishment is essential in comforting society about its norms and in raising a collective conscience, as Durkheim suggested. The counterpart to the American culture of violence and to the freedom of criminal entrepreneurs in the no-go areas of a vast country is strong punishment. Yielding to the tyranny of majoritarianism, elites accept that violence "as a legitimate monopoly of the state" be exerted as a counterpart of freedom and at the expense of solicitude.

An interactive process

Stuart Scheingold warns us to be cautious in our judgment of politicians' behavior. He admits that

> national politicians . . . have strong incentives to politicize street crime. For them, it provides a unifying theme and thus a valence issue. While victimization is experienced differentially according to class, race, gender, and geography, the *threat* it poses to property and person evokes comparable fears throughout the society. National and gubernatorial political leaders can,

> therefore, deploy the fear of crime to unify the public against the criminal. (1991: 178)

Yet, he also says:

> it is an interactive process *combining elements of responsiveness with elements of manipulation.* Politicians do not so much "expropriate our consciousness" as take advantage of punitive predispositions about crime that are rooted in American culture. The public engages and disengages from the politicization process for reasons that have at least as much to do with the place of crime in the culture as with the impact of criminal victimization in our lives. (1984: 54)

In other words, politicization is a reciprocal process, with political leaders taking the initiative as much as responding to the public.

In an August 1994 interview with journalist Jeff Plungis, New York Governor George Pataki illustrated the whole process: "The governor reads the polls. He knows that his economic tax-and-spend policies are out of touch, so he tells [the people] he's going to do what they want. What he does and what he says are two different things. It's the same thing with welfare reforms. It's the same thing with criminal justice policy."

Law and order political candidates take advantage of opportunities to create genuine moral panic, as they target a racially stigmatized urban poor without any political clout and hardly able to get other social groups to identify and sympathize with them. But politicization has only an indirect and unpredictable impact on policy. For instance, while 62 percent of Americans in 1992 supported a stricter control of guns – firearms being the second most common cause of death for Americans between the ages of 10 and 24, and the leading cause for young blacks – and while four out of five in 1998 wished to make gun manufacturers liable for injuries linked to guns, only a few politicians took measures on this issue (the mayor of Chicago, among others: see Chapter 4). Power games begin long before demands surface and debates take place. Alternative sentencing is another example – it is in popular demand, but difficult to enact in the USA and France because it requires more work from judges, probation officers, educators, community leaders, and city bureaucracies, etc.

Finally, which groups support the policies of punishment? The objective distribution of risk does not provide any clue. Those who can reduce personal risks by privately investing in suburban homes, sophisticated security systems, private police, and so on, pay millions of dollars in taxes to alleviate their fears and frustration. Higher income Americans who are much less likely to be victimized than those on lower incomes are also the more vocal and expressive on the crime issue.[27]

A Trickle-Down Phenomenon

Some governors have used more or less deliberately wrong information to appeal to the public's demand for action. When "truth in sentencing" was passed in Virginia, the governor claimed that it was necessitated by the "rapid rise of violent crime" in the state and that it targeted "violent career criminals." Data show that violent crime fell in the two years preceding the introduction of the plan and that determinate sentencing would affect almost four times more non-violent than violent offenders. But the governor nevertheless went on with his discourse and claimed that "putting dangerous predators back on the streets" was a leading cause of victimization (Governor's Commission on Parole Abolition, August 1994).

Some elected officials in southern states have tried even harder to "get tough," reinstituting striped prison suits with the word "convict" printed on the back in Mississippi and prison chain gangs in Alabama in 1995. The state of Alabama, however, was required to halt this practice, after numerous safety hazards involved in the chaining together of inmates were exposed. Other examples can be drawn from the Texas, Florida, and Maryland gubernatorial campaigns. In Maryland in 1994, six out of seven candidates for governor promised to cut parole to keep thousands of people in prison, the very year when the legislature had already doubled the minimum time that a violent offender had to serve before being eligible for parole (Donzinger, 1996: 81).

At the beginning of this chapter, I asked whether the federal government has had an impact on the social control that cities choose to exercise. No one can deny that the six major crime bills signed into law by presidents since 1968 have either influenced subnational levels or been used by them to convince their electorate that the repressive trend was majoritarian.[28] At the federal level, the President is in charge of order all over the national territory. If necessary, he may send in federal police and the Marines in response to tensions in a state, as presidents did in the 1960s and, more recently, during the LA riots. But in contrast to the centralization of police and justice processes in France, movements of linear contagion also occur between states, in addition to the pyramidal bottom-up and top-down dynamics. There is a process of diffusion that may explain the success of ideas such as "three strikes" or drug-testing. It travels initially from state to state on a restricted scale. Then it is amplified federally and diffused down again to the remaining states. Unlike France, there is little counterweight to these ideas as they spread, sociologist J. Fagan remarks.[29] The intellectual forces to oppose them often develop long after their diffusion has occurred. The media play a prominent role in this process of diffusion.

What is disquieting is an evolution connecting segregative processes, the criminalization of "dangerous classes," restrictive tax policies, and managerial logics of containment. In a *Blade Runner*-like scenario, inner cities could, in the future, become low-intensity prisons.[30]

Surveillance in No Man's Land

According to the following scenario, the political system will continue to dwell on the "risks" emanating from distressed urban areas. A *fin de siècle* revanchism will appropriate deep-seated public insecurities (Smith, 1997) and policies of care will be transformed into policies of containment in the urban landscape. In a way, this scenario has already been turned into reality in some places. In New York City, for instance, after criminality had been spatialized, streets and public spaces were spectacularly reclaimed by the carte blanche and expanded powers that the police precinct commanders received from the mayor and by the withdrawal of certain legal and bureaucratic constraints, granted so that the police could reverse the decline in public order (see Chapter 4). As early as 1991, the city launched a campaign to remove the homeless from bridges, tunnels, train stations, parks, and vacant lots where they gathered. At a "confidential" meeting with newspaper editors, Mayor Giuliani explained that the shrinkage of the poor population, in general, was "not an unspoken part of our strategy . . . That is our strategy" (quoted in Barrett, 1995). Of course, some New Yorkers would admit the mayor was somewhat "dictatorial" and "unfair to the poor," but "what politician isn't?" and as a result, "the streets are safer" and "people want to live in New York City" (quoted in Smith, 1999).

The obsession with security transforms "cities of quartz" into fortresses and in the new militarized space the police are mandated to pursue criminalized categories (Davis, 1992). The affluent, reluctant to pay more taxes for punitive facilities that they do not use, "brazilianize" themselves by locking themselves into gated communities. They buy more and more sophisticated weapons, security devices, and services, and their police and justice systems are privatized (an already observable trend). As penitentiaries become too costly, the decarceration of many offenders and the growth of parole and probation populations represent a lower level of investment and of operating costs, thanks to the new technologies of surveillance by satellite.

Distressed neighborhoods, geographically sequestered, are then absorbed into general crime containment strategies. Gary Marx calls these large urban areas "the functional alternative to prison," which act as low-intensity prisons (1988: 221). In this scenario of a motionless city, silences and communal solitude exacerbate the neighborhoods' fragmentation. According to Cohen (1985), who was influenced by Foucault, we already see such operations at work within families. Foster homes, family placements, and substitute homes are required to act as surrogate parents. Families with one house member under arrest become "treatment resources" and sites for surveillance penetration. Schools supervised by the police weed out potential delinquents; some sign behavior-contract agreements with criminal justice agencies and become part of the machine, installing video surveillance, metal-detectors, hot lines to the police, and redesigning their space. They become "converted into closed-security prisons" (Cohen, 1985: 81). Police, prosecutors, and probation officers infiltrate the neighborhoods, requiring cooperative "intelligence" on the part of the residents, whose details are put on file, in addition, by welfare agencies. A safety net surrounds such neighborhoods, with helicopters and police squads specially trained by the army to intervene against this new "internal" enemy. Then, obviously, neighborhood-watch groups cooperate with the police to protect their streets, their spaces, and their buildings. Eventually, the surveillance of individual residents by others, the exchange of information with the police, and raids on apartments are encouraged by city hall and are meant to give residents a sense of "empowerment." Such neighborhoods are no longer communities sharing communal values with the mainstream.

Diana Gordon suggests that an electronic panopticon already infiltrates the inner city, beyond the prison walls. With the presence of probationers, drug-treatment centers, halfway houses, and other quasi-prisons in the inner city, the success of these new surveillance techniques may produce a lower-cost criminal justice system, she says, that will make an exclusion strategy for the poor more politically palatable (Gordon, 1991).

This second scenario differs from my earlier scenario, outlined at the end of Chapter 1, in that the world is still holist, based on Durkheimian assumptions. The law-abiders of the revanchist city "contain" the marginals via a sophisticated logic of management, but trespassers and law-abiders share the same institutions and references to justice. The geographical containment of this punitive city within a city allow American taxpayers to develop a greater indifference and short-sightedness to the causes of crime.

A third scenario will be discussed towards the end of Chapter 5, after the demonstration that cities have their own capacities of innovation and that local resistance to repression matters. It is now time to examine the national choices made in France to meet the challenges of urban disorders.

Notes

1 This form of violence was regarded by some as a means of access to the system which had remained deaf to the peaceful demands of the civil rights movement (Body-Gendrot et al., 1984: ch. 3).

2 By "actuarial justice," they refer to preventative strategies like selective incapacitation, criminal profiling, risk of custody scales, preventative intervention with "at-risk groups" and community-based initiatives. See also Crawford, 1997: 86.

3 I follow Massey (1995) on the issue of segregation.

4 The section 8 housing measures initiated by the federal government during the Great Society programs of the 1960s imply that when poor tenants moved to a better neighborhood, the government would pay the landlord the difference in rent between what he required and what the tenant could afford. This policy was re-enacted in 1995. By enabling tenants to vote with their feet and by giving them vouchers, the Department of Housing and Urban Development (HUD) was hoping to force local housing authorities to compete with private landlords for tenants (a housing version of school choice).

5 US Dept of Justice, 1992, FBI, Crime in the US 1991, Uniform Crime Reports.

6 A French official study of 1997 from the CSA observed a major difference on this topic: if French TV programs grant four times less space to violence than their American counterparts, they also distinguish the good guys who do not resort to violence and the bad guys who do – in the USA violence and weapons are frequently presented as a legitimate solution of self-help and freedom of expression is protected by the first amendment to the constitution.

7 Zones of social distress contain families with severe hardship: large numbers of single-parent families on welfare and of school drop-outs, for instance, whereas in zones of extreme poverty the focus is on income: 40% of the households in the census track are under the poverty line. See Kasarda, 1993.

8 President Clinton had this measure passed, according to which a tenant breaking the public housing rules will be expelled. Sometimes, mothers are expelled for problems raised by their older children.

9 A racist film called *Choice*, aired during the 1964 campaign and amalgamating black rioters and looters, a white victim beaten by a gang, and helpless policemen, anticipated George Bush's Willie Horton spot on the breakdown of law and order. Cf. Anderson, 1995.

10 Scheingold (1984) points to similar vibrant appeals to the restoration of law and order. For instance, President Eisenhower's contribution to Goldwater's campaign went in these terms: "Let us not be guilty of maudlin sympathy for the criminal who, roaming the streets with switchblade knife and illegal firearms seeking a helpless prey, suddenly becomes upon apprehension a poor, underprivileged person who counts upon the compassion of our society and the laxness or weakness of too many courts to forgive his offense" (quoted in Scheingold, 1984: 78).

11 In retrospect, the Kennedy Mobilization for Youth programs look very similar to those launched by the French social democrat government twenty years later, and they met with the same criticisms of ineffectiveness, corruption, and waste.

12 Although women were less frequently victimized, they were also less punitive. Young males, the most frequent crime victims, were not the most fearful nor the most punitive (Currie, 1985: 227).

13 Though they bear the same name, the laws are quite different from one state to another according to the definition of a "strike."

14 "What makes these youths predators and not just killers? Do they eat their victims? What solution can be imagined to 'predators' other than to shoot them or at least cage them forever?" asks Simon (1996: 9).

15 In 1996, only 19,000 more police officers were on the streets and they were spread throughout the country and not specifically in the crime-plagued areas.

16 At the time of writing, June 1999, immediate background checking by gun sellers has been made mandatory by Congress, but loopholes allow all kinds of buyers to purchase their guns in fairs.

17 The first federal penitentiary, Fort Leavenworth in Kansas, was not built until 1895. During the eighteenth and nineteenth centuries delinquents were locked up mostly in southern state prisons, and until 1930 there were only 5 federal penitentiaries. Less than 50 years later, the Federal Bureau of Prisons ran 44 jails. And, in 1994, that number had climbed to over to 100, with 80,358 inmates, 34% more than the space was designed for.

18 Most of the figures given below come from the two-year-long work of the National Criminal Justice Commission in a report published in 1996 under the title *The Real War on Crime*. A group of 34 citizens relied on the testimonies of police chiefs and prosecutors, defense attorneys and educators, doctors and other participants, and used official data to support their analysis. The existence of such a commission and the publishing of its critical report are only possible in a democracy where citizens expect to be informed about the functioning of their institutions and the use of their taxes. It is far less likely that such a process would have been initiated in developing countries where the democratic process is in transition. But the lack of success of the commission's report seems to indicate that the American public is not ready to hear disquieting facts about law and order.

19 In 1986, House Speaker Tip O'Neill introduced a bill that established stiff mandatory sentences for possession of crack cocaine. The bill made such sentences 100 times greater than those for powder cocaine, even though only the race of the people using them differed. There was virtually no debate in Congress and in public, though it could be anticipated that the bill would reshape the racial composition of federal prisons. African-American leaders held ambivalent positions about the crack-related sentences which caused such carnage in their communities, and they were hardly heard nationally.

20 According to recent research, the allegation that crack breeds wanton violence has never been substantiated (Reinarman et al., 1989). Violence comes from turf battles between police and crack dealers and among crack dealers

fighting to control lucrative markets. The physical effects of crack, medical research points out, are not 100 times stronger than those of powder cocaine, as was claimed, but probably just twice as strong.

21 Take, for instance, boot camps. They have been established in many states to give youth offenders shock treatment in rigorous discipline modelled directly on military training (Kane, 1989). The public is fond of them because they evoke order, physical fitness, punishment, and rigid discipline. Yet the recidivism rates suggest the technique is a failure.

22 For instance, the Justice Department claimed that the imprisonment of one inmate would save society $430,000 a year. This implied that each inmate would commit 187 street crimes per year at a cost of $2,300 per crime!

23 The analysis of "No answer" is important, as well as the intensity of divergent answers. In other words, polls give access to a limited and instantaneous knowledge, they must be carefully used and completed by in-depth studies.

24 Only 28% of the respondents were then in favor of the death penalty for those under 21 in 1936. In 1957, the proportion dropped to 11%, rose to 21% in 1965, and in 1989 reached over a majority (57%) for those aged 16 and 17.

25 While four Americans out of five support paying inmates for their work, they would also require them to return two-thirds of this amount to their victims or to the state for the cost of keeping them in prison. Public opinion on that issue was identical 13 years ago, but this is not similar in France, where prison is unpopular as a solution to crime (Flanagan and Longmire, 1996: 86).

26 That belief is supported in part by some experimental research, including the results of a weekend forum in Manchester, UK, which assembled a focus group of 350 people, representative of a national sample. Attitudes toward crime were tested before the experience, then for two days the participants were submitted to contradictory information from experts, politicians, prison directors, etc. Between each session, small groups in workshops would have the opportunity to exchange their opinions. After the experience, most views had entirely changed, one way or the other. A similar experiment was conducted in Delaware. Support for imprisonment among the participants declined dramatically. The number of people who wanted a prison sentence would drop from 80% to 38% in the case of a juvenile with no prior conviction committing an armed robbery. In Alabama, incarceration was favored in 18 out of 23 cases before the experience and in only 4 afterwards (Doble and Klein, 1989; Doble et al., 1991). This experience has its limits and dangers. What if the "experts" were biased and the opinion-makers fierce racists?

27 The chances of a white woman, aged 65 or older, becoming the victim of a serious violent crime (murder, rape, robbery, or assault) are one-sixtieth the odds of an African-American male teenager.

28 Prohibition in the 1920s considerably enlarged the police powers of the federal state in the fight against organized crime. While the severe repression

of alcohol consumption exemplified federal crime control of that time, today, federal law-enforcement is exemplified by the number of prisons being built with federal money.

29 Interview with the author, spring 1998.

30 This scenario is inspired by Simon's (1993) own scenario.

3
Solidarity and Social Prevention in France

This chapter intends to show that, for nearly two decades (1980–97), a gap has separated discourses and practices of solidarity towards problematic neighborhoods in France and the mounting urban unrest in such areas. Policies of social prevention developed throughout the 1980s and the beginning of the 1990s revealed their limits and yielded to concerns for public safety expressed by mayors and local constituencies. But, urged to take action, elites have been confronted by internal dissents as well as by dysfunctions of national institutions.

The National Context

From a cross-national point of view, the urban crisis in France seems less severe than the one experienced in the USA, for several reasons. The physical, social, and racial fragmentations isolating the suburbs from the core of the city are less extreme, qualitatively and quantitatively, and inner cities are not ghettos (Brun and Rhein, 1994). The French are less likely than Americans to move in order to follow jobs and, as a result, demographic stability is a positive factor ensuring social and cultural continuities in cities. Privatization, deregulation, and the denationalizing of space are also less intense trends than in the USA; there are fewer weapons circulating and families are less fragmented. Whether one agrees with it or not, one could add that the process of cultural and social assimilation that stimulates differences gives way to a more cohesive society, on the whole, which acknowledges its dissensus.

A powerful centralized administration still dreams of regulating 60 million clones. Agents of resistance to the market trends within the state apparatus can count on all kinds of regulation inherited from the past to block rapid economic turnovers. Expectations that they will do so remain strong in the country. In a quickly changing world and at a time of European challenges, a majority of the French still takes the intervention of the state in the solution of problems for granted. They rely on public services, never question their costs, and the image of civil servants and especially of mayors remains positive. In brief, the national culture does not encourage people to resort to self-help.

Yet, as in other Western countries and cities, destabilizing trends are observed, including the growth of inequalities, spatial polarization, the flow of newcomers, the long-term unemployment of unskilled workers, the concentration of socially handicapped families in public housing, and the failure of mass education to promote social mobility for lower classes. The diffusion of anxieties added to individualistic tendencies leads to a retreat from the public space (housing, jobs, schools, transportation) by the better-off. Fear of urban unrest appeared as the second largest concern of the French in November 1998. Three-quarters of those polled said that the situation is either extremely or "very" worrying. The mayors of large cities rank public safety immediately after unemployment as their major concern. National elites seem unable to appease public anxiety, based on rates for juvenile crime, which have increased by 50 percent since 1990 (juvenile delinquency represented 21 percent of all delinquency in 1998, as opposed to 12 percent in 1989) (Bousquet, 1998), and on urban violence, both of which are widely commented on by the media.

How is "urban violence" constructed? From the 1950s to the 1970s, large apartment complexes were built rapidly and cheaply in the *banlieues* to accommodate population growth and alleviate the pressure on city centers. Some 10 million housing units were thus built, most frequently – but not always – at the periphery of cities. Among other things, the problems with these urban spaces were environmental (the first oil crisis in 1973 prevented the development of adequate public transportation, of social amenities, and of commercial facilities), social (the arrival of lower-income categories, including working-class immigrants and their families, followed the exodus of former, upwardly mobile tenants), and political (a change in policies favored aid to ownership over the improvement of public-housing units). The crisis was exacerbated by rising rates of long-term unemployment and under-employment which effectively trapped people in these spaces as well as by a difficult mixing of cultures, and by the accelerated decay of the buildings (Body-Gendrot, 1993a). The widespread vandalism of buildings and of public amenities may be interpreted, on the one hand, as a protest by residents against the way the projects were designed for the poor, furthering their social stigmatization. On the other, it can be seen as their response to the state intervention that placed all public-housing residents in nearly identical mortar and concrete boxes, denying them the right to choose how they wanted to live (Bachmann and Le Guennec, 1996).

A new development refers to a delinquency of exclusion from youths frequently of immigrant background, but not always, who grew up in these anonymous areas, without reference points, in a normative blur, and with an identity disorder. In pre-industrial and industrial societies, individuals' identities were defined, sometimes negatively, according to age, gender, and class. A control – too much social control, as Foucault suggested – was exercised by institutions and by a community where everyone knew everyone. The major problem for the youths of today "is less one of social control than of social exclusion, less an issue of regulation than of how to learn to live without norms, less an appeal to freedom than to disenfranchisement, less a resistance to social pressure than an acceptance of a vacuum" (Garapon, 1996: 121). The mobility of positions and their precariousness generate a new freedom but they also bring about pain from a lack of barriers. "The price of modernity is the difficulty in being oneself" (Gauchet, 1985: 201). Delinquency and urban violence today reveal this identity disorder. If delinquency is usually linked to a denial of social order as-it-is and to socialization problems, violence is also a sort of language to be understood.

Juvenile delinquency is initiated by groups of adolescents attempting to forge collective identities and searching for boundaries in a world that has been deserted by adults. It may be an attempt to force others – the victim, the family, the social group – to react. Action then becomes the last resort for communication, a negative dialogue, a way to say that they are alone and lost (Body-Gendrot, 1993a). Their working-class fathers have often had little to transmit to them, while mothers generally occupied a subordinate position. Interviews reveal that immigrant fathers have been disrespected and humiliated throughout their work and residential experiences. They remained silent because they did not feel that they "belonged" to the receiving country. As a result their sons and daughters received from an early age a negative experience of socialization. "Acting out" can appear to observers as a social act by default; as if their *passage à l'acte* became a resource in their quest for constructing an identity. Confrontation with the police and the legal system may give some of these youths acting collectively an opportunity to derive some meaning for their action in a world that is largely indifferent or which attempts to de-dramatize rites of entry to adulthood.

Violence may then be interpreted as a way to construct new and more dangerous rites, or to return the violence that a family, and, more generally, a society, has inflicted on them and their families. The difficulty for social research comes from the clash between ethics of conviction and ethics of responsibility, between sociology and law. Ethics of conviction lead to see these youths as victims. Yet they are also offenders against those who threaten the self-contained world that they attempt to construct and, as such, they challenge ethics of responsibility. Moreover, an extreme fluidity in the youths' motivations and actions forbids any reification. Physicians note the growth in the number of minors either in the emergency room or in custody for violating public order. They are two sides of the same phenomenon: they are both actors and victims.

Some of these youths experience a frustration without object, which they turn against themselves or against weaker and marginalized elements perceived as outsiders. Most often, this violence is bundled up in the very exhibition of violence. It can be invisible, as it is with drugs and the drug-related underground economy, or hyper-visible in neighborhood supermarkets, in stadiums, in confrontations with institutions, the police, firemen, schools, and sports centers. Counting on TV crews to exhibit their actions, these youths act as if they only had violence to prove their existence.

Urban Violence in French Problematic Neighborhoods

What is behind urban violence? Are statistics a reliable indicator? Urban violence could be defined as weakly organized actions, from youths acting collectively against goods and persons linked to institutions in neighborhoods regarded as "sensitive" by authorities, but also against other young and older residents. This definition is biased, since it does not include other categories also resorting to violence in cities to express their discontent. But both the public and the authorities have chosen to dissociate the two sets of actors, even if the results of their actions and their motivations are similar, using violence as a prelude to negotiation.

Statistically, urban disorders are difficult to measure[1] and statistics must be interpreted cautiously since they emanate from police forces and point to the necessity of reinforcing public order. Other national surveys indicate that only 5 percent of the population has been seriously victimized (in 1997). The statistics of Commissaire L. Bui-Trong[2] at the head of the Department of Cities and Neighborhoods at the Central Service of General Intelligence (Renseignements généraux) constitute one indicator among others, the interest of which is to point out an evolution.[3] Every day, Bui-Trong collected all of the information emanating from field officers from all over the country. She describes urban violence as "collective, open and provoking; it is both destructive (arson of schools and gymnasiums, rodeos, excessive noise in public space), emotional (hostile gatherings, riots), expressive, sometimes playful, frequently criminal (assaults, rackets, mugging, thefts in stores, razzias), always juvenile" (Bui-Trong, 1993: 235–6). Among the characteristics of urban violence, as presented above, it is noticeable that, as in other countries, the offenders are younger and younger, live in tight circles, and goad each other into lethal "games."

Data from Renseignements généraux grade urban violence on a 1–8 scale (see table 3.1). In 1995, 684 neighborhoods were classified

Table 3.1 Indicators of violence in problematic neighborhoods, France

	Characteristics of violence from low to high intensity
1	Vandalism with no anti-institutional connotation raids in shops rodeo of stolen (and then burnt) cars villainous delinquency in gangs against individuals (racketing, spoils) fights, gangs setting scores
2	Collective provocation against vigils or guards verbal and gestured insults towards adults in the neighborhood, uniformed police officers, and teachers furtive anti-institutional vandalism (schools, post-offices, teachers' cars, public spaces)
3	Anti-institutional assaults: uniformed police officers (inspectors, firemen, soldiers, vigils) and teachers
4	Hostile crowds gathering during police intervention threats to policemen on the phone police patrol cars being stoned demonstrations in front of police stations drug-dealer hunting
5	Vindictive crowd gathering, hindering police intervention police station invasion diverse visible traffics (drugs, receiving)
6	Assaults of police officers open attack on police stations ambush of police officers, fender bangs
7	Open and massive vandalism: shop-window havoc, car-breaking Molotov cocktails rapid climax, intensive actions
8	Havoc and looting, assaults of individuals, fights with law and order forces guerrilla warfare, riots

Source: Bui-Trong, 1993

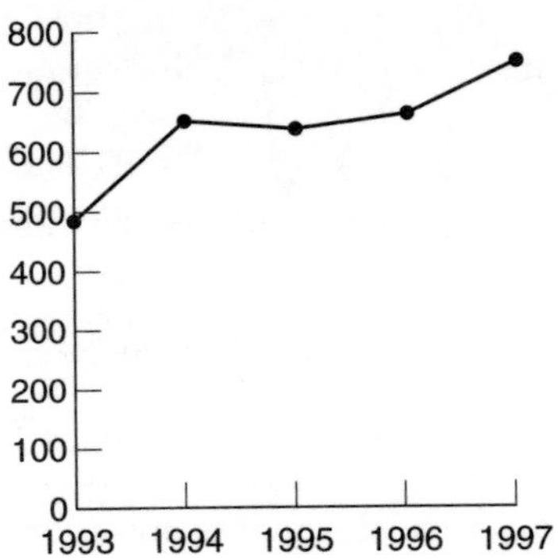

Figure 3.1 Evolution of urban violence in France, 1993–1997
Source: Bui-Trong, 1998

scale 1 (incivilities) and 18 as scale 8 (ambush of police forces, gun-related shots, riots). In 1999, more than 1,200 neighborhoods have been classified scale 1 and the number of those grouped in categories 7 and 8 has increased (ambushes are carefully prepared). In six years, the volume of urban violence has been multiplied by five (see figure 3.1). However, many urban areas in France – around 100,000 – do not experience disorders. The *départements* in which more than 1,000 violent events occurred in 1997 are Seine Maritime (Rouen) followed by Seine St-Denis near Paris, Rhône (the Lyons region), Essonne (near Paris), and Nord (the Lille region).

According to a second source of statistics, the central Direction of Public Safety, during 1997, in fifteen geographical departments, more than 2,000 policemen were injured following assaults, and could not return to work (an increase of 25 percent since 1996). Attacks against shopkeepers have increased by 14 percent. Confrontations with youths, with or without the actual use of weapons, were the cause of 61 deaths in 1998. Arson and vandalism registered an increase of 44 percent against utilities' infrastructures and post offices. Hundreds of attacks involving cars with reinforced bumpers have occurred (Body-Gendrot et al., 1998). Police statistics may be questioned, but their showings are disquieting. Moreover, they indicate a deterioration of collective life. As one prosecutor asks:

> What can three, five, ten beat policemen do in the midst of 35 high-rise projects, twelve thousand souls, 25 of whom are jobless? How do you erase the fear which has put iron curtains on day-care center windows and which forces the police to accompany the postman on his deliveries? (Davenas, 1998: 124)

National Urban Appeasement Policies

The national discourse of the struggle against exclusion

For a long time, elites have refused to associate disorders with urban dysfunctions in working-class and immigrant areas where a lot of public housing is concentrated. Gradually, the "urban question" imposed itself on the debate as a substitute for the "social question" of former days. It referred to the major changes experienced by the post-industrial society, but focussed more on problematic places than on disenfranchised subjects.

National elites, who frequently come from the same background and from the same educational molds, share a consensus: causes of urban disorders are structural and offenders are not to be blamed, since they have been victimized at birth. Inequalities and social "exclusion" are to be redressed through governmental action (the problem of institutional racism is usually sidestepped). Terms such as "sanctions" and "repression" are avoided. From the 1980s onwards, governmental elites have excelled at diffusing discourses of solidarity and they were given opportunities to do so by the launching and consolidation of a national *politique de la ville* (urban policy) during the 1980s and early 1990s.

In his famous speech in Bron, near Lyons, in December 1990, for instance, after some minor rioting had occurred in the inner cities, President F. Mitterrand warned that if solidarity was missing and if society did not care about inner cities,

> tomorrow there could be horrible tragedies of the type which occurred in this very region . . . It will be necessary to take from those who have much to give something to those who have nothing, lest the poorest localities flounder . . . I have learned all my life that whoever owns something hates giving it back. [We should not] fear the hostility of those who own.

Prime Minister Rocard then added that "the emergency was solidarity . . . To fight social segregation, we must build and organize this solidarity." He evoked the law, making housing mandatory for the poorest (the Besson Act of May 1990) and (in 1992) redistributing part of the taxes of the wealthiest 400 localities to the 400 poorest ones. We have here two examples of discourses based on historical traditions and values differentiating countries from one another.

Strangely enough, whether the governments were right or left wing, the discourse of solidarity and of social integration was unaltered, despite increases in unemployment rates and an explosion in juvenile crime rates. In 1993, Minister Simone Weil, who was responsible for urban policy and social affairs as part of a right-wing government, added the immigration question to the analysis and emphasized the necessity to fight segregation:

> Today, foreigners are concentrated in specific neighborhoods because other localities refuse them. It is a disastrous result, the opposite of what we look for: segregation instead of integration . . . Our very society is accountable. The social fabric is to be restructured . . . It is my role as Minister of Social Affairs to defend the rights of the most deprived to the right benefits which are not just "emergency" benefits. (*Le Monde*, 5 June 1993)

The welfare state's function of cementing society via an equal distribution of benefits was reaffirmed. President Jacques Chirac, the former head of the conservative party, RPR, also emphasized solidarity in October 1995:

> The origin of social fracture is unemployment. It is not the only one, but it is the major one . . . Inner cities are zones to which we must offer our support [*tendre la main*] . . . reinserting economic dynamism, work, jobs with free enterprise zones. Then the best public services. Then, by massively helping all those admirable community organizations, then public order will gradually be restored. (*Le Monde*, 28 October 1995)

Such discourses have been accompanied by symbolic acts of solidarity. In a spectacular gesture expressing the refusal of society to accept violence, for instance, Mitterand threw a rose in the Seine where skinheads had drowned a young Moroccan immigrant during a street demonstration. Then after an adolescent was stabbed in Marseilles by a youth of Algerian origin in 1996, Education Minister F. Bayrou required all high school students to stage a debate on violence on a certain day, at a certain hour, renewing the Third Republic's regalian and "virtuous" approach.

Urban politics of inclusion

Four chronological stages distinguish the evolution of politics of prevention known as "*la politique de la ville.*"

The "Bonnemaison" model of social prevention In 1981, the issue of social exclusion erupted in the political arena after youths from a housing project (Les Minguettes) on the outskirts of Lyons caused widespread disturbances. For several days, in their neighborhoods, these youths organized "rodeos" with stolen cars and burnt them in front of the TV cameras. The newly elected left-wing government had been used to urban struggles against renewal and removal during the previous decade. Some of these new representatives had participated in those struggles. Beginning in the 1970s, a succession of government working groups had addressed urban housing issues, after it had been revealed that large concentrations of people in public housing, including immigrant families, experienced a vast array of daily life problems. In 1977, an inter-ministerial committee for housing and social conditions, Habitat et vie sociale (HVS) was established, aimed at giving lower-middle-class residents who then lived in public-housing units more voice in the management of the neighborhoods in which they lived and at funding projects for housing refurbishment. But the disturbances at Les Minguettes were of another nature (Donzelot and Estèbe, 1994: 38). These youths were not political: they revolted without any specific claims, they expressed their *mal de vivre*. A series of governmental taskforces on social housing (Dubedout, 1983), on the prevention of delinquency (Bonnemaison, 1987), and on the incorporation of low-skilled youths in the labor market (Schwartz, B., 1982) appeared then as adequate structural preventative responses to an emergency situation. The taskforces were meant to introduce innovation into the handling of new urban and societal problems and to connect a variety of actors, specifically local elected officials, with the services of central ministries. The local sphere was to become the new terrain of political action, with the hyper-localization of physical, technical, and social problems analyzed in their complexity. Soon, the taskforce on the prevention of delinquency was supported by a commission of 800 mayors and revealed its ambition to bring more flexible, rapid, and adjusted responses to urban disturbances (Commission des maires, 1982).

This topic vividly illustrates some of the subsequent ambiguities of these urban policies based on decentralization, caught between a will of devolution and of modernization from the central state. Moreover, on the part of sectorial ministries (education, interior, justice, equipment, social affairs, health), there was a reluctance at granting too much leverage to local elected officials and losing control.

A national council to control juvenile delinquency and vandalism, Conseil national de prévention de la délinquance (CNPD) was chaired by Bonnemaison, the mayor of a Parisian suburb, with four mayors with different partisan labels and common goals as vice-presidents. At the departmental (CDPD) and local (CCPD) levels, councils were created, numbering as many as 400 in 1985, in order to establish partnerships among diversified public actors. These partnerships were meant to enable all parties dealing with delinquency to express their views and be given a hearing. Their conclusions were the result of a confrontation between contrasting interpretations and points of view. Rumors were taken into account as much as scientific surveys; elected officials intervened side by side with experts, citizens, and decision-makers. Their involvement was based on three conditions: equal status for all participants, respect for deontological rules and individual rights, and the definition of a process of agreement to fix goals and programs (Vourc'h and Marcus, 1994: 139). A more ambitious goal of reforming the macro by means of the micro, reforming the state and transforming the periphery, was also operating throughout the discussions.

The notion of contract prevailed, as French civil society aspired to emancipate itself from hierarchies and to prove it could offer its own perspective. Yet everywhere there was a search for confederates, rather than for a unified social body. There was an explosion of initiatives and a multiplication of actors connecting the central sphere and the regional, departmental, and local levels. The Commission nationale pour le développement social des quartiers (CNDSQ) was created to promote innovative regeneration schemes in eighteen pilot neighborhoods. That the urban policy was global and transversal meant that it was supposed to cut through the sectorial administrations, detour the inertia generated by routinized practices, and explicitly force them to work together in the problematic urban areas. Donzelot and Estèbe (1994) emphasize the lessons drawn from a 1986 survey of emergency (911) police calls as a successful example of such an approach. The survey showed that 60 percent of those calls did not refer to any criminal activity. Rather, they revealed the dysfunctions of other administrative services: people would call the police by default. Following the survey, the local prevention councils in a number of cities took the initiative to conduct a debate among the concerned services to help them respond to the citizens' needs.

Despite these initiatives, the dysfunctions of the urban space became more visible and more difficult to redress, especially when they concerned general problems related to poor urban planning. What had been in the 1950s neighborhoods, urban villages, places where people encountered each other, such as stores, cafés, or cinémas, had given way to areas of flux, connections, transit exchange places between functional spaces where people either worked, or relaxed, or slept. The unified space of the ancient city, which encouraged links between people, the celebration of common values, and the staging of events, was vanishing, replaced by a multiplicity of abstract nodal points. "The urban has killed the city," urbanist F. Choay dryly remarked (1988), and some would add that it did so under the influence of capitalism (Castells, 1972), of merchandizing (Lefèvre, 1974), of the state (Coing, 1966), of planners and technocrats such as Le Corbusier and/or of utopianism. An "anxious" city (*la ville inquiète*) was to be reassured by the political elites and regenerated with a globalized/localized structural approach.

At one point, in the early 1990s, it became widely acknowledged that the initiatives from the early 1980s had underestimated the severity of the urban crisis and of social disintegration. Minor tools had been used for a major trauma (Linhart, 1992). Elites had vainly waited for the return of a fully salaried society and for an economic boom to take care of social problems. But with unemployment continuing to rise, especially for the youths of immigrant origin, social relations becoming more conflictual in stressed areas, and urban disorders threatening the social peace of the whole community, more comprehensive economic, social, and physical strategies were put forward. The large question of social marginalization had become the new social question.

To fight the growth of inequalities, the government offered benefits for "insertion" (Revenu minimum d'insertion: RMI), a form of welfare for those who had exhausted all their benefits and unemployment allowances, requiring from them a form of workfare (Paugam, 1993). This approach was based on the belief, dating from the Revolution, that a society of equals could not tolerate poverty and should provide jobs to all to fight pauperization. If jobs could not be provided for everyone, then allowances, based on a national effort of solidarity, would take care of the deserving poor's subsistence (Procacci, 1993). Along with the RMI redistribution process, the central government mobilized further to address the deindustrialized territories in crisis. This time, the aid was to go beyond the problem-ridden neighborhoods, estimated to be on too small a geographic scale, and be more comprehensive.

The hypertrophy of prevention policies During the second stage of the *politique de la ville*, a Délégation interministérielle à la ville (DIV) was created in 1988 to provide a clearer administrative focus for urban policy. The DIV brought under its control the CNDSQ and the CNPD, and its budget rose to 3 billion francs by 1990. At that time, the local councils of crime prevention numbered 600, with 100 at the departmental level. It was thought that policies of urban regeneration and crime prevention were complementary and could be combined and coordinated to better address general problems of housing, education, and job qualification. Prevention was understood, from the start, as social action. The DSQ, the neighborhood regeneration programme, was expanded further into 450 well-funded state programmes, Développement social urbain (DSU), committed to urban renewal, training, and job placement, and community initiatives.[4] Special education programs, called ZEP, (Zones d'éducation prioritaire) were targetted at students from distressed urban zones and more flexibility and resources were granted to teachers and administrators. This approach represents a territorial affirmative action, based on equity and departing from universalistic principles of equality. It fails to address either the problems of ethnic discrimination preventing youths of immigrant origin from having the same structures of opportunities as others and generating urban disorders, or institutions' dysfunctions in the areas concerned.

The contract model for the city was the basis of this policy. Its purpose was to tackle specific social and economic problems in thirteen pilot areas (a number later to be expanded) through an agreed-upon inter-agency partnership designed to encourage coherence between the goals of various ministries and local authorities. The "urban solidarity" approach called for the participation of all sections of the community and areas of the city, collaborating beyond local boundaries in order to tackle metropolitan problems. This was the ambition, but unfortunately not the reality.

In 1990, a series of riots at Vaulx-en-Velin, near Les Minguettes in the region of Lyons, indeed showed the limits of such initiatives. Once again, sectorial administrations had not followed the global and transversal approach. Instead, they had pursued their own internal logic and the youths from the plagued neighborhoods, confronted with intractable problems, expressed their discontent.

On 4–5 December 1990, Président Mitterand, as described above, used these dramatic events to draw lessons from the failure of previous urban policies. As long as actions are redundant and efforts dispersed, "the activists of today will remain in a blind alley. They cannot on their own, gather, catch the threads and create a coherent work," he said. A new ministerial position, Ministre de la ville, was then created to coordinate the various ministries' approaches to urgent urban tasks, to provide a strategic thrust to the urban policy, and to ensure that the DIV fulfilled its missions in the problematic neighborhoods with the support of local elected officials. In addition, the financial contributions of the different ministries were consolidated in a single block budget for urban issues. The phrase "*État de solidarité*" refers to "the focus of administrative objectives and methods on the fight against exclusion" (Donzelot and Estèbe, 1994: 107). Actually, once the ministries lost control over their budgets for urban policy, they not only lost interest in the policy itself (which isolated the DIV), but also tried to regain at the local level – via the Préfet and the elected officials – the control they had lost at the top (Le Galès and Mawson, 1994: 29).

A moderate follow-up In the third phase of the *politique de la ville,* starting with the national electoral victory of a right of center administration in 1993, the preventative urban policy was conducted without enthusiasm. The DIV, no longer under the jurisdiction of the Prime Minister, was transferred to the Ministry of Social Affairs. Interdepartmental wrangling occurred over sources of funding, with the Ministry of Finance trying to scale down the experimental urban initiatives taking place in more than 800 neighborhoods. Some politicians thought that the decentralization policies had gone too far, while others wished to strengthen the leverage of mayors to "experiment," along the lines of the British model. The policy reflected a centrist approach and a desire to attract a maximum of voters. A budget of 9 billion francs over five years was allocated to urban programs in 1993.

A change occurred, starting in 1994, when a pact involving the relaunch of urban policy took form. It was active in three areas. First, a policy of free enterprise zones, inspired by the USA and the UK, was adopted but the drawing up of boundaries created problems of equity for mayors. This program, initiated by the Minister of the City, attempted to add an economic component to the traditional socio-urban perspective (Body-Gendrot, 1996). Second, some zones received city jobs that were created specifically for troubled youths by the Minister of Social Affairs.[5] Finally, contracts addressed the zones formerly targeted (Donzelot and Estèbe, 1994: 55). The success and failure of such processes depended partly on the human factor.

From the start, the role of the Préfet, the formal appointed representative of the state at the *département* level, was central to the negotiation that went on between the national and the local arenas. Traditionally, the Préfet coordinated the field services of the ministries, receiving his orders from Paris, expressing the local needs, and adjusting the resources to the local situation. Decentralization measures strengthened his role, particularly in relation to urban policies. The various ministries and the local players tested his negotiating skills. In the case of the *politique de la ville*, the Préfet had to establish priorities for funding and convince various mayors and administrations to share a common agenda. A few strong Préfets filled that role successfully, including the Préfet in the Lyons region, for instance. He was helped by the fact that the mayor of central Lyons was surrounded by powerful political partners from smaller communes at the periphery. This allowed him to exploit divisions between the local mayors, act as a mediator, and, in return, press for "solidarity" on various aspects of the *politique de la ville* (Le Galès and Mawson, 1994: 46). At the *département* level, the elected assembly's president performed the same conciliatory role within the CNPD. Thanks to the charisma of a few political leaders, the reality approached the ideal of the "elastic city" (Rusk, 1993), at least temporarily.[6]

In contrast to the British and American model, where cities bid competitively for assistance, distressed areas in France were selected and then state and locals negotiated their intervention (Tabard, 1993). However, in most cases, the cross-cutting networks of power and influence extending from the center to local areas made the process of negotiation extremely difficult and protracted (Le Galès and Mawson, 1994: 91). To avoid protest and because they lacked commitment, political support, or technical capacity, the Préfets frequently chose to redistribute the funding for urban problems to all elected local officials, regardless of whether they came from poorer or richer communities. Only after serious riots erupted did any significant innovations take place.

Urban policy in the fog Strangely enough, there was little activity related to the *politique de la ville* following the return of a left coalition to power in 1997. At the time, the DIV's chairperson wrote in *Le Monde* that he was at the head of a ship, with all its lights turned down, and in the fog. Would the left actually let down its natural supporters, i.e. the militants involved in fighting exclusion and building up neighborhoods?

Figure 3.2 shows the spectacular expansion of jobs and of accompanying bureaucracies that have been given to people administering urban and social problems. Would the "militants" in social prevention become "municipalized," losing their intermediary position between the central and the local spheres? Was the government's prioritization of job creation efforts going to solve the neighborhoods' problems and their crisis of identity? Eventually, during the spring of 1998, urban policies of the third type (elevated rhetoric and small budgets) were confirmed. In 1998 the DSU received 2 percent of the revenue spent at the national level. The Ministries' funding for the local needs of the *politique de la ville* came to 11 billion francs (Sueur, 1998). (It is worth remembering that the French spend more than 35 billion francs each year on their pets.)

New initiatives? After 15 years of urban policy, the diagnosis is somber. Though one could argue that without these efforts, things would have been much worse for cities, there were numerous failures. Rivalries between the ministries involved never abated, and there were also feuds with the civil servants in charge of urban policies. The routines of sectorial corporatism won out over globalized and transversal programs. For instance, with few exceptions, it remained difficult to get the police and justice services to work with social workers on specific cases. Moreover, the field services of central administrations and the mayors continued to connive to grab subsidies from one ministry after another. Elected local officials became accustomed to using the grants earmarked for the *politique de la ville*, not for innovation (too little time, they claim, too many zones getting specific grants), but to finance their routine operations,[7] a phenomenon observed in the USA at the time of the Model Cities, in 1966, after President L. B. Johnson launched his Great Society (a policy of distribution directed at cities). Then, the federal grants allotted to ghettoized neighborhoods were frequently used by mayors for middle-income areas, from where their votes came.

On the other hand, public partnerships obliged mayors to operate visibly with an objective of solidarity and equity, whereas, according to the logic of elections, each was supposed to be autonomous.[8] With a coordination of partners at the local level, the mayors' power of secretly negotiating with one actor after another was weakened.

In an 800-page report issued in the spring of 1998, former Minister J. P. Sueur described a strategy for addressing urban problems and recommended budgeting 35 billion francs a year over ten years. Titled *Demain la Ville,* the report defends the contracting mechanisms by which both the state and localities express their solidarity and share responsibilities for policy and programs. The principle behind the transfers of funds to the jobless (RMI) also justifies, it says, transfers to problem-ridden territories in order to bring back social mixing and public services and make life as liveable as elsewhere.

A number of experts involved in this report defend elections of representatives at the metropolitan level. And they support a levy of city taxes in order to fund "contracts" addressing the more general problems of public transportation, urban renewal, and urban landscapes, intended to reduce the stigmatization at the edges of cities. The advocated reforms would also reintroduce more equity in the distribution of public services. In areas in difficulty, there are on average relatively fewer police, welfare agencies, and mail deliverers than at the national level. Judges, overrepresented in rural areas, are urgently needed in the recently urbanized zones. Too much zoning for the allocation of funds and too many partnerships have become dysfunctional. But integrating the metropolitan space and sharing functions with less rural areas would raise enormous corporatist opposition, not least from elected officials from smaller localities who are overrepresented in the Senate. It was that sort of opposition that led the right to dismantle the 1992 *Loi d'orientation sur la ville* – whose redistribution strategies included forcing the 400 wealthiest localities to share part of their taxes with 400 poorer localities. In the USA, the blatant refusal of suburbs to bear their share of cities' problems represents a similar obstacle.

Once again, it should be noted that the reforms advocated to repair the urban fabric are designed at the top. They require experts' ambitious schemes and national mobilization initiated by the central state, in the same way as projects such as high-speed trains, the Ariane rocket, and the highway networks (Sueur, 1998: 47). At the local level, the mayor keeps his regalian privileges and participation is limited to the input of just a few private firms and residents' organizations.

Police and Justice within Urban Policies

When the left returned to power in 1981 and handed the Bonnemaison taskforce a mandate to think over the question of delinquency and of social

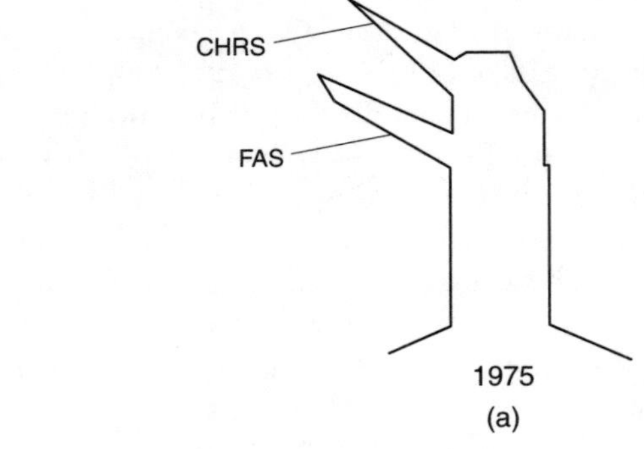

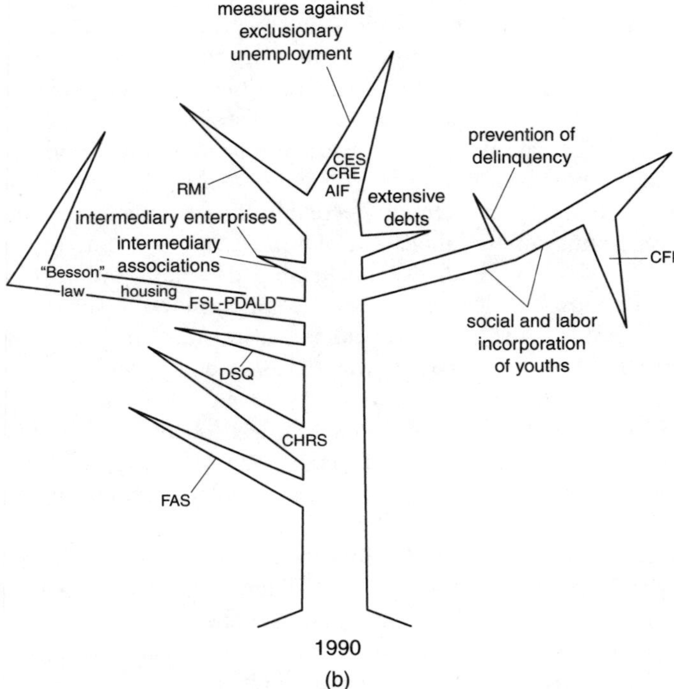

Figure 3.2 The growth of the tree of struggle against exclusion, France, 1975–1997
Source: Hardy, 1999

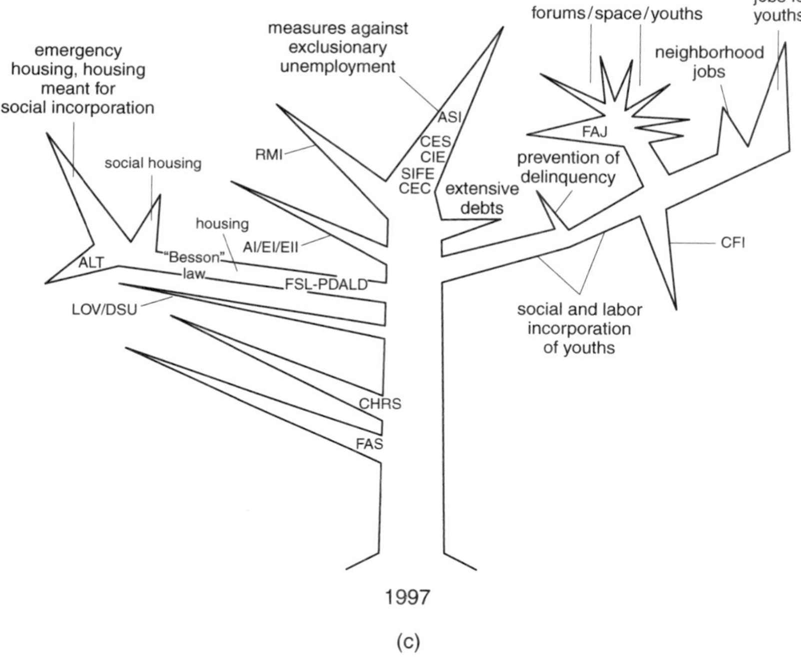

(c)

AI/EI/EII Association intermédiaire/entreprise insertion/entreprise interim d'insertion (individual support to job training)
AIF Actions formation insertion (job training)
ALT Allocation logement temporaire (temporary housing support)
ASI Appui social individualisé (individual social support)
CEC Contrat emploi consolidé (consolidated job contract)
CES Contrat emploi solidarité (job contract in a solidarity program)
CFI Crédit formation individualisée (individual job training)
CIE Contrat initiative emploi (job initiative contract)
CHRS Centre d'hébergement et de réadaptation sociale (shelter and social readjustment)
CRE Contrat retour emploi (back to work contract)
FAJ Fonds d'aide aux jeunes (youth action fund)
FAS Fonds d'action sociale (social action fund)
FSL Fonds solidarité logement (housing fund in a solitary program)
HVS Habitat et vie sociale (commission from the 1970s)
DSQ Développement social des quartiers (social support to problematic neighborhoods)
DSU Développement social urbain (urban aid program)
LOV Loi d'orientation sur la ville (1992 tax transfer from rich to poor localities among other measures)
PDALPD Plan départemental pour le logement des personnes défavorisées (regional program for the housing of disadvantaged persons)
RMI Revenu minimum d'insertion (minimum subsistence benefit)
SIFE Structure insertion formation emploi (job training program)

Figure 3.2 Continued

disintegration in poor urban areas, the page was not entirely blank. In 1977, the Peyrefitte Commission had clearly identified the law and order strategies necessary to restore order. While many of the Peyrefitte Commission's proposals, such as the refusal to distinguish preventative from repressive approaches, were to be adopted by the mayors working with Bonnemaison, this later commission was anxious to differentiate itself from the conservative positions always associated with security. It dismissed demands for law and order as alienating, and as attempts by the previous government to strengthen its regalian power. Claiming to present a pragmatic strategy for dealing with delinquency and to remove the negative influence of politics, the Bonnemaison Commission laid out policies focussing on local tools to fight local dysfunctions and involving the mayors in issues of safety (which do not strictly fall under their jurisdiction: the police and justice system are centralized in France, and the police became a tool of the central state after 1941[9]).

This implied that the national institutions of repression – the police and the justice systems – would not be involved in the policy of social prevention beyond simple representative functions. Moreover, their opponents were given full expression:

> In 1982 . . . a whole series of terms were progressively banished, as shameful, from the vocabulary of people who wanted to develop a positive approach towards problems of street crime. The result was that, only those who had a negative vision of words like sanction, prison, violence, delinquency . . . used them to reinforce their negative meaning. (Bonnemaison quoted in Dourlens and Vidal-Naquet, 1996: 13)

The very title of the 1982 report on safety from the mayors (Commission des maires sur la sécurité) was heavily debated and when the term repression was inserted, it was placed between prevention and solidarity: "The major role of the repressive apparatus is . . . to maintain the national consensus around a certain number of values on which the commonalities of life are based and to consolidate national solidarity." The majority of the proposals then concerned prevention and ignored repression. It was not seen as the object of a national or local debate (quoted in Dourlens and Vidal-Naquet, 1996: 14).

Most mayors at that time also expressed the greatest difficulties in imagining that repression could exist at the local level. For many militants on the left, the police were seen as "a necessary evil, obscure and disquieting" (Monjardet, 1988). Their report stated that:

> "The police has to remain under the authority of the state for public order and under the control of justice for judicial matters . . . Means must be found to manage conflicts between the strategies of the police and

> city hall, without having the mayor become the police chief or the police chief in charge of prevention." . . . If repression is to be used, it should target less *the delinquents, who will be handled by preventative measures*, than hard-core criminals. (Dourlens and Vidal-Naquet, 1996: 15, 16; emphasis added)

Since the time when *politique de la ville* began focussing on delinquency, there was a gap between the growing social disintegration taking place in former working-class areas and the tools used to slow down its consequences. The very terms "prevention" and "delinquency" were defined quite differently, according to places and situations (Estèbe, 1990). In the meantime, the geography of social risks expanded year after year, as in other Western countries.

After the early 1980s, disorders plagued the *banlieues* and the fear of crime increased among all social categories. The police, justice, and prison systems became more and more costly, and were less and less efficient. Confronted with social and cultural mutations, the police hierarchy was too slow at reforming its structures and its methods of intervention. In part, this dilemma was created by policy-makers who were not able to solve the controversy within the police over whether they were most concerned with maintaining order (targetted intervention) or public safety (reassuring the population), as was remarked by P. Cardo, the mayor of Chanteloup-les Vignes, a problem-ridden Parisian suburb, and the author of a report on urban violence (1991). Maintaining order was most often given the highest priority, a bias inherited from the post-World War II economic boom time when security was not a priority. Consequently, small- and medium-scale delinquency and disorders in inner cities were neglected.

Since the 1970s, the growth of urban violence has been interpreted by public opinion as a crisis of authority and a lack of vision from the elites. "A source of controversy between France and its police comes from the perceived gap between what the police should do and what they do. This feeling may differ according to categories of users or of observers, but it is shared by the whole population," Belorgey wrote in a report on the reform of the police, which was ignored (1991). The feeling that the police do not care about the deteriorating quality of daily life resulting from increased delinquency is still with us. According to the Peyrefitte Report of 1977, there were twice as many crimes and offenses in 1976 as in 1967, five times more robberies with assaults, and twenty times more hold-ups (1977: 31). While reported crimes against property continued to double through the 1980s, crimes against persons remained stable. But it was obvious that petty delinquency was not a priority for the police: to this day, only 15 percent of thefts (two-thirds of all criminal cases) are solved in

France (Lévy, 1996); 80 percent of cases are closed with no follow-up. Many victims are thus discouraged to report crimes to the police, and if they do, it is for insurance requirements. This lack of accountability has other consequences. Take the following example. An overburdened local prosecutor decided to close all cases of thefts by youths at a local supermarket. As a consequence, the police chief gave up registering the complaints from the supermarket's manager. The youths then carried out even more thefts causing the exasperated manager to hire more repressive private guards. Violent confrontations occurred between the youths and the new guards, forcing elected officials to intervene to stop the riots after four youths were arrested.

If the police benefit, on the whole, from a positive image (people are convinced that police are needed), negative opinions increase when the efficacy of their work is questioned (Zauberman and Robert, 1995: 63–4, 71).

The attitudes of left-wing elites, this "culture" of social prevention which is the fundamental distinction between France and the USA, reveal a difficulty in sorting out what is within the jurisdiction of the central state and what comes under the responsibility of other actors. The *politique de la ville* aimed, indeed, at restoring the socio-urban fabric, at promoting the participation of residents, at reinvigorating community policing, and giving more autonomy to the local governments. Moreover, if no convincing correlation can be established between institutional vandalism, riots, and the fear of crime, this fear is supposed to disappear when the "social link" (another favorite French expression referring to trust between people) is re-established. In the rhetoric of the left, law and order policies must come second after social prevention (*politique de la ville*). As a consequence, the initial pragmatism of the local councils of prevention was overdetermined by a sort of doctrine, almost a culture, based on a distrust of repression, even a renovated one (Dourlens and Vidal-Naquet, 1994: 16–17).

After 1986, the majority of local councils opted for minimalist attitudes and "soft" lines rather than a debate on the local methods of police and justice (Estèbe, 1990). One must remember that in the French system the mayors can, at best, rely on the limited capacity of the municipal police. They are not informed of national police operations designed to maintain order in their locality, even major ones, unless they have clout in national political channels. One of their wishes is to be better informed. In the local councils of prevention created by Bonnemaison, the participation of the police remained marginal and often limited to the sharing of statistics on delinquency. As will be seen, it was only after the preventative methods showed their inefficacy at reducing urban unrest and juvenile delinquency and became the targets of public and media criticism that the repressive police and justice apparatus found a new legitimacy in the eyes of mayors. This happened at the end of the 1990s.

New approaches to policing in urban policies

Some background information may help to clarify why the French national police appear dysfunctional when it comes to solving moral panics.[10] The French police and gendarmerie system is both centralized and plural. That is to say, it is composed of several distinct forces, themselves centralized. In total, there are around 220,000 police under the authority of the executive power. The country is the most policed in Europe, after Spain and Italy.[11] The police are divided into several sections: public safety, judiciary police, intelligence, counter-espionage, and immigration. But inside their operations, there are three distinct functions, according to Monjardet (1996): law and order, the fight against organized crime, and community policing.

The government's emphasis on linking the social question to the local arena after 1980 may be interpreted as a challenge to the first two regalian functions of the police. It was nearly impossible to develop the third function more because it required the police to adjust its working hours to local conditions and to the local characteristics of the phenomena that generate moral panics. This had to be ruled out due to the closure and centralization of the police. For instance, it is extremely difficult to adjust the working hours of the police (each individual works on average three days a week and there are more police operating during the day than at night) to the local needs of the population. Citizens want the police to be always accessible and nearby, as they frequently are in other North European countries. But the police unions, which effectively determine policies alongside the Ministry, are opposed to local negotiations and adjustments, painting them as unnecessary strategies of "Balkanization." This is not a new story.

Community policing When attempts at implementing community policing (*ilôtage*) were made in the mid-1980s following the Bonnemaison proposals, observers were struck by the gap between the discourse and the practices, a gap that still has not been filled (Monjardet, 1996: 240ff). For twenty years or so, successive governments tried to get the police closer to the community. Yet, to this day, police on the beat represent 3 percent – or less – of all police. Why? As Monjardet observed, community policing is burdened by a triple handicap. It is a policy designed outside of the Ministry of Interior and no administration likes to be told what to do by outsiders. It is not elaborated by the state, but by elected local officials who already benefit from decentralization laws. And, it is marked by measures of prevention, as opposed to toughness (Monjardet, 1976: 245). A majority of the police that I interviewed told me that community policing is not "real police work." I have often heard this remark: "We did not get in the police to be janitors or nannies." The police think that community policing should be left to those who are motivated (militants or marginal elements). This kind of work cannot be evaluated, they say, and would harm their careers. Since their work is not integrated with the national service, beat police are isolated and information does not circulate among them. As a consequence, community policing is often left to inexperienced police adjuncts or to those whose actions are not coordinated with their colleagues' activities. Unlike the British intelligence service, which relies on information from police officers on the beat to design strategies for the whole force,[12] the French police system has marginalized community policing. Even when it has the political will to do so, as in 1999, the central state seems incapable of reforming a corporation in which most actors firmly resist accountability, that is to say, the possibility of inserting the citizens between them and the state and of listening to their demands. For the police chiefs, decentralized urban policies have meant "more (central) state" intervention in order to avoid fiefdoms, the parochial vision of mayors, and a link with the territory (Lévy, 1996: 11). Conversely, with the *politique de la ville* of the 1980s, mayors refused to let community policy evolve into repression. If repression is to take place, they wrote in their intermediary report, community police should not participate (Dourlens and Vidal Naquet, 1994: 112).

In France, the police have never had legitimacy from the citizens of a defined territory, as they do to a certain extent in the UK, the Netherlands, and Germany (Body-Gendrot, 1998b), but from the central state. The consequence is a mutual ignorance and avoidance between police and citizens, especially in areas at risk. Tenured British police officers frequently look forward to going where the "action" is (so they told me) and are asked to leave their district after ten years. But, in contrast, young French police assigned to the *banlieues* never live there (unlike the gendarmes) and will most likely commute to their provincial home town for the three days a week they are on leave. And most will immediately ask to be transferred. There are some exceptions, however. The discussion on French urban practices in Chapter 5 will show that, here and there, local space effects have allowed the introduction of police innovations and genuine partnerships.

Emergency police measures In response to emergency demands for law and order emanating from hundreds of neighborhoods at the beginning of the 1990s, the Ministry of the Interior initiated three changes to previous policies:[13] a build-up in forces and an increase in police resources (supported by both left- and right-wing governments), reforms to the policing structure, and a redesign of police strategies.

The first response falls within a continuity of law-enforcement policies. In 1981, Interior Minister Gaston Deferre (who had also been the mayor of Marseilles) recruited 10,000 new and better-trained policemen. From 1986 to 1990, his successor, Pierre Joxe, increased the police budget by 5 billion francs for the development of infrastructures. Between 1994 and 1995, the new minister, Charles Pasqua, authorized 5,000 additional recruits and 7 billion francs to keep modernizing the police structures. Thousands of young men were allowed to do their mandatory draft service in the police. At the same time, some police tasks were transferred to other services (identity papers, lost property, etc.). Yet, amidst all this, police officers' work hours have been reduced. (For one full-time police officer's day, at least seven people are needed, when one takes into consideration sick leave, holidays, training, etc.).

In 1998, it was decided that the gendarmes and police forces would be redeployed: 96 police stations in small and quiet localities would close, leaving those districts to the care of the gendarmes and allowing more police to move into "sensitive" urban areas. This decision was so fiercely opposed by the mayors of the small localities and by the police unions that it was postponed. The police are meant to do more community policing. But is it a good idea to send armed young adjuncts into the field after just a few months of training? Who will supervise them? These questions have been the source of numerous debates on the media. The second initiative concerns territorialization "to better handle crime" and emphasizes action rather than organization. It seemed a pertinent response, due to the growth of delinquency. Yet it appears that this structural reform (later called "*départementalization*") was most of all a new management technique. It was terminated eighteen months after its launch in 1993.[14]

As for the third response, the "local programs of law and order," Monjardet calls them "revolutionary" (1996: 243). Indeed, they challenged the imposition of uniform rules all over the country according to the French model of administrative centralization. Under the traditional model, policing was done similarly in Marseilles and in Lyons, in Neuilly and in Vaulx-en-Velin, according to demographic statistics. The new programs admitted a differentiation in the modes of operation and the granting of diversified resources, according to the idiosyncrasies of the places and their problems. Under that policy, the police are proactive and they focus on problem-solving. The mayor and the district police chief have to agree on a contract, meaning that a third party – the local arena – is inserted between the central state and the police. This excessive innovation did not work at first – indeed, it could not work. In other words, the police corporation "efficiently showed its capacity of resistance to social pressures and demands. Neither the growth of delinquency nor urban riots shook it" (Monjardet, 1996: 245).

Partnerships were established, however, between the police and public transportation managers, public-housing management, private security services, and the national education structure. According to the culture of places and the charisma of the actors, they were more or less successful within the bounds of the *politique de la ville*. Increasingly, the central and local levels used situational prevention, based on video cameras and high-tech devices, and management strategies proved effective for law-enforcement authorities.

In 1997, the new Minister of the Interior launched 400 new "local contracts of law and order" as a move to take the lead on urban policies. The local contract defined what was to be done to restore safety. In the most dubious cases, private consultants diagnosed the local problems. In others, the population and the various partners were consulted. The mayor, the prosecutor, the school administration, and the préfet all signed on. It is obvious that these contracts will not be efficient everywhere. In some areas, magistrates and educators have been reluctant to abandon their preventative savoir-faire and be dictated to by the police over what to do. The Ministry of the Interior plans to train 13,000 police officers (600 adjuncts a month) and one-third of the police should be renewed within five years. This commitment requires clear objectives and evaluation, and results in a "co-production of security," a concept prevalent in Britain twenty years ago. Ten teams of inspectors are to supervise the state services involved in the contracts and to pass along security-related information to the Higher Council on Safety.[15]

New approaches to justice in urban policies

The justice system is one of the most criticized administrations in France.[16] Yet it has to play its part in the responses given to moral panics. How can justice be made a partner in the *politique de la ville*? Not only is there a dramatically insufficient number of infrastructures, magistrates, and staff, but justice is itself subject to a moral crisis relative to its role in society.

There are three models for justice in France. First, there is a traditional model, according to which the rule of law is autonomous, the judge is sacred, and the rites are dominant (as I described in the Introduction). Second, there is an activist model in which an all-providing judge interprets the rule of law and adjusts it to the social circumstances of the disadvantaged. Finally, there is a public service model in which the judge mediates conflicts and acknowledges multiple contracts from negotiated rules of law. Sanctions gradually give way to goals, regulations, nonprescriptive norms, etc. (Dourlens and Vidal-Naquet, 1994: 24). From a penological perspective, it is striking to see that retribution to society is being overtaken by the goal of social peace. Justice, then, is invested with a heavy task. As former juvenile court judge, A. Garapon, remarks (1996: 177), it is supposed to bring authority to democracy, to legitimate political action, to structure the subject, to organize social bonding, to produce symbols, and to ensure truth.

The link between justice and the *politique de la ville* evolved in four stages: a strictly formal presence in the local councils of prevention of delinquency at the beginning of the 1980s; after 1988, the development of mediating processes at the local level; the extension of Maisons de justice (community courts); and more legitimation given to repressive actions against delinquents and their parents.

Left-wing policy-makers have largely ignored victims during their attempts to strengthen social ties between problematic areas and the rest of the city, in contrast to the American experience. Consequently, the issue has been manipulated by the far right's denunciation of soft judges and the neglect of victims. But the moral left (also called the Old Left) claims that some victims are more "worthy" than others and that, in its eyes, this is the case with young offenders. Local procedures of criminal mediation are more or less successful depending on individual magistrates. A broad range of views on the efficiency of mediation is indeed found among judges. Some see mediating processes as a means of pacifying urban areas where social relations have disintegrated. Others use them as a support for reasserting the law in no-go areas.

In 1990, the decentralized Maisons de justice et du droit in problematic areas reinforced the goal of a justice closer to people's concerns. Their aim is less to help the residents solve their conflicts than to handle petty delinquency which disturbs communal life and gives the victims the feeling that nobody cares. The judge mediates between conflicting parties, but uses the law and the threat of prosecution to enforce an agreement after negociation has taken place. Between closed cases and criminal prosecution, this method represents a third way of addressing disorders in distressed neighborhoods. In principle, social workers, educators, probation officers, associations working with victims, etc. follow up the judges' decisions. A further forty-one Maisons will be added to the existing thirty-four in the early years of the twenty-first century.[17]

At the same time, a few criminal justice courts have chosen to put cases resulting from delinquency and urban violence on the fast track. The police phone the prosecutor, after speaking with the offender (and hopefully the victim). The prosecutor asks questions and, if satisfied, can immediately start a judicial response (mediation, restitution, or warning, for instance) instead of setting the case aside. The location of the Maison de justice within the problematic neighborhood allows the prosecutor to appreciate the context of the case, as well. In the best cases, the partnership with the police will be founded on trust.

The crisis of the justice system in France has been brought about by several overlapping logics: the regalian logic based on the rule of law, the logic of regulation, and that based on unsolved social and territorial conflicts. This system has thus become increasingly fragmented. In the regalian system that operates as a pyramid, with the Chancellor at the top directing policy, law, which comes above society, has to be enforced. The Maisons de justice, for their part, represent minor innovations linking justice to territorial units and to the participation of civil society. But this justice is then seen as "second-hand," on the side, designed for marginalized social categories. In practice, we are close to the scenario described at the end of Chapter 1. The sites of the justice authority are dispersed throughout a hyper-complex system, with each level and each actor claiming their autonomy and not referring to common references. Uncertainty and rivalries govern the relationships between prosecutors and judges, between them and experts, educators, and probation officers, who act as satellites, linking the justice system to the problematic areas through unstable ties. Control of the policies addressing disenfranchised areas' problems constantly oscillates between the police and the local prosecutor. Prosecutors "interpret" criminal justice policies and adapt them to local situations. This means that it is not the national level which dictates their personal evaluations of cases. But then, as a matter of professional self-protection, magistrates frequently refuse to work alongside mayors and other city partners, marking a break in what is supposed to be local governance.

No one can predict what form change will take at this point. The evolution of the French justice system is subterranean, horizontal, with ramifications, linkages, and partnerships here, and individual punctual actions, initiatives, and extensions there (Dourlens and Vidal-Naquet, 1994: 93).

One thing is certain: this change will take a lot of time. Currently, the justice system continues to be accused of being too slow, inefficient, opaque, and costly.

In a number of large cities, the number of magistrates is the same as in 1912. In 1998, the overall budget for justice represented 1.5 percent of the national budget. Yet, the consciousness of a social emergency is real. Immediate appearance before the judge and fast-track treatment influence the delinquents more than a decision which will take effect one or two years after a crime. But this "judiciarization" process is denounced by those defending the young offenders' education and socialization as a priority, while at the same time fighting against the devalorization of their work. As former juvenile court judge D. Salas argues:

> The multiplicity of rapid procedures . . . causes a loss of judiciary substance and an evolution towards an administrative expression of justice. Why should we alert public opinion to the danger emanating from some youths while, at the same time, we must affirm that any society should tolerate fooling around at the margins of the law, as most teenagers do? (1997: 85)

Juvenile justice needs confidentiality, time, and doubt, he adds, reflecting the views of numerous juvenile court judges.

Since the early 1980s a serious vocational crisis among educators and probation officers has stymied the goal of efficiently resocializing delinquents. Currently, a juvenile court judge must wait six months at Bobigny, a poor suburb of Paris, to have a young delinquent meet his educator for the first time. There are simply not enough educators and probation officers. Four thousand judicial opinions were handed down in 1997 related to such youths, 10 percent of them in Seine St-Denis, which were not enforced for lack of staff at the Judiciary Protection of Youth (whose function is to assist judges and accompany youths at risk). In eleven localities in Seine St-Denis, with 400,000 residents, one counts only ten educators, one psychologist and one or two social workers in the educational services of the court in 1998. In the 1990s the budget has been cut by 30 percent and three educative centers out of four were closed.[18]

The justice system is accused of being lax by the police, by the victims, and by public opinion. That is probably wrong. But it is easy to understand the criticism, granted that magistrates do not communicate information to the police and to the victims, and rarely to the public during an investigation (which may be lengthy). Few people know that between 1994 and 1996 the number of juveniles sent to judges increased by 46 percent (from 18,100 to 26,500). Out of 144,000 procedures related to juveniles, 54,000 received warnings or were closed (Body-Gendrot et al., 1998).

The popular mediation and restitution processes described above are largely limited to Bobigny, Créteil, and Bordeaux (their mediation cases are 50 percent of the national total and 8 percent of all criminal justice activities nationally). Incarcerated juveniles comprise 1 percent of the prison population (645 out of a flux of 3,600 juveniles entering and leaving prison, on 1 January 1998) (Lazerges and Balduyck, 1998: 135, 193–4). Two-thirds of them serve sentences requiring less than two months of incarceration. In 1998, an important report to the Prime Minister on juvenile delinquency recommended that an end be put to juvenile quarters, which is another major difference from the USA. This opinion caused a controversy with the Minister of the Interior, who is symbolically in charge of law and order.

Controversial prisons There is a general consensus in France among elites and a great number of professionals that incarceration is a short-sighted solution for people who do not endanger society. According to Kensey and Tournier (1998), French prison numbers have been stable since 1988: in 1998 there were 50,744 inmates, as opposed to 49,328 ten years earlier. France comes in tenth position in Europe, with 89.9 inmates per 100,000 inhabitants (the figure for England and Wales is 106.8). However, this stability hides longer detention stays (4 months on average in 1997 compared with 2 months in 1975). That is a minor phenomenon, however, in comparison with the USA, where the average sentence is 26 months (Mazetier and Portelli, 1998).

The first cause of detention in 1998 was from drug-related crimes, immediately followed by rape and sexual offenses, then by theft. Are more offenses being committed, or are more being reported to the police, or are sentences getting heavier? It is very likely that all three factors overlap. Take foreign inmates, whose number rose from 18 percent in 1975 to 29 percent of the total in 1995. For criminal offenses, their increase is similar to that of French nationals, but it has jumped to 180 percent in cases of breach of immigration regulations. On the whole, when arrested, a foreigner's chance of conviction is twice as high as for French nationals. The same is true in other European countries.

Opponents to prisons state that they "contain people whose place is in psychiatric institutions or in drug-treatment centers but who are not, because no one handled their case . . . Too frequently, prisons are places of arbitrary decisions and where rights are ignored. Inmates cannot choose their cell, their co-inmates, their future" (Vertet, quoted by Garapon, 1996: 211). Since discriminating practices within prisons are not denounced enough, inmates' rights are highjacked.[19] Prisons are criminogenic, with an over-proportion of suicides, physical degradation, and brutal assaults, especially among juveniles.

French reformers also claim that penitentiaries should remain the exception and be used only for the most dangerous offenders. In extreme cases of recidivism, short periods of incarceration should give professionals time to find alternative means of punishment. This could be applied, for instance, to non-violent undocumented immigrants. Instead of being spaces where rights are ignored, prisons should be viewed as an extension of public space containing subjects with rights, belonging to the same world as others and with common references. While reformers at the juvenile court prison in Chicago assert that incarcerated juveniles can later succeed in life (in part, because the school system at the prison is excellent, so they say), French experts think that the simple fact of having been socialized in a penitentiary will weigh negatively on the formation of young people's identities and on the risks that they will take or not take later on. It is a good thing, in my view, for French elites to share these negative views about prison, but inasmuch, should they continue to deny the responsibility of individuals who commit disturbances in the collective space? Can it be tolerated that poor residents have their cars burnt and the buses vandalized, preventing them from access to work? that parents have to keep their children home after a school has been ransacked and teachers have gone on strike? Again, this is when the two types of ethics clash. But there is no doubt that the laws applied to adolescents need to be understood and to be accepted by them as fair – that is, punishing other social categories (including policemen) when they transgress the laws with the same treatment as the youths from inner cities.

The bone of contention comes from the 1945 ordinance which still inspires most juvenile court judges, also called "social" judges emphasizing protection, assistance, surveillance, and education. "One must trust the educative potential inside each young person" and understand that their violence is only the expression of fear and of their own pain, one such judge explains. Yet that, acting collectively, they might be a threat to a community is rarely admitted and all the possibilities of sanction that are offered by the ordinance are not fully exploited.

Public Discontent

Many French people share the opinion that the *politique de la ville* has failed to buy social peace. Overburdened taxpayers[20] have become reluctant to support schemes of solidarity which, after so many years, have not proven to be efficient in terms of social peace. As in most Western countries, fear of crime has become a major concern for the French, ranking second after unemployment (according to an August 1998 poll). Let me reassert the fact that polls offer a simple picture of the moment, that some answers are brought by the questions, that giving no response may be as important an indicator of unmeasured intensity as positive answers. On such a topic, many answers are predictable because of the continuous and dramatic media coverage of urban violence. This being said, comparisons in the structure of answers point to a malaise concealed in the package "urban fears." According to polls of 1998, 82 percent of the French think that acts of violence in cities and suburbs have reached a dramatic, unprecedented level. Of those under 25 and those over 65, 70 and 93 percent, respectively, agree with that idea. Of those interviewed, 66 percent think that police presence must be massively reinforced to improve safety in "sensitive" neighborhoods. Only one-third of them (but 41 percent of those under 35) expect the governing left coalition to do better in this matter than previous right-wing governments (Institut français d'opinion publique

Table 3.2 "What should be done to improve public safety?" Poll Credoc/IHESI, France, 1990–1998 (%)

	June 1990	*Jan 1993*	*Jan 1996*	*Jan 1998*
Harsher sentences	33.0	42.0	45.0	47.8
Alternative punishments to prison	49.6	43.3	43.5	42.0
More police and gendarmes	26.1	28.9	24.5	35.6
Fight undocumented immigration	32.8	34.8	36.5	28.3
Collaborate with police and gendarmes	20.7	19.6	19.2	19.0
Learn to be less exposed to risks	28.2	21.2	21.8	17.6
Self-defense	3.6	4.6	3.6	3.8
Call for private services	3.2	3.3	3.1	3.0

Source: Le Parisien, 25 November 1998

(IFOP) – *Libération*, 5 January 1998). A study for the Institut des hautes études sur la sécurité intérieure carried out by Credoc over eight years and published in November 1998 shows that 47.8 percent of the French (compared with 33 percent in June 1990) wanted harsher sentences and more protection. This last poll indicates that, unlike Americans, although the French require more repression, they are not ready to involve themselves in self-help actions nor to "collaborate" (a loaded term since the Second World War) with police forces, or to turn to private safety services (see table 3.2). A poll conducted by SOFRES in September 1998 revealed that of those interviewed, 58 percent thought that "politicians pay almost no attention or little attention to what people like us think" (a 16 percent increase since 1977), 57 percent felt a distrust in politics and 30 percent felt hope. Less than a third felt that they are well represented by a political party, 27 percent by a political leader, and 16 percent by a trade-unionist, representing a sharp decline over ten years. This does not mean that the French are not interested in public affairs, but that individual assertions are accompanied by civic desertion. Citizens require genuine collective commitments, and when they do not find them, they retreat from the public space (Perrineau, 1999). This double trend expresses a deep democratic malaise.

The Credoc poll reflects the evolution of attitudes over time. It says something about the French state of mind at the end of the twentieth century. People want harsher sentences as if the country was under siege, yet only 5 percent of those over 25 admit that they have been the victims of acts of violence, including verbal violence (nine times out of ten their aggressor was known or identified[21]). Currently, only one French person out of five feels unsafe. Exasperation comes from the gap between representations, expectations, and the perception of institutional inaction. The police do not give the impression that they care about repressing delinquency. It has become exceptional for police to visit a home where a burglary has taken place. A "technical investigation" occurs in fewer than 10 percent of burglaries in the Parisian region.[22] Disenchanted with their police, their justice, and governmental elites in general, more French abstain from voting or vote for marginal parties, both forms of "exit" or "voice," but not of "loyalty" (Hirschman, 1970).

The media and the creation of insecurity

As in the USA, the media play a role which cannot be underestimated. The collective fate of society is questioned through singular incidents (*faits divers*) or riots. Political elites react hastily to such disorders under the scrutiny of the media. Such haste annihilates long-term political elaboration and the promotion of collective political goals.[23]

First, the French media, as in America, launch populist appeals, thereby weakening credibility in institutions. In a problematic move, they amplify the most archaic mechanisms related to the circus games in which a public of voyeurs contemplate the victimization of a few by cruel adolescents. The media excite voyeurism and reify those that they display to the audience. In building up feelings of fear, victimization, and revenge, they reintroduce the mechanisms of scapegoating and lynching of former days. They do not distinguish between legitimate and illegitimate violence, they simply unroll one report after another, as if only moral outrage counted (Garapon, 1996: 112 *passim*). The mass media rarely emphasize what works and what solutions are appealing. In talk shows, they frequently channel fears (people calling the radio, not being representative of the general public's views) and amplify stereotypes. Moderation and nuances are inhibited by the unease fanned by the media. While promising to make reality more transparent and to suppress social distance, the mass media actually distort it and annihilate the sense of shared citizenship to which national elites vainly appeal.

Second, in generating doubt, cynicism, and pessimism among the public, the media provoke a paralysis and an inaction among elites. They activate frustrations in a democracy besieged with conflicts. This has consequences for public opinion, which "is more ready to identify with the victim than with the umpire, with the governed than with the government, with counter-powers than with power, with the vigilante than with the lawmaker" (Garapon, 1996: 96). Opinion yields to emotional and manichean views. The mayhem exported by multinational entertainment companies to TV and cinema screens and in video games does not spare French society or its youth.

Some researchers contest the idea that the mass media have a real impact on the genesis of public fear. As early as 1985, Robert suggested abandoning this idea that the media have an impact for France. The media's sphere is a scene run by its own rules, he said, and actors with different means attempt to transmit their message more or less efficiently: "That the play on 'fear of crime' partly takes place on the media-political scene does not mean that it is devoid of social substance but rather that, first of all, it concerns the place and role of the state in social relations" (Robert and Pottier, 1997: 623). For instance, when in a talk show a radio invites police union leaders to share their alarm with the listeners, it is obviously to the state that the police is talking, using citizens and the media as tools to exert pressure for more resources.

The Punitive Trend

Public authorities attempt to respond, sometimes symbolically, sometimes pragmatically, to demands for repression. For example, undocumented immigrants were used as scapegoats during the tenure of the right-wing Minister of the Interior Charles Pasqua, after 1993. Immigrants' children were required to express visibly their wish to become French, a move sometimes made difficult by colonial and post-colonial ties. This was an easy way to gain a national consensus at the expense of vulnerable categories who have little power of sanction.

The Debré laws in 1995 pursued the repressive trend started by Pasqua. Several bloody acts of terrorism provided an excuse for the new Minister of the Interior Debré to increase the surveillance of the public space under programmes called "Vigipirate," requiring the presence of the repressive force CRS (Compagnies républicaines de sécurité) and of the army. Meanwhile, the conservative Minister of Justice Jacques Toubon (1993–7) was launching several boot camps called Unités d'encadrement et d'éducation renforcée (UEER) to take care of multirecidivist juvenile delinquents.

Within the left in power in the late 1990s, tougher rhetoric and policies have begun to emerge from the top due to electoral considerations, pressures from mayors, and street-level civil servants, as well as to disquieting rates of delinquency. Interior Minister J. P. Chevènement himself alluded to "les petits sauvageons" (interpreted as the little barbarians or as badly grafted trees) who do not attend enough school and who watch too much television. Not as hostile as "superpredator," the term evokes a nostalgic quest for restored order. It is probably as far as he may go, due to a massive resistance to toughness within the moral left. The French state council – the highest administrative court – forbids, for instance, mayors to take curfew ordinances and the law denies the pursuit of vagrants and of panhandlers (including those young beggars who squat in the collective space with unfriendly dogs). The constitution of a file, allowing the police to stock data on offenders and victims, is hotly contested by organizations protecting civil liberties. In the spring of 1998 the Minister of Justice conceded the suspension of family benefits for delinquent parents and the Prime Minister supported this idea in the fall. Numerous acts of intimidation against bus and train drivers had indeed led to general strikes that disrupted the life of millions of commuters.

The crisis of authority is real and there is a genuine institutional powerlessness at erasing internal fractures. In the 1950s an American observer, Jesse Pitts, called France "a community of delinquents." Looking at classrooms, he noticed that school teachers never hesitated to ridicule bad students, putting the class on their side with sneers, shaming the bad students with posters that they carried on their backs, or threatening them with all kinds of punishment. But if the teachers lost authority for some reason, they became, in turn, the object of sneers and acts of delinquency from all the students (Pitts, 1963). Today, the major cultural change concerns the crisis of the vertical model of authority, the difficulty of asserting the law, "the cultural incapacity of enforcing the old techniques of the disciplinarian society, the crisis of our capacity of surveillance and sanction," as Brazilian sociologist Peralva comments, after observing French classrooms in poor suburbs (1997: x). Intergenerational communication is blurred. Young people frequently complain of being "dissed" by adults, employees, policemen, storekeepers. Violence is interactive: the state, instead of "producing social [links]" seems to have abandoned its global project of integration and resorted to a mere social maintenance by the utterance – not the substance – of necessary regulation rules (Dourlens and Vidal Naquet, 1994: 188).

Thirty years after he first uttered them, we could apply to the French situation the words of Kenneth Clark, the famous African-American sociologist, when he addressed the Kerner Commission after the 1960s racial riots in the USA:

> I read this report . . . on the Chicago riots of 1919 and it is as if I were reading the report of the investigation commission on disorders in Harlem in 1935, the report of the investigation commission on disorders in Harlem in 1943, the report of the McCone commission on riots in Watts. I must sincerely tell you, Members of the Commission, that we think we are in Alice in Wonderland, with the same film, over and over again: same analysis, same proposals, same inaction. (National Advisory Commission on Civil Disorders, 1968: 483)

To conclude Part I, it appears that the policy of suppression in the USA and the policy of social prevention in France mark changes which can be theoretically explained in different ways, following the model elaborated by Cohen (1985):

1. According to a *linear narrative*, reforms take place one after the other when old practices seem to be outmoded. Following a consensus that "nothing works," managerialism has taken hold of the criminal justice system in the USA. Meanwhile, in France at the end of the 1970s, the left defended social prevention policies to tackle the roots of urban delinquency. The end results were more or less as envisaged: the system became insulated from criticism in a narrative found in official commissions, inquiries, and reports. A constant leitmotiv is that "things would have been much worse without these efforts."

2. According to an *organizational convenience*, reforms are blocked or are welcome when absorbed and co-opted. Thus, community policing and justice reforms in France cannot develop because of a litany of impediments: the resistance of institutions – police, justice, teachers, educators – undermining the reforms, the lack of governmental determination, inappropriate selected client groups, inappropriate service modalities, inappropriate settings, etc. The struggle between the social demand of the public and the resistance of corporations is unequal.

3. French mayors who are alarmed when they are in the field turn into distant actors when, as Parliament members, they pass national "soft" laws. Conversely, in the USA, the preference of the public for alternative sentencing, mediation, and job creation for youths is ignored by policy-makers, who yield to tough symbolic politics in Congress or state legislatures. The policy arena is a site for *contradictions in ideology*. It questions the status of ideas, theories, knowledge, intentions and ideals. Contradictions, distortions, paradoxes, and anomalies are internal to the ideology. It explains both the appeal of prevention for French law-makers and the way the patterns work themselves out at the national level, while, when they are accountable at the local level, the same persons cannot ignore massive demands for urban safety.

4. With the *political economy explanation*, economic rationality explains why business as usual prevails. State expenditures are less reduced than redirected. In the USA, the hard side of the repressive system grows (prison-building is a lucrative business), while the soft preventative side contracts in response to monetarist objectives or budgetary balancing goals. What remains of the soft edge are symbolic exercises of legitimation. In France, symbolic politics redirect resources after each emergency crisis. But the slow process of larger reforms, such as transferring more police to urban zones where they are needed, hides the fact that the state does not have the authority nor the necessary resources to build more police or gendarmerie stations at this stage.

The current phase betrays uncertainties and hesitations at the national level in France. Successive governments defended a policy of social prevention without ever giving it the resources and the experts it needed in the field to bring disqualified areas in line with the rest of society. Incentives were never created to tempt the most experienced teachers, police, judges, and educators to these areas. The urban violence question conceals a lack of political will in favor of more equality in French society. Lawmakers and governmental elites fear corporatist and clienteles' reactions more than the discontent of public opinion requiring more authority. As aptly remarked by Bourdieu, the left hand of the state is ignored by the right hand of the state elites. These observations on national contexts lead us, therefore, to examine the solutions which have been adopted by cities and how they may differ or not from one country to another and from one city to another.

Notes

1 Collective urban violence cannot be confused with juvenile delinquency, which makes it difficult to sort out the statistics. Driving drunk in a stolen car is not an act resulting from urban violence, for example. Many incivilities, as defined by Skogan (1990), are not punished by law, yet they are perceived as urban violence. That is the case, for example, with residents who are frightened by youths gathering in the stairs of their public building with dogs or spitting when they pass by.

2 L. Bui-Trong's personal trajectory is quite unusual: she was a philosophy teacher (*agrégée*) in a secondary high school before she resigned to raise her children. At 42, she passed a difficult civil service competition to become a police chief (*commissaire*) and gained notoriety after she created a national observatory for urban violence in 1991 and often talked to the media.

3 This "political police" service comprises 3,000 mainly plain-clothes officers. It has a considerable automy and is supposed to exert infiltration and surveillance. It repositioned itself on *banlieues* after the threat from the far left declined in the 1980s.

4 This discussion relies on the Le Galès and Mawson (1994) report for the local government management board.

5 The youths were bypassed in a number of cases, for clientelist reasons, and here and there the jobs went first to the public servants' children and their connections.

6 In *Cities Without Suburbs*, Rusk suggests that the problems of inner cities could be reduced by cutting new administrative units incorporating part of the city and part of the suburbs in order to address urban problems with more resources.

7 Abuses caught national attention when in Reims, for instance, such funding was used to restore the Clovis baptismal stone in the cathedral instead of regenerating poor neighborhoods (which do not vote).

8 In the same vein, Crozier (1963) described how face-to-face meetings were avoided at all costs by administrators secretly protecting their scope for maneuver.

9 Following World War II, the police were seen as a force of repression and linked to the maintenance of order. During the Algerian war, they persecuted Liberation Front militants, fought against union demonstrators, and were called up in response to fears of revolutionary subversion and political violence in May 1968. But their latest mobilization was linked to the rise of delinquent actions and to efforts to appease moral panics, especially among senior citizens, shop-owners, and farmers, the traditional constituents of the right, but also those experiencing precariousness in their status.

10 This section relies on a development from Lévy (1996).

11 We will not examine the national gendarmerie, except to say that it appears as a good "model" when compared with the national police. The gendarmerie is under the authority of the Ministry of Defense. It is the oldest police force and served as a model for numerous European systems. It is formed of 90,000 individuals operating in rural and semi-rural areas. Due to excellent investigative methods and to its professionalism, 66 percent of its cases are solved. The gendarme is compared with the sheriff by Brodeur (1990: 203ff) in terms of legitimacy, community policing, and accountability.

12 Interviews with Metropolitan police inspectors, commander, and superintendent by the author at Scotland Yard, spring 1998. In France, symbolic gestures will, however, be made in favor of community policing, such as in 1989, with the creation of a subdirection of prevention and social action within the central management of urban police forces, and after 1998.

13 I follow Monjardet, 1996: 240ff.

14 The reform failed partly because of the expressed hostility of the General Intelligence service, which would have lost its autonomy.

15 This Higher Council created in the fall of 1997 is made up of the Prime Minister, the Minister of Justice, and the Minister of the Interior.

16 This section relies on ideas expressed in the 1994 report by Dourlens and Vidal-Naquet.

17 One goal of this decentralized process is also to reduce the gap between the police in charge of the investigation and the justice system. Frequently, the police are accused of not providing enough written evidence to help the judge decide whether to follow the case or not. The files get lost amongst thousands of other files, and the public and police are convinced that delinquents enjoy impunity, while the victims are abandoned.

18 Interviews with specialized educators by the author, fall 1998.

19 These remarks come from discussions at the Committee on Prisons of the League of Human Rights in Paris and from my own observations when I participated in a one-day debate on violence with inmates in the maximum security penitentary of St-Maur, at the invitation of the composer Nicolas Frize.

20 Only one household out of two in France pays direct income taxes.

21 *Le Monde*, 18 December 1996.

22 *Evénement du jeudi*, 2 November 1995.

23 In 1791, in the debates relating to the elaboration of the first national crimi-

nal code, Representative Duport had warned: "it is not always by a punctual and servile obedience to public opinion's orders that law-makers pass the most useful laws for the country; frequently such laws relate to temporary needs and bring relief only to impacts: successful and broad decisions generating the happiness of people are processed through meditation and design" (quoted in Garapon, 1996: 99).

Part II
The Politics of Reconciliation

Part I has shown that globalization matters. We now need to evaluate the impact that globalization has had on specific cities. Appropriate data will support the idea that, cities being the sites where globalization is played out, local policies must address its impact. I also showed in Chapters 2 and 3 that national values and choices and governmental and parliamentary processes mold political decisions at a time of uncertainty. It is also appropriate to estimate their impact, if any, on cities. The question could then be formulated in the following ways: Which national and urban conditions allow the adoption of certain types of local policies? Do national politics matter? Do they create expectations among anxious voters? Do they give birth to vocal and disruptive lobbies?

In Part II, I will attempt to track down innovative practices in the treatment of marginalization and delinquency. I will consider whether the structural position of local elites and their links to other institutional strata matter. It is indeed essential to distinguish between those who formulate policies and those who implement them and to make a difference between the true believers, the opportunists, and the resisters. If, at the highest levels, law-makers are judged on the validity of their ideas and on the strength of their convictions, at the local level there is a dissonance between declarations in principle and constituents' concrete expectations. Relativism, accommodations, negotiations, and space effects all define local politics. Debates are less politicized than at the national level, where crime and violence remain abstractions and where political responsibilities are more diluted.

The issue at stake is pragmatism. How do mayors and their partners work to slow down the exodus of the middle and working classes from central cities? Is making neighborhoods safer a central requirement for job creation (for instance, in the tourism sector) and stability? How is this being done? Is it by upgrading education, as in Chicago, and fighting criminalization, as in New York, in order to give a city a more business-friendly image? How do French cities control potential threats of disorder from distressed areas? How do law-enforcement officials get their information about "dangerous classes and areas"? Do they suggest policies to national and/or local officials based on this information? How is local governance functioning? Are the civil society and the private sector associated in the search for solutions? How do political actors in charge of the city establish ties to local residents? In our study of particular cities, are the authorities in charge (mayors, policemen, care-providers, community center staff, etc.) listening to the residents' needs and working with them? Those are all questions to which the case studies which follow should help in providing some answers.

Since we are dealing only with large cities, we may wonder whether the size of a city is actually of benefit. Large cities have a lot of resources and wealth, but they face colossal social problems, visibly seen in some segregated areas. Cities are often impoverished when they are faced with the flight of wealthier populations. Metropolitan areas are socially and spatially split. Do they have adequate resources to meet the challenges of polarization and segregation? How much do outside forces and demographic factors influence the social well-being of urban areas? Do local stabilizing factors counterbalance cyclical market trends? How does local governance express itself in the search for social control?

Part II brings forth arguments, if not forcible proofs, that the historical and cultural specificity of a place and, in particular, its tradition of community activism, do matter. But it will be shown that the concept of "coproducing" safety is ambiguous. A discussion on the local governance of security cannot be avoided. Examples drawn from field research will reveal the complexity of local governance, its subtle dosages of prevention and repression, the importance of civic cultures and of repertoires for problem resolution, and they will tentatively offer an understanding of local social control and of the nature of cities.

4
Managing Polarization: New York and Chicago

In this chapter, New York and Chicago parameters support our hypothesis, according to which there is a link between globalization, loss of income for lower social categories, institutional disinvestment and racial discrimination (welfare cuts, low-income housing destructions) in poor neighborhoods, family disruptions, and high levels of crime, gang violence, and drug-trafficking. Governing through crime seems to describe the political regime of Mayor Giuliani who has also discovered the backlashes of zero tolerance.

New York, a Global City

New York has changed a great deal since the 1960s, when the Ocean Hill or Canarsie controversies (Body-Gendrot, 1993a; Reider, 1985; Zukin and Zwerman, 1988; Gregory, 1998; Sanjek, 1998) pitted many white ethnics against racial minorities on issues of school control, social housing, and public jobs, and aligned all of them against City Hall. Such scenarios are now repeating themselves thirty years later in France's troubled regions. Not that America anticipates European situations, but urban conflicts frequently have a déjà vu look. Under Mayor Lindsay (1966–73), New York City experienced difficulties in fulfilling the promises of social justice for all, just as a left-wing government now does in France, torn between balancing the budget and helping cities with their crises of social integration. But the comparison stops there, as the context in which disturbances take place is so widely different.

The demographic context

Of all of the so-called "global cities," New York perhaps most deserves that title. According to the 1990 census, 18 million people were living in the New York metropolitan consolidated area, a population as large as that of São Paulo. The city of New York (with 7.5 million residents) is famous for its cosmopolitan character: one resident out of three is foreign-born; more than 25 million people visit the city each year. It is a segmented city, compared with more homogeneous European cities; minorities represent a majority of its population. In the 1970s, an exodus of whites resulted in a 26 percent shrinkage in the population, and in the 1980s another 14 percent left. Immigration of blacks increased their numbers by 11 percent in the 1970s and 9 percent in the 1980s, and the non-Puerto Rican Latino population grew 14 percent in the 1970s and 63 percent in the following decade.

The economic context

Industrial jobs have been disappearing from the city for more than a generation and jobs in other sectors, like FAIR services (finance, advertising and communication, insurance, real estate), have not been numerous enough to compensate for the losses. Following the 1987 stock-market crash, 600,000 jobs were lost (between 1989 and 1992), while 133,000 others were created. Moreover, the manufacturing jobs which remain in the region are increasingly found in suburban locations. For inner-city residents this creates an obvious problem, which Kasarda (1993) has described as "the spatial mismatch hypothesis." There is some controversy among labor economists as to whether the hypothesis is correct or important. It is not clear how mobile disadvantaged categories are within metropolitan areas or how neutral suburban employers are as to where their labor supply comes from.

Mollenkopf's (1997) analysis of Current Population Surveys since 1977 points out that inequality in New York in recent decades has become more striking, as the income of those on the middle and top rungs improved. The growth of social polarization in the city and region over time is well documented. On average,[1] it is greater in the New York region than in the urban north-east. As a consequence, New York has a disproportionate share at the low end of income distribution, while maintaining notable strength at the top end (see figure 4.1). In 1987, there were more than a million people below the poverty level, some thousands of them homeless (Chevigny, 1995: 60). "The income gradient from the Upper East Side of Manhattan around 80–89th Streets to East Harlem north of 96th Street must be one of the steepest and longest anywhere in the world," Mollenkopf points out (1997: 14).

In 1990, the 300,000 households at the top received fifty times the income of the 300,000 households at the bottom of the economic scale. The ratio of the top tenth's total income to that of the bottom tenth

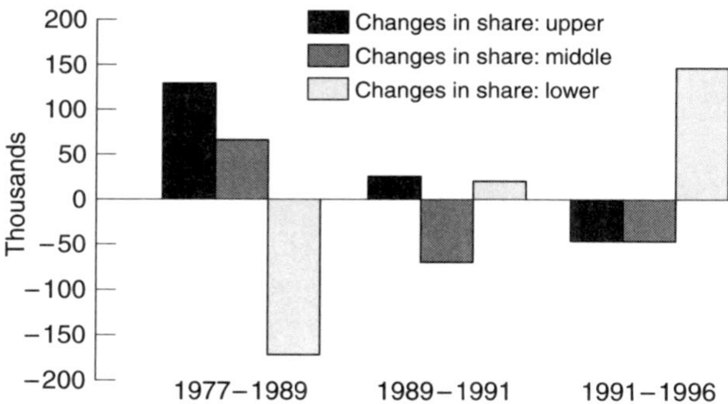

Figure 4.1 Changes in share – upper-, middle-, and lower-income groups, New York, 1977–1996
Source: Mollenkopf, 1997: 19

increased by 45.5 percent in the 1980s. Among the reasons put forth for the growth of inequality, as pointed out by Sassen (1991), the status of New York as a global city means that a large, advanced corporate sector requires both highly trained and unskilled workers. By the end of the 1980s, the New York region was home to the headquarters of 138 of the 500 largest corporations in the USA. Its dominance over transnational corporations was even more pronounced: 40 percent of them were located in the region (Abu-Lughud, 1995: 180), as were 54 percent of service providers to corporations. International investment and financial markets allowed the city to exert a function of control over world transactions and to retain enormous assets, but at a price.

Of those who work in New York's financial industry, 77 percent are not professionals but hold clerical jobs or jobs that require even fewer qualifications. The various ethnic groups (foreign-born and American-born) occupy diversified economic niches according to their capabilities and to their employers' perceptions (Waldinger, 1996). An August 1998 report by the Federal Reserve Bank of New York, which looked at wages from 1979 to 1996, has shown that at a time of huge gains on Wall Street, full-time workers near the bottom of the region's income spectrum have seen their paychecks shrink, after inflation. The number of working poor in the service sector has increased in the city.[2]

Middle incomes grew during the 1980s, Mollenkopf notes, but this was not due to globalization. Part of the manufacturing sector continued to provide steady and well-paid jobs. Segments of the black and Latino populations shifted into the top half of income distribution, increasing income inequalities in their communities (this was more true for blacks than for Latinos.) As the maps of New York show (see figures 4.2 and 4.3)

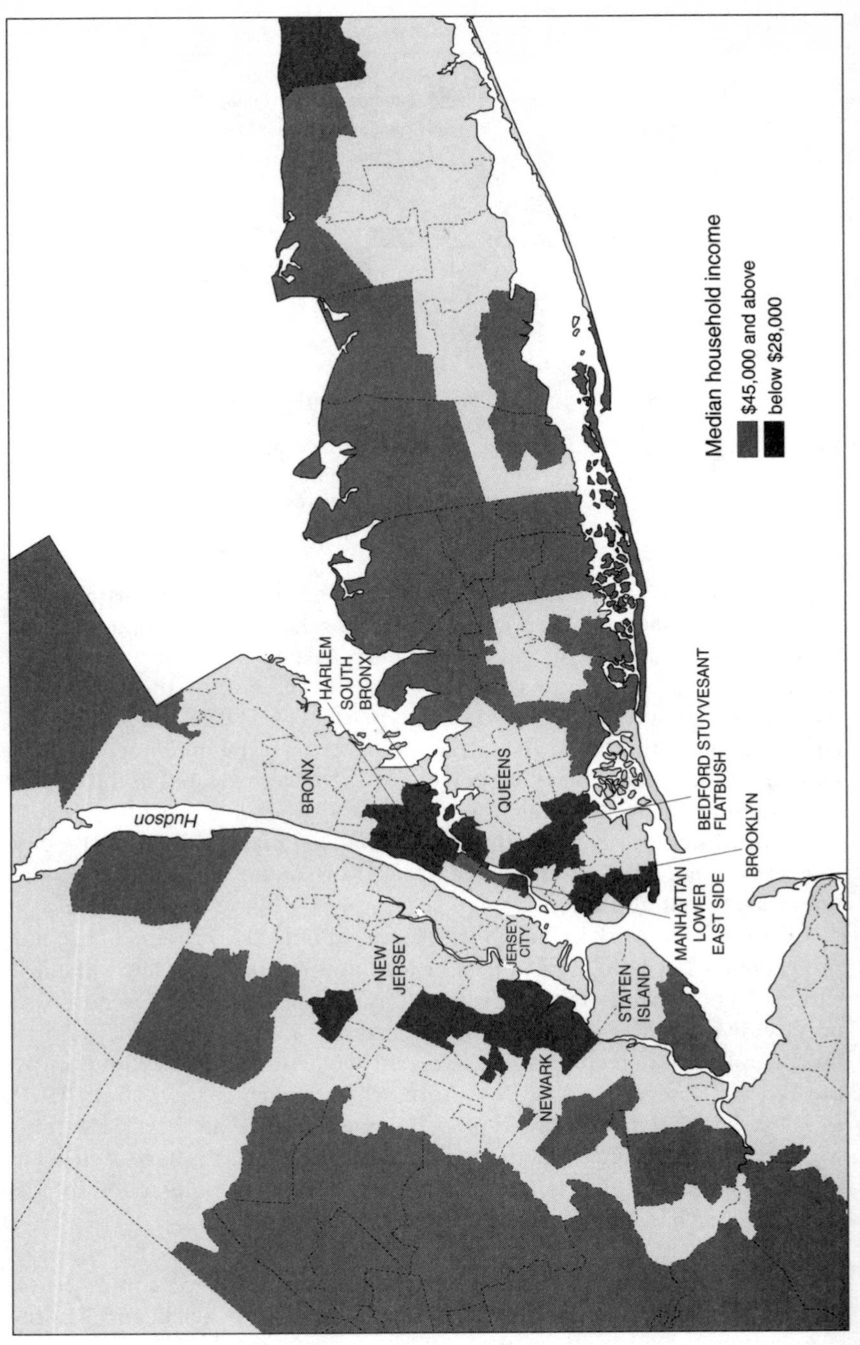

Figure 4.2 Poor and rich neighborhoods in New York, 1990
(Courtesy of Professor John Logan, State University of New York at Albany)

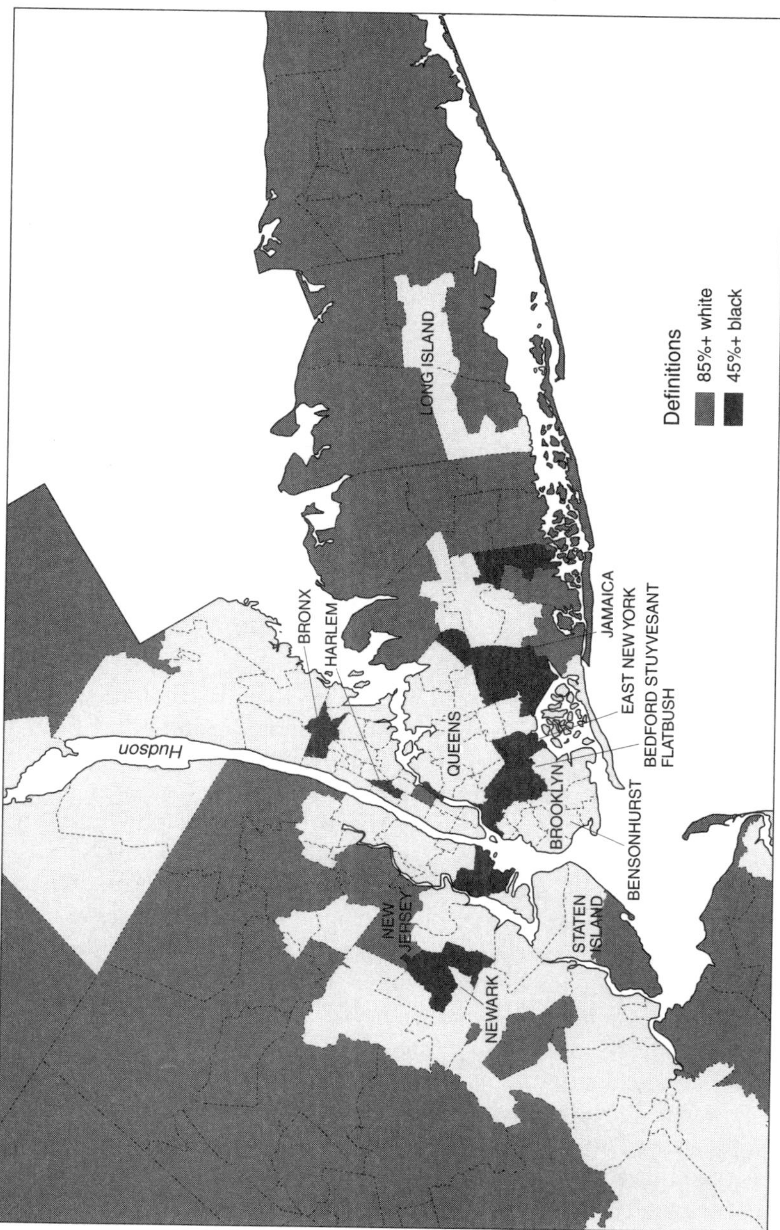

Figure 4.3 Black and white neighborhoods in New York, 1990
(Courtesy of Professor John Logan, State University of New York at Albany)

disparaties by race and by income do not coincide. There are more immigrant neighborhoods amongst the poorest in Brooklyn, the Bronx, and Spanish Harlem. At the same time, neighborhoods where more than 45 percent of the population is African-American have shrunk and the census of 2000 will show whether this trend has been reinforced. Three sectors experienced internal upward mobility: health, education, and social services, followed by business and personal services. Moreover, family economic strategies allowed job-holders (45.6 percent of the total population) to support non-job-holders, who still represented 28.8 percent of the adult population in the late 1990s. Labor-force participation increased in each decile, except for the lowest, where it deteriorated. Economic prosperity did not pick the worst off up from the bottom (Mollenkopf, 1997: 7).

The Impact of National Repressive Politics on Welfare

In New York, a long series of mayors from Fiorella La Guardia (1934–45) to Robert Wagner (1954–65), John Lindsay (1966–73), Edward Koch (1978–89), and Rudolph Giuliani (1994–2000) have been effective at capturing national changes of mood. With their liberal Democratic past, many New York voters still supported a Democratic agenda, however, for instances in cases such as rent control and support programs for the homeless. Mayor Giuliani started his career as a prosecutor, then served as an Attorney-General. He based his campaign on the anti-crime issue. But, at the same time, he has reflected New Yorkers' mood as a pro-immigrant, pro-choice, "quality of life" mayor.

When he became Mayor of New York in 1994, Rudolph Giuliani made a symbolic attempt to distance himself from the conservative social policies of other Republicans. On the question of welfare reform, "the city of New York is really just a better place than a lot of other places in terms of caring for the people and understanding the worth of human beings," he said, capitalizing on the historical culture of the city (*NYT*, 5 August 1996). He evoked a tradition of social generosity in the city and state that goes back to the nineteenth century, when the Tammany Hall political machine won working-class immigrants' voting loyalty with generous helpings of public relief. Providing public jobs and easing bureaucratic requirements for immigrant families was a task the machine performed even during the Depression. It accomplished that through the support of labor unions and fusion mayors, such as Mayor Fiorello La Guardia, a non-Irish mayor who enlarged the profile of ethnic constituencies benefitting from the city's generosity.

The politics of austerity enforced by Mayor Koch in the 1980s following New York's severe fiscal crisis of 1975 provoked ethnic and racial tensions among groups competing for meager resources. Planned shrinkage of services disproportionately hurt the disadvantaged. In recent years, the trend has continued but not as fast as in other large cities. The state of New York did not cut its benefits or drop its home relief program as much as some other states. Between 600,000 and one million tenants still live in public housing, which has a better reputation than it does elsewhere. New York State was one of the few in America to enshrine the principle of relief for the poor in its constitution. The state's monthly welfare grant of $577 (in 1997) for a family of three, though hardly generous relative to the high cost of living, is still one of the four highest in the country. The intensity of the need is much greater in New York, which has the highest population of welfare recipients, 1.2 million in the family aid program (one-tenth of the national total), and 287,000 on home relief, 70 percent of whom live in the city.

Two years after national welfare reform became law, a quarter of the New York recipients – 100,000 people, mostly single mothers – were supposed to work in exchange for benefits. By 2002, this number could have doubled. The mayor says that the law deliberately shifts costs from federal to local taxpayers without providing the funding to take care of workers' children "in a humane or a decent way" (*NYT*, 13 August 1996). According to the chairman of the City Council's General Welfare Committee, the kind of work offered by the city – cleaning parks and courthouses, answering telephones, supervising school lunchrooms: 20 hours a week at a minimum wage of $6.66 – offers little more than dead-end jobs, "creating a permanent underclass of people pushing brooms around the city" (*NYT*, 13 August 1996).

The size of the city makes a difference, as does the culture of providing home relief. Liberal experts anticipate, nevertheless, that, were a recession to occur, cuts in welfare would likely push more families out onto the street, into public hospital beds, toward domestic violence and crime. In 1995, in the state of New York, federal, state, and local governments spent nearly $4 billion on welfare (about one-sixth of what is spent nationally) and $20 billion on Medicaid. "There is a commitment among a substantial part of the population to maintain a basic level of decency for their neighbors as if they understood the consequences of the social collapse that would otherwise result," the director of the Center on Social Welfare Policy and Law explained (*NYT*, 2 and 6 August 1996).

Whether his discursive concern for welfare recipients is founded on principles or not, the mayor's actions differ from his words. Constrained by federal requirements (stricter rules and the five-year lifetime limit) and by welfare costs, he has removed 155,000 recipients from the welfare rolls since taking office. Such a move is contentious.

Unlike the situation in many European countries, the general withdrawal of social support in the USA that was supposed to alleviate dependency did not in fact lift the lowest stratum out of its poverty. As a result, the distance between this stratum and others has worsened (Carnoy, 1994). Racial discrimination and spatial entrapment penalize those already in a situation of poverty and of racial exclusion. In that respect, the number of non-working poor female-headed households with children and no other worker in the household jumped in the poverty data (Edelman, 1999). Institutions, choosing the planned shrinkage option, invested insufficiently in health care, housing, and education for these households. It is within this context of extreme inequality and institutional withdrawal that policy analysts make a socio-spatial connection with patterns of crime.

Crime Patterns in the City

The neighborhoods with the highest crime rates are also the poorest: Bedford Stuyvesant, East New York, Brownsville, Harlem, Washington Heights, Soundview, and the South Bronx, according to the United Hospital Fund data (see figure 4.4). The explanations for crime fluctuations are numerous. By 1982, crime was declining for the first time in more than twenty years. At the time, as has continued to be the case, credit for the fall-off was given to the vigorous actions of the police, to harsher sentencing policies, and incarceration, and to the declining proportion of males in the crime-prone late teens and early twenties age brackets. When crime climbed again after 1986, the consumption of crack was blamed. At the end of the 1990s now the consumption of heroin as a substitute for crack is viewed as part of the explanation for the decline in crime (Beiser, 1995; Krauss, 1995).

Explaining racial disparities among offenders is complex. Victims and offenders are often the same people in the USA, as in France. Sullivan carried out an ethnographic survey of three types of neighborhood in Brooklyn, differentiated by race and income (Sullivan, 1991: 226). He established a link between race-ethnicity, criminal offenses, and arrests. Blacks and Latinos accounted for 84 percent of all male arrestees but only 48 percent of the borough's overall population. Of the arrestees, 60 percent were under 25. Taking into account the ecological dimension, the Chicago School research of Shaw and McKay (1942) showed that environment plays a major role in the socialization of youths. In the 1920s, the delinquency of black boys varied from one area to another. If it was, on the whole, higher than for whites, it could be that in the white working-class neighborhood, more employed adult men attached to households performed an informal social control function. It could also be that white

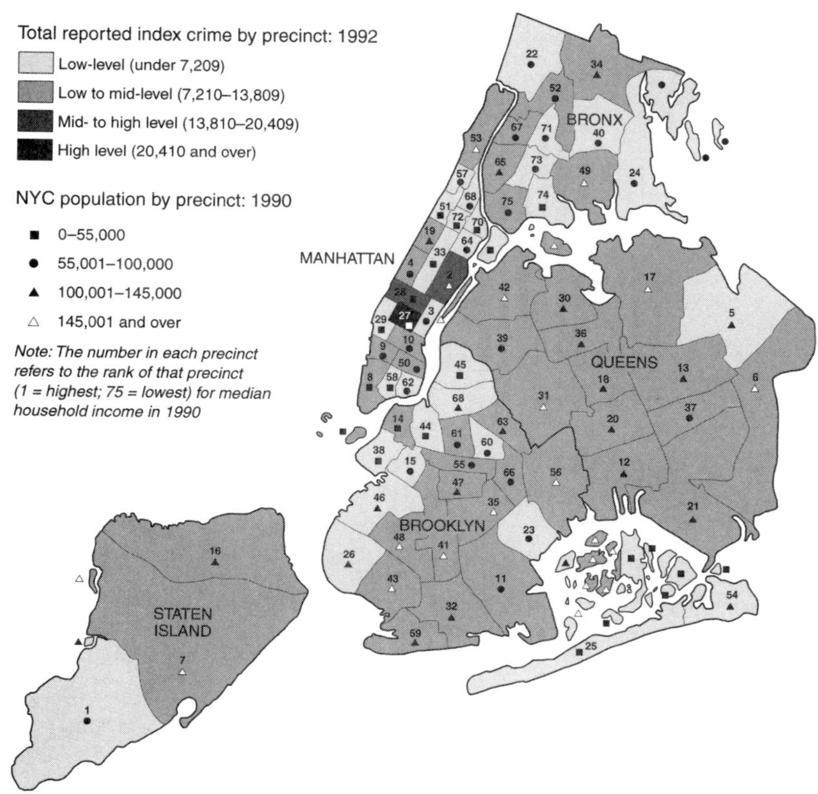

Figure 4.4 Crime, residential population, and household income, New York, 1990–1992

Source: Odubekun, 1993. This map first appeared in *The Vera Institute Atlas of Crime and Justice in New York City*, © Vera Institute of Justice, 1993. For more information, contact the Vera Institute at 377 Broadway, New York, NY 10013 (*www.vera.org*)

youths had more structures of opportunities to find jobs and join the mainstream, while youths in minority neighborhoods were more isolated, more on their own, and more able to commit offenses with impunity (Shaw and McKay, 1942; Sampson and Wilson, 1995). A criminal career was one of their options. Joblessness, related to single-parent families and family disruption, is correlated to juvenile violence and to weaker social control and guardianship, as Sampson and Groves (1989) explain. A high level of joblessness has, therefore, an indirect impact on family disruptions, as does a lack of social organization in neighborhoods, and consequently these factors can contribute to high levels of crime, gang violence, and drug-trafficking.

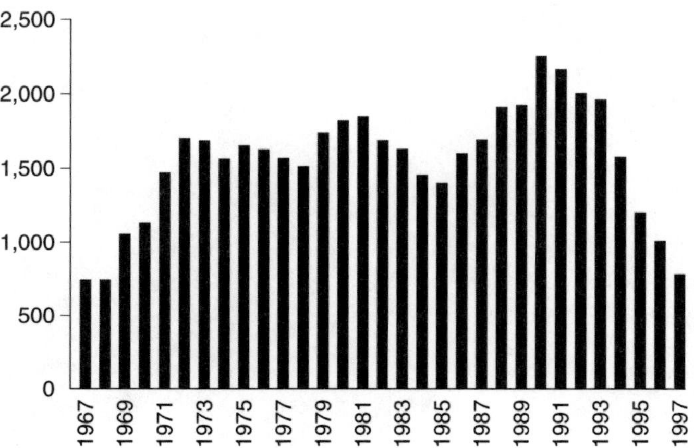

Figure 4.5 Homicide rates in New York, 1967–1997
Source: New York Police Department

The maps of poverty and lawlessness are very similar. They reflect the dynamics of social conditions and the weight of ecological factors. Homicides rose from 1,000 in 1940 to over 2,000 in 1991. In 1992, when crime was at its peak in the city, New York ranked sixth (or seventh, if Chicago is included) among the ten largest cities in terms of homicide. The drop in crime in New York has been all the more spectacular and rapid, since it had been particularly out of control in the 1980s (see figure 4.5). Due to three drug epidemics – heroin, cocaine, and crack – and to drive-by shootings resulting in hundreds of deaths, the social conditions of poor neighborhoods with high-rise public housing worsened. In the two previous decades, tens of thousands of mentally disturbed people were deinstitutionalized and a lot of single room occupancy units were destroyed, pushing more people onto the streets and making them more vulnerable.

The rates of homicides, burglaries, and thefts have not been lower for a quarter of a century. In March 1998, for the first time in a generation, no homicide was registered in Brooklyn for a whole week. (In each precinct, police officers report offenses on a voluntary basis.) Homicides in New York hit 2,262 in 1990 – 29 per 100,000 residents – and fell in 1996 to 984 – 13 per 100,000 – a drop of 57 percent (Lardner, 1997: 52). In 1998 New York registered around 600 homicides, down 75 percent from six years ago. Are the police underreporting or reclassifying crimes? As crime statistics fall, pressure rises on police departments to alter data (Butterfield, 1998). Felonies such as assault and burglary are downgraded and misdemeanors such as vandalism are not reported to the FBI. Police Commissioner Safir admitted that the head of the police department's Transportation Bureau had underestimated crime in the subways by about 20 percent, but he insisted that the overall crime numbers had not been affected.

The crime curve in figure 4.5 shows two downward peaks: from 1982 to 1985 murders decreased by 24 percent (from 1,832 to 1,392) and from 1991 to 1996, by 52 percent (from 2,166 to 984). One has to look back to 1943 and to 1951 to observe such substantial declines.

Repressive Policing: National Influence and Local Innovation

New York's policing policies raise several questions. Should we be influenced by the sophisticated marketing of the New York Police Department (NYPD) and the triumphalist discourse it disseminates all over the world? The NYPD invites foreign experts to observe its success *de situ*, but would applying its recipes be enough to see delinquency curves decline? Would not serious scientific examination show the shortcomings of such a discourse? Would it not be better to search for other causes for the decline of crime in New York City (NYC) and compare the situation there with other cities?

At the national level, many actors are eager to take responsibility for the drop in crime: the same is true at the local level. The election of Mayor Giuliani and of Governor Pataki in 1996 and the recent support for the death penalty mark a change of majorities among the local voters, reflecting the conservative national mood and expressing a rejection of the approaches of both Mayor Dinkins and Governor Cuomo. The incoming mayor and governor both took advantage of moral panics.

The impact of the federalization and politicization of the crime issue can be felt in New York in several ways, such as "zero tolerance" policing, the adoption of "truth in sentencing," fixed term sentencing, the increase in the number of inmates, and the enforcement of tough drug laws. However, the impact of other factors, such as the economic health of the city, demographic changes, drug-consumption changes (see Chapter 2), cannot be denied, not to mention the essential role played by the citizens. But the media and politicians downplayed these factors in order to sell good stories about the New York police's miraculous methods.

NYPD innovations

Former New York police chief, William Bratton, for example, claims that he created the New York "miracle." In 1994 he was asked by the mayor to supervise the largest police department in the country, with its 32,000 employees. Bratton devised new strategies at a time when crime was already declining. He turned to J. Maple, his assistant, who had developed original methods of policing as a police lieutenant in charge of the NYC transit system. Maple had run decoy squads in the subway which had arrested hundreds of teenage muggers working in gangs during the mid-1980s. He had drawn maps for himself, noting subway entrances, sleazy bars and places where thefts were most frequently committed. Analyzing the charts, he could anticipate delinquents' actions. Bratton himself was convinced by the "broken windows" theory, as it was developed by Wilson and Kelling (1982; Kelling and Coles, 1998). American society, they argued, had become lax in standards of public behavior. Signs of social disorder had accumulated and formerly prohibited behaviors were now tolerated: fare-beating in the subway, drinking and urinating in public spaces, playing loud boom-box radios, harassing passers-by, washing car windshields and extorting money from motorists in an intimidating way, etc. Affluent families with children were tempted to leave the cities, and those who could not were buying dogs and weapons and abandoning public transportation. An experiment carried out by a sociologist inspired Wilson and Kelling. Two cars were abandoned on purpose, one in a ghettoized area, the other in an affluent neighborhood. After 24 hours, the first car was entirely dismantled, while the second one was still intact after a week. The sociologist wondered whether things would change if he were to signal that the second car had actually been abandoned, and he shattered its windshield with a baseball bat. Within 24 hours, the car had been dismantled. He understood that inaction by law-enforcement authorities had started a spiral of delinquency. When a first broken window was not repaired, more were broken, then the building was vandalized, followed by other buildings on the same street, and other buildings on other streets, until residents, unable to tolerate what they interpreted as a frightening and unbearable break from the norm, moved out, further accelerating the decline. Such infractions against "the quality of life," such "incivilities," as Skogan (1990) termed them, provoked a deep feeling of discomfort and fear among the remaining residents. But, by adopting a preventative strategy, these feelings would be changed and serious delinquency combatted, as petty delinquents would not become hard-core ones (Wilson and Kelling, 1982). This theory has been contested, but for Bratton and Mayor Giuliani, it provided a strong rationalization for the new law-enforcement policies that they wanted to sell to the New Yorkers.

Between 1990 and 1992, Bratton and Maple worked on a program for the "reconquest of the transit system." Hundreds of police officers were stationed at turnstiles and fare-beaters were arrested. Fraud dropped dramatically: figures fell from nearly 200,000 fare-evaders per day in 1990 to 45,000 in early 1996. The number of disorders decreased in 76 police

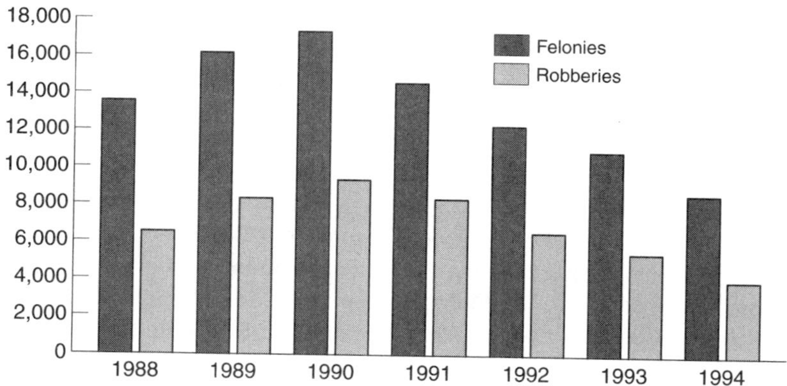

Figure 4.6 New York City subway felonies and robberies, 1988–1994
Source: Kelling and Coles, 1998: 152

precincts (Remnick, 1997: 100). Between 1988 and 1994 this strategy reduced subway felonies by 75 percent and robberies (mostly thefts in the subway corridors) by 64 percent (see figure 4.6). Graffiti disappeared from subway trains and the atmosphere of stations changed dramatically. Fare-evaders who could not show any ID were taken right away to a bus used as a mini-police station, equipped with computers for background checks. Some of the people stopped by the police had warrants for their arrest outstanding – one in seven was already wanted by the police and one in twenty-one carried a weapon (Lardner, 1997: 55). Some of the delinquents, once arrested, gave information to the police that allowed them to expand their files and target their actions in order to seize more weapons or disrupt drug markets.

The mayor, Rudolph Giuliani, born in 1944, is himself the nephew of four police officers and a fireman. As federal prosecutor in the southern district of Manhattan, he pursued organized crime and fought corruption on Wall Street. Defeated in the 1989 mayoral elections by African-American David Dinkins, he won four years later on a platform of anti-street crime. During the years between those two elections, his activities included a stay at the Police Foundation in Washington, participation in a think-tank financed by the Ford Foundation, and active support of the Wilson/Kelling approach. Once elected, he continued to encourage the police who had contributed heavily to his campaign. Bratton's crime-cutting innovations (see below) helped ensure the mayor's re-election in 1997 although Giuliani had, in fact, removed Commissioner Bratton from office in 1996 in favor of a less colorful figure, Howard Safir. The Bratton reforms were as follows:

- While crime rates were declining, Bratton increased the police force to 40,000 officers for a population of 7.5 million residents.
- In line with the "broken window theory," a zero tolerance approach was developed towards "quality of life" crimes.
- New training programs were developed for pursuing gun- and drug-traffickers and those involved in committing crimes.
- The discretion of officers at the lower end of the hierarchy was expanded and patrol officers were empowered. Day-to-day operations, managed at the precinct level, afforded commanders greater responsibility and control. Merit bonuses were awarded for outstanding officers and disciplinary action was taken against those who failed to meet high standards (the Patrolmen's Benevolent Association is opposed to such systems).
- Uniformed officers were authorized to arrest drug-dealers and seize guns, cash, drugs, and cars instead of letting the specialized narcotic investigation units operate alone. A notorious street crime unit proceeded to make immediate arrests.
- Undercover operations, aggressive crackdowns on drugs, ID checks, searches of car drivers and transit users, and raids on suspicious apartments were launched.
- In order to curb youth violence in the schools and reduce gangs' influence, School Security Plans were elaborated for every public school. Police cooperating with security guards located and returned truants to their schools. The number of youth officers was increased.
- Police attempted to break the cycle of domestic violence. More violence officers and investigators began to work with extensive databases to intervene proactively and put the victims in contact with shelters and social service agencies.
- Police were directed to reclaim public spaces and address minor public disorders.
- Reforms were instituted to improve managerial efficiency and computerize as much information as possible. These efforts led to the merging of the transit police and the public housing police into the NYPD.
- Weekly statistics were produced by the precincts for the Compstat (Computer statistics) meetings.

The Compstat process In New York, information has always been a form of power. Until the reforms came along, it would take several months for the police to gather operating statistics. When the crack epidemic hit the city in 1986, the police reacted very slowly. Crack-dealers carried large sums of money and weapons to protect themselves from being assaulted. If a dealer owns a gun, then the other "homeboys" would want guns too, for self-protection (Canada, 1996). By chance, a police officer would notice that gun shots were always exchanged on the same street corner and that drug transactions and arrests were recorded in the same "hot spot." All complaints, arrests, and summons activities are compiled by police officers on the beat on the local computer system, and then forwarded to the Chief of Police's Compstat (Computer Statistics) Unit, where the data is collated and loaded into a city-wide database.[3] A computer analyzes the data and a weekly Compstat report is generated, broken down by precinct, patrol, borough, and city-wide levels. The computerized system developed by Jack Maple allows the police to access this information instantaneously and to examine it visually. Relying on this system, they draw up strategies and analyze their impact. These data are presented on a "week-to-date, prior 28-day, and year to date" basis with comparisons to previous years' activity.

Compstat represents a technological revolution for city policing. Weekly crime-control strategy meetings began in April 1994 as a means of increasing the flow of information between officers and commanders of operational units. The briefings are referred to as Comcon meetings, since many of the discussions are based upon statistical analyses. The Commander Profile Reports also permit executives to scrutinize commanders' performances on non-crime issues, such as the amount of overtime generated by members of the command, absence rates, civilian complaints, etc.

From 7 to 10 a.m., twice weekly, Comcon (Command and Control) meetings examine the districts of NYC one by one (it works out that each district is discussed once a month) in an intensive interactive management strategy. During the meetings, held at 1 Police Plaza in Manhattan, huge maps produced by the *GIS* (*Geographic Information Systems*) appear on the walls depicting the sub-neighborhood under discussion. The database produces the date and time of the incident, the address, the jurisdiction, the penal sanction, etc., showing the evolution of crime by week, month, and year. Electronic pins mark incidents on the street maps. The message conveyed by the Compstat is that any means seems justified to reach the desired end of reduced crime statistics.

Police executives have 36 hours to study the files before a Compstat meeting. The relevant precinct and operational unit commanders discuss their successes and problems and are questioned on their failures. They speak in front of an audience that includes representatives from the District Attorney's office, federal and state law-enforcement agencies, crime strategy coordinators from other boroughs, internal affairs bureau personnel, and ranking officers. Obviously, their presence exerts a tremendous pressure on the precinct brigades. The processes are driven by four principles: accurate and timely intelligence on crime, effective tactics adapted to shifting trends, rapid deployment of personnel and resources, and relentless follow-up and assessment. Direct contact between the police commissionner and precinct commanders helps valorize rank-and-file officers. The feeling that a large machine has been set in motion, that the operation is tightly run, that the police form one block, that they are supported strongly by the mayor, that results are expected week after week, that feedback mechanisms ensure that local commanders remain accountable for the discretion they have been given, all represents a threatening police regime within the city. The vocabulary of the "war room" is military, the headquarters feel that they are waging a war against criminals and that tolerance is not an option. The idea is to get results. A binary logic of "us/them" is used to mobilize the police, the consequences of which are at times measured in terms of abuses and excessive police brutality.

The limits of police success

It is worth noting a number of points concerning crime rates in NYC in the late 1990s:

(1) It has to be emphasized again that the drop in crime followed a spectacular rise a few years earlier.
(2) In cities that did not enforce the Bratton/Safir strategies, crime also declined (in Houston by 48 percent, in San Diego by 46 percent, and in Dallas by 45 percent).
(3) Attributing the decrease in crime solely to new police technologies and to situational crime prevention would be scientifically dubious. Such approaches dealt less with the roots of crime than harassing the delinquents and forcing them to constantly redefine their own strategies.
(4) The computerized information may be problematic. The maps produced by *GIS* juxtapose social, economic, and demographic factors such as single-parent families, race, social housing, and unemployment rates, as if such factors had a deterministic correlation with crime. Numerous studies have shown that crime is more unpredictable than statisticians think. The collection of such data by the police would not be allowed in France.
(5) Are homicides committed by the hot-spot residents or by outsiders? What is the density of the population on the maps? There may also be little understanding from pin points that represent one or ten crimes on a map.
(6) Police strategies need to be constantly reviewed and updated. Otherwise, in time, their efficiency declines. Between 1993 and 1995 the number of weapons seized by the police declined by 9 percent, and dropped by 15 percent in 1996. The reason? Teenagers adjusted

their behavior and left their weapons at home. In 1993, one fare-evader out of 438 was arrested with a gun, one out of 904 the following year, and one out of 1,034 at the beginning of 1995 (Lardner, 1997).

(7) Not one, but several strategies were implemented. And some of them had already been implemented under Mayor Dinkins (he increased the police force by 4,000 officers, for instance).

(8) There was a co-occurence of factors (economic health, demographic decline in the number of teenagers, reduction in drug wars, more aggressive policing, massive incarceration, etc.) rather than a causal link between the police reforms and the drop in crime (Moran, 1995).

(9) The greatest limitation of this police strategy comes from its heavy human cost and from citizens' mobilizations of protest.

It is in poor minority areas that arrests are the most numerous. The situation in NYC reveals a heavy repression whose abuses have been covered up. In 1992, Mayor Dinkins appointed a commission headed by Judge Milton Mollen to investigate corruption in the police department. The commission found shocking evidence that numerous police officers treated citizens brutally, committed thefts, protected drug-traffickers, sold and used drugs, falsified police reports, and lied in court. It concluded that

> it was a multi-faceted problem that has flourished in parts of our city not only because of opportunity and greed, but because of a police culture that exalts loyalty over integrity; because of the silence of honest officers who fear the consequences of "ratting" on another cop, no matter how grave the crime; because of willfully blind supervisors who fear the consequences of a corruption scandal more than corruption itself; because of the demise of the principle of accountability that makes all commanders responsible for fighting corruption in their commands; because of a hostility and alienation between the police and community in certain precincts which breeds an "Us versus Them" mentality. (Mollen Report, 1994: 1–2; Human Rights Watch, 1998: 268–313)

At the same time, a report by Amnesty International about NYC in 1992 showed that the city ranked last in terms of ethnic police representation (African-Americans represent 28 percent of city residents, but just 11 percent of the police force; Latinos are 24 percent of NYC's population, but 13 percent of police officers). The command staff of the NYPD is 90 percent white. Efforts to increase diversity in the police force compete with family traditions that push police officers' children into entering the profession. Half the white recruits in 1994 had a relative currently or formerly with the NYPD (*NYT*, 10 October 94). For a long time, the police and fire fighters made up an Irish fiefdom in New York (Waldinger, 1996). But the

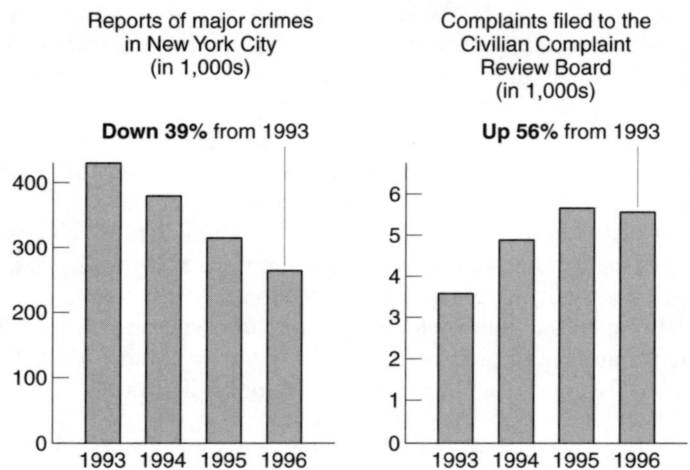

Figure 4.7 Crime rates versus police complaints, New York, 1993–1996
Source: J. Lii, *NYT*, 16 May 1997

police have become estranged from the city residents whom they are there to serve. Many policemen and women have grown up in the suburbs and a majority do not live in the city. This socialization does not help them understand minorities – a similar problem to that in France. Moreover, many of the officers sent out to interact with the public did not receive enough training or close supervision.

The report also pointed at numerous abuses.[4] As David Dinkins learned with the Crown Heights episode, aggressive policing cuts both ways for a politician. In the past, police brutality cost NYC $87 million in court judgments. In 1987, 81 percent of polled New Yorkers judged their police to be too brutal, and the rate of complaints increased by 56 percent between 1993 and 1996[5] (see figure 4.7). Yet, as observed by Paul Chevigny, "the NYPD is not a notably abusive department in the big cities of the United States" (1995: 85).

Since the Amadou Diallo episode, which involved a street peddler immigrant from West Africa who was gunned down in his doorway with 41 bullets fired by four frightened officers from the elite street crime unit in March 1999, a spotlight has been turned on the serious and systematic police abuses accompanying the crackdown on crime. The abuses were overlooked because of the ambivalent attitudes of New Yorkers, including low-income minorities such as those I have also observed in the São Paulo *favelas*: on the one hand, everyone appreciates safety and order in a neighborhood, especially those who cannot buy private services of security as other categories do and who are the first targets of crime, but, on the other hand, mistrust and fear of the police, especially by the youth of poor minority neighborhoods, is very strong. Nearly nine out of ten black residents in New York think that the police often engage in brutality against blacks (*NYT*, 17 March 1999). Other statistics show that individuals who come through the criminal justice system are usually drawn from the communities that are the most victimized by violent crimes. Young African-American males are 23 times more frequently incarcerated than young whites (11 percent of Latinos between the ages of 20 and 29 are in jail).[6] Are they 23 times more guilty? Police arrests lead them to be imprisoned for traffic violations, drinking in public, riding a bicycle on the sidewalk, etc., and it is unlikely that they will be better socialized and integrated when their time is up.

Policing is a "tainted occupation" (Bittner, 1990), and it means almost exclusively working with the poor. Almost all street work by the police is conducted among poor people. Policemen and women have enormous discretion and are almost always on their own when on patrol. An iron-clad code of silence protects them from being accused or testified against by other officers.[7] The Mollen Commission found that there was virtually no effort to punish policemen for lying when they got caught, that supervisors taught them how to prepare convincing false testimony, and that nothing is really wrong with some unnecessary force. "Brutality strengthened the bonds of loyalty and silence among officers and thus fostered corruption tolerance," the Commission found (Chevigny, 1995: 63, 80). A young Dominican told me that he was so scared of being "framed" by police officers in Washington Heights that he feared going to work at a bodega nearby. He was on the verge of a nervous breakdown when we met and his friends were advising him to move out of the neighborhood.

In the spring of 1999, Major Giuliani discovered that governing through police with a "go-get-'em" mentality could backfire. At the time, only 43 percent of New Yorkers thought that the police were doing an excellent or good job and the popularity of the mayor had sharply declined to 42 percent from 63 percent five months before (*NYT*, 17 March, 4 April 1999). The Diallo case resulted in several demonstrations in New York City during the spring of 1999. The incident occurred after numerous other incidents of police abuse. Thanks to action orchestrated by – among others – the Revd Al Sharpton and his Harlem-based organization, more than 1,200 people – including the two former mayors, local and national public figures, a number of actors, radical black and white citizens, Jewish rabbis, and hundreds of ordinary moderate residents – submitted themselves to arrest by the police in an act of disobedience and protest against illegal searches and ethnic profiling. In other words, in the defense of social justice and respect, civil society stood up against a dysfunctional institution, compelling it to alter its conduct. Even Republican Governor Pataki criticized the mayor for insensitivity and a poor handling of the situation. The Revd Al Sharpton admitted to me that the black leadership was to blame for having been passive for too long. Some of the leaders became part of the establishment. But with a new generation, remarkable changes are taking place, he added. The minority leadership has had to find a new vigor in the issue of social injustice and resistance, mobilizing a diverse and wide constituency that outreaches beyond the city limits. After continuous protest demonstrations during spring 1999, the police modified their behavior, Sharpton observed, and fewer acts of brutality and abuse were reported, more rules were enforced. Sharpton's message is clear: cosmetic changes will not be tolerated; what is required is better police training and supervision, residence requirement in the city, and more psychology training given to police officers. If enough is done to transform police culture, then pressures from civil society will gradually be alleviated (interview, June 1999).

The ambiguous position of criminal judges

Does justice contribute to the drop in crime? This issue is of interest from a comparative point of view. In France, tensions between the police and the justice systems are strong, each body accusing the other of hampering its efforts. The police ask why they should be efficient if juveniles are released from court an hour after their arrest and start trouble again, sneering at the same police officers who just arrested them. Juvenile court judges ask how they can indict youths if the cases are poorly managed by the police and there is insufficient evidence. If 98 percent of penal cases are closed in the UK, the situation is hardly better in France. Things are dramatically different in the USA in this respect.

The laws of New York State have changed a lot in the last twenty-five years and determinate sentencing has been imposed by the federal government.[8] A classification of offenses from A to E (A being the most serious) and managerial goals have turned justice "into an assembly line," Judge Fried remarks.

In 1972, New York State passed the nation's "toughest drug laws" at the behest of Governor Nelson Rockefeller in the aftermath of Nixon's war on drugs. The governor had national ambitions and wanted to be remembered for his policies. The laws instituted a mandatory life sentence for anyone convicted of the sale or possession of narcotics. It prohibited plea-bargaining and eliminated parole. Criminologists Feeley and Kamin say this is an example of how moral panics related to crime led to draconian policies in the local criminal justice systems, followed by adaptive behavior and decisions which reduced the laws' effect. They show that even if judges, prosecutors, and other court personnel vigorously opposed the Rockefeller laws in New York, they nonetheless found it useful to embrace the exaggerated claims of the law-makers to secure more resources for themselves. Then they demonstrate that after a while, officials employed a host of discretionary devices to adapt to the laws in ways that reinstituted long-standing operating procedures. Finally, court officials quietly but publicly pressed to amend the harshest provisions of these laws and to rectify inequities (Feeley and Kamin, 1997: 136).

On the one hand, we have here an example of the politicization and dramatization of the crime issue, for the benefit of the governor who then promised that the law would be "brutal" and that it would provide "the strongest possible tools to protect our law-abiding citizens from drug pushers" (Goldstein, 1973: 1). On the other, it also illustrates the capacity of local actors to resist and adapt the law in a more peaceful and reconciliatory way.

As a result of the law, the number of cases increased so much that fifty additional judges had to be added at the lower level of the criminal justice system and extra prosecutors were hired and courtroom space was expanded – all measures that justice officials had long waited for.[9] In the three years following the passage of the law, there were fewer drug convictions and fewer offenders sent to prison than expected, probably because judges dismissed charges more often than they had previously done (the same behavior is observed with the "three strikes" laws). In other words, the system adjusted and the laws were normalized (Feeley and Kamin, 1997: 140). In 1976, when the chief drug prosecutor said he would no longer enforce the prohibition against plea-bargaining, there was almost no public response: the moral crusade had passed. In 1979, sentences related to marijuana were reduced and plea-bargaining and flexibility in sentencing restored. Sentences, however, remained harsh for other drugs and someone convicted of selling two ounces or possessing four ounces of narcotics would be sentenced to 15 years or more.

The judges whom I talked to complain that their discretion has been severely limited by conservative politicians. They face pressures similar to those put on defense attorneys. They are allowed limited time to look at mitigating circumstances, regardless of human consequences, and feel they have few options. Judges remark that they frequently meet juveniles whose lives could easily figure in a novel by Charles Dickens. As judges, nonetheless, their hands are tied by law-makers. Legal assistance to the poor has been decimated by budget cuts. Alternative sentencing is not used enough, as a result of political pressures. Moreover, public emotion is constantly fueled by the media calling for prosecution in defense of victims, as demonstrated by the Bernhard Goetz case in 1991 (Rubin, 1986).[10]

The prison explosion From 1983 to 1989, under Democratic governor Mario Cuomo, the number of new inmates serving state time for sale or possession of drugs increased over 600 percent. Governor Cuomo initiated a prison-building policy, in order, he said, to avoid inflicting the death penalty. (A "three strikes and you're out" policy was supported by Cuomo in 1994, along with thirty other governors.) The death penalty was passed by the state legislature after Cuomo's departure in 1995, however, and the prison-building policy remained. In 1994, inmates from poor neighborhoods were overrepresented among the 64,500 inmates in the state prisons from NYC (they made up 69 percent) and in the 16 jails containing about 18,000 defendants, (a slight decrease since 1992) and among the 80,000 offenders on probation, not including the 4,000 juvenile offenders in various institutions. These numbers are not surprising: they confirm that the neighborhood context influences the willingness of police to use coercive authority against suspects; it influences their evaluation of the danger of the suspects to the community (according to their employment, marital status, and length of residence), and it affects the decision to detain them or not. The poor are unable to afford bail and remain incarcerated in pretrial detention, then meet their court-appointed lawyer only for a few minutes in the court pens. Fewer than 5 percent of NYC's felony defendants and fewer than 1 percent of those facing misdemeanor charges ever go to trial (Page, 1993: 616). They are offered a plea-bargain, though they themselves are not present at the conference between the judge and the defense attorney. The pleas are based on very scanty information, and plea-bargaining negotiations, which last four minutes or less, can result in years of incarceration.[11]

Research shows that there may be some racial discrimination, some of the time in some places, but, according to criminologists Sampson and Lauritsen, race discrimination is not as pervasive as commonly thought (1997: 348). Again, crime is what society decides it is. Crime rates and incarceration rates are independent.

When the "toughest laws" in the nation were passed in the state of New York, over 90 percent of the individuals arrested for drugs were black or Latino (Page, 1993: 611).[12] One reason for this is that it is easier for the police to arrest them on the street than their white counterparts. On the whole, as in most of the urbanized states, the number of incarcerations since 1980 has doubled (*NYT*, 8 December 1991). In the late 1990s, prison-building cost New York State $850 million, adding to its debt.[13]

More judges resort to sending juveniles over the age of 13 to adult courts. The death penalty is applied to those of 18 and over in the state.

Boot camps experience mixed results, according to Judge Bingham. Incarceration time runs from 90 to 180 days. The quality of the staff and the care given after the camp explain success or failure. Some camps emphasize discipline, others education, others labor. Drug treatment is prioritized.

On the whole, the judiciary seems to have been colonized by politics. Whenever possible, some judges choose education and probation over incarceration. But in the last few years, such judges have been accused of giving a "phony" justice and have been called "idiots" by the mayor and the governor.[14] They reacted by appealing to citizens through public forums and trying to justify themselves, as will be seen with the case of Robert Johnson.

The capacity of resistance of a Democratic city to a national conservative mood is limited by the subnational state, which determines judicial orientations. The city's funding from the state of New York is allocated by the legislature, after heated negotiations between Democrats from the city (who are in the minority) and Democrats and Republicans from the suburbs and the rural areas upstate. The current Republican governor relies on the politicization of crime for obvious political gains and aims at reducing preventative measures. He has constantly pushed for more prison space ($810 million was allocated in 1997 for a plan to build 7,000 double-bunking cells over three years, even after 1,500 additional beds and cells had been built in 1996), while offering to cut funding for retirement homes, medical care, universities, youth centers, public housing, mental health services, probation services, legal aid, alternative sentencing, health in prison, drug treatment, etc.

In March 1995, the governor signed legislation reinstating the death penalty in New York for anyone convicted of first-degree murder. As he said in his message to the legislature in January 1995:

> It is time to hold criminals responsible for their individual actions . . . New York State must toughen its laws to put violent criminals behind bars and keep them there . . . We must say: no more second or third chances for violent criminals . . . Young violent felons must be treated the same way we treat adult violent felons . . . Your criminal acts are your responsibility and you will be held accountable.

An example of local resistance and culture: the Johnson case The Johnson case demonstrates the importance of a local (in this example, progressive) culture and the resistance which can be mounted by actors within the system in the face of a national conservative mood. Robert Johnson, the African-American district attorney (DA) in the Bronx area, opposed Pataki over a criminal prosecution under the so-called "Murder One Death

Penalty Statute." For obvious electoral reasons and "visceral gratification," the governor tried to force the prosecutor by executive order to introduce the death penalty into murder sentencing. Johson explained his resistance to the governor in a press release on 9 March 1995:

> I was raised by loving parents who instilled in me an intense respect for the value and sanctity of human life . . . As a result, I have devoted my life to the criminal justice system. During more than 20 years, I have seen the devastation inflicted by those guilty of horrible crimes. I have felt the rage and thirst for vengeance which all but consume the victims and their families . . . But I have also personally witnessed the devastation of those wrongfully accused . . . Under our system of justice, the death penalty neither can nor should be mandatory.
>
> Consequently, it is highly uncertain that the penalty actually will be imposed by a jury in a given case, that its application will be fair, that the sentence will be upheld, on appeal, that the defendant will be executed and that others will be deterred . . . Moreover, the price of this uncertainty is enormous, given the costs in time and resources of trials and appeals in death penalty cases. Clearly, this money could be better spent on providing more judges and courtrooms so that more defendants could be brought to trial more quickly. The money could also be better spent on valuable and broadly based crime fighting and crime prevention programs, including reducing the flow of illegal guns, incarcerating more violent criminals and providing more assistance for crime victims. While these programs may not provide the visceral gratification of the death penalty, they will do a lot more to improve the quality of our lives.
>
> For all these reasons, while I will exercise my discretion to aggressively pursue life without parole in every appropriate case, it is my present intention not to utilize the death penalty provisions of the statute. (quoted in Silver, 1996)

It was difficult for the governor to claim that life without parole was not an exemplary sentence. And on 2 November 1995, the people of the Bronx re-elected Johnson, giving him an 89 percent majority of the vote.

But just over a month later, when a man was accused of killing five people in a shoe store in the Bronx, the conflict arose again. Pataki immediately wrote the DA to inquire whether his approach to the case would be based on "your professional discretion or . . . on a policy decision not to seek the death penalty in any case in Bronx county." The governor demanded an answer by 5.30 p.m. that very day. Johnson replied that he would seek life imprisonment without the possibility of parole and the governor indicated that he accepted this decision "with grave reservations." "I am concerned," Pataki said, "that the laws of the State of New York may not be executed . . . [The people] have determined that the death penalty should be an option available to jurors in capital murder cases" (quoted in Silver, 1996).

In March 1996, a police officer, Kevin Gillespie, was murdered in a shoot-out in the Bronx. Three men were arrested. The governor put pressure on the DA again: "the case is one in which the death penalty seems particularly warranted . . . The people have spoken. As governor, I cannot permit any DA's personal opposition to a law to stand in the way of its enforcement. No one, including a DA, can substitute his or her sense of right and wrong for that of a Legislature." Johnson said that the death penalty was no more the law of New York than is the penalty of life imprisonment without parole: "The statute in no way suggests that a sentence of death is 'the better' or 'presumptive choice' . . . Yet you are demanding the death penalty within hours of the killing of Gillespie" (quoted in Silver, 1996). (The DA has 120 days after an indictment to exercise his statutory option to seek death in any given case.)

One day after Johnson's answer, Pataki issued Executive Order 27, substituting the Attorney General, Dennis Vacco, for the DA in the prosecution of the individuals linked to Gillespie's murder. Johnson decided to petition the Court, claiming that the Executive Order was unconstitutional and illegal, that it went against the people of the Bronx who had elected him, and that this power was unprecedented: the governor was creating the law rather than enforcing it. The DA's independence was no longer protected. Throughout the controversy, Johnson retained the unfaltering support of his constituents. And the infringement on the free speech and discretion of the DA had a chilling effect. But his request that the Executive Order be ruled unconstitutional was dismissed.

On the one hand, the incident illustrates the extremism of some politicians eager to flatter some public thirst for vengeance and their emotions for electoral purposes. On the other, it reveals the local culture of New York, individual capacities of resistance, and the Bronx voters' clear heads. That voters would support their representative against state domination is also a phenomenon of the late 1990s.

Prison interactions Former Commissioner of Correction Facilities, Michael Jacobsen, suggests that the number of inmates in the state is not excessive "granted the current repressive trend."[15] At the end of 1997, there were 69,500 people imprisoned in the state, as opposed to 12,500 in 1973. Broken down, these figures include 46,600 inmates in city jails, 7,238 in suburban jails, 7,971 in upstate cities, and 5,412 in rural zones. While the national average is 354.3 inmates per 100,000 citizens, New York State is slightly higher, at 367.6 (compared with 645.6 in Texas and 529.4 in Louisiana).

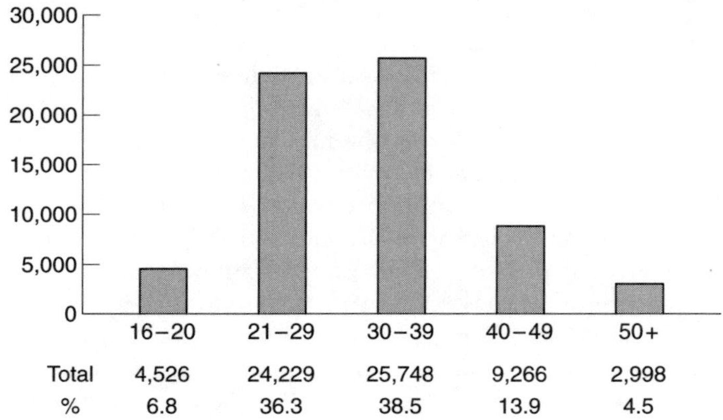

	16–20	21–29	30–39	40–49	50+
Total	4,526	24,229	25,748	9,266	2,998
%	6.8	36.3	38.5	13.9	4.5

Figure 4.8 Age of inmates in New York state prisons, 31 December 1994
Source: Annual Report on Crimes and Offenses, 1994, New York State Division of Criminal Justice Services

The Rockefeller laws impacted on the number of inmates. As already stated, from 1983 to 1989, during Cuomo's tenure as governor, the number of drug-related detainees increased by more than 600 percent. Over 90 percent of them were minorities (Harries, 1996). African-Americans represented 12.4 percent of the state's population and 50 percent of its inmates (Latinos accounted for 10.8 percent of the population and 33 percent of all prisoners.) Of prisoners in New York City jails, 91 percent are minorities and two-thirds of them come from the city. Juveniles represent 6.8 percent of all inmates (see figures 4.8 and 4.9).

According to Commissioner Jacobsen, public debate is simplistic. He cites the "three strikes and you're out" law. "We are constantly looking for new ideas facing the repeated failures of the criminal justice system. But alternative sentencing yields better results. Juvenile programs of intensive care bring back teenagers to school. Yet politicians surrender to emotions triggered by the media and to a thirst for revenge" (interview with author).

A reformer with an $800 million budget in 1997, Jacobsen has relied on financial arguments to convince law-makers in Albany that alternative sentencing actually costs less than incarceration. The provision of 1,500 beds for drug treatment inside the jails allows a reduction in the number of security guards and creates jobs for social and care workers provided by non-profit organizations. Recidivism is decreasing as a result of alternative sentencing.

Since 1981, Governors Hugh Carey and Mario Cuomo added 39,561 prison beds at a cost of $100,000 out of a budget of $4 billion for the justice system. Governors and mayors do not have the same financial con-

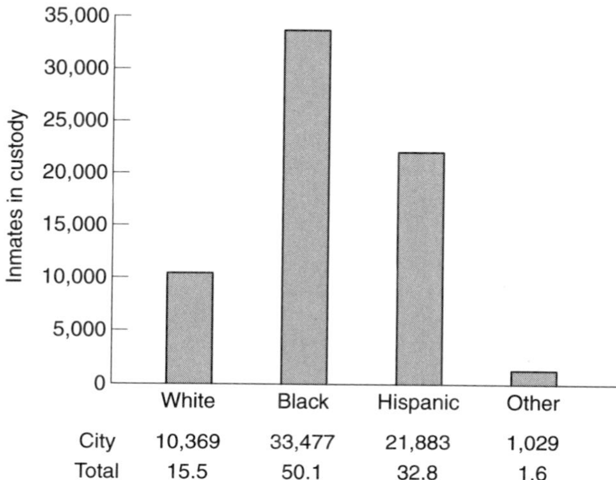

Figure 4.9 Ethnic profile of inmates in New York State prisons, 1994
Source: Annual Report on Crimes and Offenses, 1994, New York State Division of Criminal Justice Services

straints.[16] According to Robert Gangi,[17] the director of the Correctional Association of New York, an organization which works in defense of inmates' rights:

> One of the differences between the past and the current situation is that, 70 years ago, Italians were routinely sent to prison. Then families would take care of them because the communities needed them. This is no longer the case. Communities get organized without a male presence. Our current economy does not seek to incorporate those unskilled young men who feel useless, thus devalorized. We know that 10 percent of men between the ages of 24 and 54, the "discouraged" workers, have disappeared from unemployment figures . . . Every day, a quarter of young African-American males are in the hands of the justice system, that is, twice the number of those on the campuses of New York.

Joanne Page, director of the Fortune Society, a New York non-profit organization which aims at resocializing ex-convicts, believes that the criminal justice system reflects the values of society and the resources it wants to allocate to it. The offenses of the poorest are punished, she said, while those on Wall Street go free.

The Rikers Island prison Rikers Island is located about 12 miles from NYC. The penitentiary there stretches over 415 acres. The facility includes ten four-story-high prison buildings, each with fifteen cells. The prison holds 19,000 pre-trial inmates: 92 percent are minorities, 90 percent did not finish high school. It has been said that the "correlation between high school dropout and prisoner rates is a trifle higher than the correlation between smoking and lung cancer."

There is one guard per 75 inmates. Space is informally divided according to ethnicity; even the use of the public phones is distributed among Puerto Ricans and African-Americans. Because the environment is dangerous, because they have enemies, inmates try to conceal weapons. Violent clashes occur regularly. Inmates sometimes attack security guards in order to be moved out of Rikers. Drug-taking is high: 70 percent of inmates use drugs, 25 percent are HIV+, 25 percent are mentally ill, and 20 percent have tuberculosis. The 8,200 guards are paid $160 a day to supervise Rikers. There are 300 admissions a day. Each inmate costs the city $58,000 per year – drug treatment alone costs $17,000. For the city, the overall cost for Rikers runs at $765 million annually. In 1993, 73 inmates died.

Rikers officials approve of drug treatment as a cost-effective way of reducing violence, an easier way of managing offenders with a drug problem. In all, 15,000 beds are dedicated to the substance-abuse program managed by the Vera Institute (see below). As a result, more civilian employees are involved, the number of penal officers is reduced, and the service becomes privatized. If violence diminishes, fewer penal officers are needed. Easing the stress on management and budgets is an argument the administration cannot ignore. Evaluation of the program done by the National Institute of Justice reveals that drug treatment reduces costs.

Inmates who test HIV+ are frequently homeless, unable to post bail, or even pay a legal aid defense lawyer. They remain in Rikers for six months on average.

When I interviewed a number of juveniles in the prison at Rikers Island, it seemed to me that their hatred and their rage could only worsen in jail. Thanks to the Fortune program for HIV+ inmates (see below) the youths I met were very tense. "The first time you are in Rikers, it is scary and people around us want to jump on us, beat us, take what we have. If we don't fight for our possessions, we appear soft (a 'pussy') . . . if we fight, we get 'respect,'" they observed. Those youths were convinced that using a gun was the only way to resolve conflicts. They refused to admit that such logic could work against them, that another homeboy could disrespect them because "they looked at them wrong" or bumped into them. They said that they were nervous when they were outside of their housing projects. They thought that everyone carried a gun in their neighborhood.

They were ready to fight for "stupid reasons," for girls, in order to "feel better." They had to act when they felt people disrespected them, they could memorize a face for a long time. They did not expect anything from life. "There is no way out," one youth repeated several times. "Either you get shot or jailed." They were certain to be targeted by racist police. They chose to join a gang, and meeting the homeboys at a hot spot would be a rite of passage. The proliferation of guns marks a difference from the situation in French housing projects. Guns are still rarely used to solve conflicts in France.

The debate that day focussed on guns. The teenagers could not remember the last time they had solved a confict with their fists (one to one). They would not admit that changing their attitude and talking their way out of a conflict could be efficient. "Guns are the solution to defend a territory" and to ease the tension. "It is so easy to press the trigger," one said. "It may seem strange. Although it is not natural to kill someone, it is not so difficult to learn." In the war zones, the oldest teach the younger ones how to survive and to fight for their honor, to defend a street corner, a gang color, or a leather jacket. Stanley and Wyllis, the mentors from Fortune whom I was with, tried to explain that the cycle of vengeance is endless and that for five minutes of gratification everybody is in jeopardy: they are in jail and their opponent is free to go. What is left?

But the highly strung teenagers answered that they lived in killing fields ("New York, it is like the Third World War"), saying they were just a target for other guys (they are more afraid of one another than of the police, it seemed): "It is always a game. Getting shot or jailed, there is no way out." "We kill everyone, Blacks on Blacks, everybody in my family has been stabbed, robbed, shot." "The pain in my leg is gone but in my heart, the anxiety increases." "It started when I was 12 years old; an idiot ran to my mother's room to rob her." To the question: would you like to move? "No, you've to stay where you have to stay." Is it an "age" thing? "No, violence affects everybody, violence is there."

Then racial tensions inflamed the African-Americans against the Puerto Ricans. The blacks accused the latter of dating white girls and the Puerto Ricans accused them back. Insults poured out as a result of sexual frustrations. The words used by the Rikers teenagers were echoing those I had heard during my visit to a maximum security penitentiary in France, at St-Maur. "Violence is everywhere," "Violence is power in a powerless life," French inmates would say.

All the incarcerated American teenagers I spoke with seemed to belong to gangs: the Brownies, the Dead Squads, the Beasts, etc. During the fall of 1997, an African-American gang called The Bloods (unrelated to the California gang with the same name) attracted the media's attention on Rikers. To distance themselves from the Latinos on Rikers, the gang began attacking Latino prisoners and cutting three stripes into their faces. Soon this "fashion" spread outside the prison walls and reached the poor neighborhoods to which the inmates belonged. The media dramatized the phenomenon for TV viewers. Metal detectors were put into action and correctional officers were required to constantly search for the razor blades that inmates hid on themselves.

Schools, delinquency, and violence

The 126 public schools in New York form a pyramid.[18] Four schools at the top are very much in demand. They are specialized, their students are selected by competitive entrance examinations, and their graduates go to the best colleges. The intermediate stratum is made up of most of the other schools, which are trying to join the top. About 47 schools remain at the bottom of the pyramid. Hundreds of violent incidents occur each year (129 gun-related incidents in 1991–2). These schools appear to be in a chronic state of emergency.[19] In these 47 overcrowded schools, located in racially segregated, deteriorated areas where the physical environment is falling apart (vacant lots filled with refuse, crackhouses, potholes, etc.), the Board of Education has installed high-tech public safety infrastructures after being pressured by parents and insurance companies. There are 3,200 guards (SSO – School Safety Officers) who operate in these schools. This type of protection cost New York taxpayers around $4.5 million in 1990 (Devine, 1996).

Two contrasted field observations illustrate how the city can choose to innovate or not, when confronted with the intractable problems of inner-city areas.

The intensive paramilitary management of spaces and bodies In a secondary school with 2,500 students in Prospect Heights (an area of Brooklyn known for its clashes between Hasidic Jews and African-Americans, its poverty, its high-rise public housing projects, and its high crime rates), the principal's first concern is safety.[20] When I went to the school with John Devine, author of *Maximum Security: The Culture in Violence in Inner City Schools*, I was subjected to several security measures: a visitors' sign-in desk run by security guards, ID card swiped through a guard-operated computerized machine which grants access if the light is green (the machine records entrances, absences, and disciplinary data for each student; other controls are done electronically during the classes), a metal detector, magnetic scanner – all of these operations amidst the loud noise of walkie-talkies, alarms, magnetic locks, etc. The students are sometimes half an hour late to their classes because of all the scanning.

That morning, with vociferous shouts, six security guards, two armed police officers, and a hall supervisor attempted to prevent a suspended student from getting in without an accompanying adult. "Verbal exchanges and demeanors . . . introduce students to police culture and introduce the language of the street into the school environment" (Devine, 1996: 27). The long walk along the school corridors reminded me of St-Maur, the French maximum security penitentiary. A gymnasium locked up while the students perform sport activities, a cafeteria also bolted and surrounded with gates to prevent students from wandering through the halls, a mini-police station inside the school, close to the nursery in charge of 24 infants so that their mothers can attempt to graduate.

The education professionals have delegated the installation and operation of all of these security devices to experts. And ultimately it is less about technology than an indicator of suspicion, a lack of respect, and a distrust of students, John Devine explains. They show that policing supersedes education. Public opinion demands, in terms of instant results and out-of-the-box solutions, contribute to a deterioration of the democratic public space. One example should illustrate this idea. When leaving the school, I saw seven teenagers handcuffed to one another. They were truants being brought back to school, handcuffed at the demand of the mayor and the request of nearby shopkeepers. In the fall of 1998, after two years of controversy, the city council authorized the NYPD to have direct supervision over the security guards and to be involved in training them at the police academy.

The principal's discourse admitted violence, even while denying it. Although he knew that "half the students carry weapons," it did not seem to him that "this school was worse than others." Other adults' reactions vacillated from complete pessimism to forced optimism. Yet violence cannot be denied; it is not a fantasy. In fact, several homicides had occurred in that school. Guns, razor blades, and other weapons circulate. Teachers know that students need to carry razor blades to reassure themselves. The scripts of their lives show that violence is an integral part of their school environment. Serious incidents – cut faces, injured bodies, ethnic clashes, beatings – may erupt after weeks of calm, one teacher says. The walls of the school are porous and street violence comes in and out without warning.

Without acknowledging it, the principal is not upset when about one-third of his students miss classes. As in France, he complains about "hard-core" students, "fifty of them who turn the school upside down." This school is constantly in motion. Students come in and out because of suspensions. Some students are in school but never attend classes. They linger in the corridors and some of them deal drugs. Troublemakers are suspended and spend a day in the cell of the mini-police station. If they have been suspended for several days, they must return with an adult. But some of them live alone, until a social service from the city decides to take action. For the more serious incidents, principals from the 47 low-tier schools "exchange their rotten apples" within the school district, the principal remarks. The teachers seem to have accepted the dichotomy between the transmission of knowlege – their job – and that of security in the hands of professionals. There is a division between the management of minds and that of bodies. Even the classes focussing on conflict resolution seem unrelated to facts. The teacher talks about it before or after the incidents; he or she seems disconnected.

Conversely, focussing on high-tech security, the security guards are anything but virtuous, according to Devine. They are not accountable to the principal but to the security department of the Board of Education and now the NYPD. Although there are already 17 security personnel in the school, the security team is reinforced by 5 hall supervisors, 18 assistants, 7 teachers (rotating to take care of security), and by adjuncts, a total of 110 people. In comparison, 150 teachers work in the school. This new paramilitary culture based on safety in the school space – practices, language, rules, equipment, locker rooms, uniforms, direct emergency lines with the police station, etc. – no doubt influences the socialization and the identity formation of students. Some of the low-paid security guards from the local communities traffic all kinds of goods and services to supplement their meager salaries. Others are gang members. They exert their influence on students. Their ambiguous attitude toward the youngest can be defined by a mixture of machismo, identification, and seduction. It resembles the ambiguous relationship between inmates and prison officers.

What will be the fate of the students from such a school? The best students beg to be transferred to a better school, others drop out and become socialized by the street. Yet, there is no determinism related to a space effect, as my second example from an inner city also demonstrates.

The harmonious and co-produced management of minds and bodies Washington Heights is a poor, crime-ridden Dominican area, also known for drug-dealing. Located at the heart of the neighborhood, the school, IS 218, is the product of a partnership between the New York Board of Education and the Children's Aid Society (CAS), a non-profit organization. With the help of the school staff, both have invested in a settlement house within the school.[21] The CAS is one of the oldest non-profit organizations and one of the largest in New York. Its yearly budget averages $30 million and it operates in 26 social centers (in the sectors of health, community, homelessness, job placement, adoption, etc.) in the city. IS 218 has introduced major changes in the ways of operating a school and making it a full-time community resource. It is open night and day, on weekends and bank holidays, and it has been turned into a resource center for local residents and specifically for youths.

A private foundation invested in the renovation of the school buildings: they had to be improved, some doors needed to be closed in order to monitor the gymnasium and auditorium access for security. Specific lighting was needed for night recreation activities and air-conditioning for the summer time.

Outreach to the Dominican community started as early as the beginning of construction work. Children were invited to summer outings and day camps, an outreach van helped people to have access to health care and to come to the school for further help. A program for handicapped children was launched. Dominican residents were trained thanks to municipal funds allocated to Beacon schools, a program started under Mayor Dinkins, supporting after-school activities, particularly homework assistance, psychological support, sports and creative arts.

IS 218 opened in 1992 and now functions from 7 a.m. to 10 p.m., 365 days a year. From the entrance, one is struck by the care given to the school's appearance, by the gardens, the murals, the cleanliness of the corridors. There are security guards, but no metal detectors. A small store sells stationery and items made by the students. The store and the cafeteria are run by the students and managed by mothers from the community. Parents acting as para-professionals are visible everywhere, in the offices, in the hall, at the social center. They welcome visitors. A resource center for the families and a health and dental center on the first floor are available to local residents.

The school is divided into four units, one per floor, to facilitate management and make access easier for the 1,200 students. Each unit has five classes and five teachers who also act as tutors. Several times a week, the teachers meet with the students to discuss their careers and their problems. At the time of my visit, the principal, an energetic and dynamic Puerto Rican woman, with a Ph.D. in education, worked closely with the teams. She would select the teachers and encourage innovations. She associated the parents with the work done in school.

As soon as they get to the school in the morning, students exercise and eat a free breakfast (the Board of Education also subsidizes the other meals because of the students' poor backgrounds). Everywhere, the environment is designed to stimulate the students: there are attractive maps and charts on the walls, brand new books, nice furniture. No student is allowed to lag in the corridors and if he or she does, the adjuncts (whose uniform was designed by the students) intervene right away. Two rooms have been equipped with computers donated by private firms and used by the students after classes. Many innovations are visible. For instance, in an optional bicycle workshop, a local resident teaches students how to repair bicycles. A great number of the repaired bikes are sent to South Africa, a decision made by the students. During open sessions on Saturdays, the community police come to the school. The Dominican parents and students teach Spanish to the police officers.

Mental disorders are taken care of by a psychiatrist, a part-time psychologist, and by care and social workers. Their function is to identify the most severe cases and address them to the relevant medical services. They say that compared with the tough social environment of the nearby schools, IS 218 exerts a positive influence.

Fordham University's school of social work carried out a detailed evaluation of the school's performance. It reported that truancy was practically unheard of and all the services fully taken advantage of. The students said that they felt respected, a term often used by the principal. There is a boomerang effect which results from everyone's involvement, as the trust and concern of the school staff, community, and students echo each other. Students tend to remain in that school because it has become a place of fulfillment.[22]

Those two examples from the educational sector reveal a great diversity of attitudes and approaches by decision-makers, based on whether they believe a school has a potential for excellence or not.[23] Unlike in France, where assistance is based on universal principles, here, aid goes to those thought to be deserving. Both schools are under the supervision of the Board of Education. One of them is a fulfilling place, the impact of which is felt in the nearby community. It is filled with enthusiasm and innovation and many volunteers commit themselves to the success of the place. The other school functions as a penitentiary, which a lot of teachers and students try to quit, symbolically or physically.

Other prevention programs are financed by the city. Another among them is the Beacon Schools program, founded at the initiative of Geoffrey Canada. The program started in 1990 with the support of R. Murphy, the commissioner for youth services at city hall. The idea was to turn inner-city schools into multi-service centers open night and day, seven days a week, 365 days a year. As with IS 218, the community can be rebuilt through the school. Mayor Dinkins had to choose between renovating a barge to help ease crowding on Rikers Island or to finance the Beacon Schools in 1991. He chose to invest in children and families. If the children are not saved, the community cannot be either. Then the administration had to be convinced. Canada explains that one of his strategies was to recreate social links among terrorized residents in Central Harlem (Canada, 1996: ch. 20). The school put up a great show and, hearing laughter and applause, nearby residents came to see what was going on and talked to one another and established contacts. The school encouraged the families to share their expectations. What they wanted was a car-free street between 8 a.m. and 4 p.m. A heated debate between parents and car-owners ensued, but the former won and the environment was spectacularly improved. Canada realized that there were 5,000 idle youths in Central Harlem. The school offered all kinds of night classes, in aerobics, self-help, African dance, literacy, training, etc. Such classes have a preventative impact; children are less ready to express their violence if they know that adults, including perhaps their parents, are nearby, ready to intervene. Such a school welcomes around 1,000 adults and children each week. The link from school to work is narrowed. The spirit has changed: teenagers have offered to clean the street, to plant trees, and to campaign against violence.

The Role of Intermediate Organizations

What comes out strongly from this field investigation is that within institutions, reformers acting individually or collectively attempt to alleviate the rigor of laws, to reintroduce more social justice in favor of the disadvantaged and to pacify the city. Jacobsen is one of the links in a preventative chain linking some judges already mentioned to the directors of the Vera Institute, to those of the Fortune Society and of the Correctional Association. Their successes are fragile and, as happens with other reformers operating quietly and efficiently, they are threatened by conservative interest groups.

The Vera Institute

The Vera Institute in New York[24] works side by side with NYC agencies to combat complex social problems. It attempts, after an analysis of the bureaucratic situation, to support the administration, to make it more efficient by looking at its concerns and working with it to solve them. Once a problem is defined and a solution in sight, Vera looks for grants to launch an initial test project. Six months of planning are generally required for projects that will last four or five years. Vera may suggest creating a new agency (e.g. a new victims service), or transforming part of the agency, or turning its responsibilities over to a non-profit organization to handle, for instance, alternative sentencing.

The Center for Equal Opportunity (CEO), associated with the Vera Institute, follows up on each offender who comes out of a boot camp. Twenty-five teams work in distressed neighborhoods and keep an eye on these youths. The center finds homes for them in Single Room Occupancy (hotels or apartment buildings where indigent families are placed by the city), day-care centers for their children and, if necessary, work through private companies that maintain public housing. Once one of the youths reports to work, the employer contacts the center, and a check arrives on the worksite at 3 p.m. after the first day of work. This pilot center has been operating for seven years with two hundred employees. Subsidized by the House of Representatives via the Training Act, it is the only one of its kind.

No scientist has been able to establish a strong correlation between economic development and delinquency. But, federal and state governments are counting on a work ethic and job creation to have a stabilizing influence on individuals who want to live like "the other half." Two-thirds of the jobs offered through the CEO come with health benefits. Employers, for example, are willing to pay for drug treatment: they prefer to invest in people screened and supported by the Center. Ex-offenders who benefit from it understand that their status is going to be improved and may seize this opportunity. (In 1997, at $7, the entry-level salary offered was more than the minimum wage of $5.15).

Vera sponsors ethnographic work in a dozen middle schools to understand how firearms circulate and when and where they are used.

In family courts, parents are excluded from the decision-making process surrounding child custody cases. Judges lack the time to understand youths' living situations. When juveniles under 15 are arrested by the police in Brooklyn, they are sent to the Spofford Detention Center in the Bronx. The families are called to pick them up. But 13 percent of all youths in these centers are from foster homes and many feel abandoned. They may have to wait as long as nine days before a decision is made about their release. The Vera Institute tries to intervene immediately. It argues that a day at Spofford costs the state $200, compared with $170 for foster care. Each year, Vera has thus reduced the cost to the city for 90 youths from Brooklyn and helped move them out of detention faster. Like Jacobsen, former research director Paul Steele thinks that the financial argument is one of the most convincing that reformers can use to introduce changes counteracting rigorous laws and practices in the city.

At the penitientiary at Rikers Island, a new method of drug treatment is used in the hope of controlling violence at a lower cost. As mentioned earlier, Vera manages 1,500 beds at Rikers for drug treatment. In privatizing the service, the prison administration reduced the amount it needed to spend on security guards. And if crime declines, even fewer guards will be needed. "Issues of management and budgetary reduction are good arguments to sell the city administration a program to the benefit of the most vulnerable categories of society," Steele remarks. The evaluation of the program is led by the National Institute of Justice, a progressive branch of the Justice Department.

La Bogeda in the Loisaida, a Latino neighborhood on the Lower East Side of Manhattan, supports the families of drug addicts. Working in a 240–square-block area, it provides social services, housing, health, and ambulance services. La Bogeda has good relationships with the police and social workers, and encourages links between the families through group activities like street fairs. The program is funded by the city, the National Institute of Justice, the New York Department of Health, and the NYPD.

The Fortune Society

For twenty-five years, the Fortune Society has worked with ex-convicts to help them heal the damage that incarceration has done.[25] Fortune plays a role in several areas: supervision of youths on probation (with the judges' authorization); rehabilitation of ex-offenders; support for HIV+ inmates; and the provision of workshops in correctional institutions.

The first role involves working with the District Attorney's office and with judges to aid the juveniles and suggest alternative sentencing. Because judges are restrained by determinate sentencing, or are uwilling to lift those restraints, alternative sentencing only occurs in 200 cases each year. But when it does take place, Fortune looks after the young ex-offender, providing him or her with tutors, education, and psychological support for several months. Nearly three-quarters of Fortune clients become non-recidivist. According to Page's calculations, that works out to 108 fewer years served in prison.

The young men that I met at Fortune all shared stories similar to that of José Vasquez. Brought up in a dysfunctional, illiterate single-parent family, he was incarcerated when he was 16 years old in an upstate penitentiary. He says that he is bitter when he thinks of his own socialization: "When families break up and women are abused, they tend to put down their sons instead of talking to them. They compare them to their despised husbands. Disgusted, boys then search for their peers' company." As for him, he was lucky to have found a mentor in jail who taught him the value of education. He passed his Graduate Equivalency Degree while in prison (an opportunity that is no longer available to inmates, since the Pell grants were suppressed). When released, he registered at a community college and this experience transformed him.

While in prison at Attica, he was caught up in riots. By chance, a video camera captured images of the guards' brutality and he was offered $15,000 by the Department of Justice as compensation. Even disempowered and belonging to a "dangerous category," he was aware of his constitutional rights and of his right of appeal to the justice system.[26]

Vasquez's second chance occurred when his probation officer found a voluntary job for him and offered him work with youths on parole. This American approach of taking the risk of trusting a vulnerable person and making him/her responsible and accountable has little equivalence in French culture. Through his probation officer, Vasquez also found housing (his landlord was not told where he came from). He placed himself in the Fortune Society program, which he correctly saw as an alternative to violence. At Fortune, he has found an exceptional milieu: most of the employees are ex-convicts or drug-abusers. There is a collective conscience of the struggles that all these individuals lead each day to resist obstacles.

Fortune's genesis is an extraordinary New York story. In 1967, an off-Broadway play, "Fortune and Men's Eyes", dramatized the horror of penitentiary life. The public was fascinated and revolted. One night, at the end of the play, a spectator stood up and asked aloud whether the play was not, after all, just fiction. An ex-convict in the audience stood up then and for 40 minutes kept the public spellbound with his own story. This episode received some publicity and very soon schools, churches, and community organizations started inviting ex-convicts to share their experience. Forums, open to the public, were organized by citizens eager to introduce more humanity and solidarity in New York society (what I call "space effects" peculiar to the progressive culture of New York). In brief, it became clearly necessary to create a support organization for the ex-inmates and to give the public information about the conditions of detention. The producer of the play, David Rothenberg, created a non-profit organization for education and self-help, calling it Fortune, as an interface between the ex-convicts' wish to reconstruct their lives and a few New Yorkers' will to repair the harm done by the repressive system.

By the end of the 1990s more than 80 volunteers and employees, many of whom are themselves ex-convicts or drug-abusers, work at the Fortune Society attempting to offer alternatives to the dehumanizing consequences of incarceration. Some of them answer dozens of letters each week from inmates asking for help, others act as mentors and help their clients to learn how to read, write, and take exams. Highly qualified social workers are also present to supervise juveniles sentenced to community service. They work with justice and probation agencies and do the paperwork for the administration. They lobby for more probation and alternative sentencing.

More than 11,000 inmates receive *Fortune News*, which informs them of free services to which they will have access when released. Each week, between 25 and 30 ex-convicts go to the Fortune offices for the first time. Counselors help them express their needs (for housing, legal support, drug or HIV+ treatments). In 1995, 850 new clients received advice on improving the conditions of their reinsertion. Other rehabilitation programs such as Phoenix House in the Bronx and The Valley in Harlem do similar work.

Fortune offers crucial support to HIV+ inmates. The risk of becoming infected with AIDS is very high in prison. And the mortality rates in New

York State for inmates with AIDS are 16 times higher than for the rest of the public. Moreover, a large number of them are ill-informed about the dangers of transmission and about treatments. The first thing someone who is HIV+ needs is support and information, and this is what Fortune does.

The staff at Fortune make 8–10 visits each month to Rikers Island to inform inmates who will soon be released about available services. Fortune also provides training to probation officers in Brooklyn and Queens, informing them about the services offered by community organizations and about the rights of the people on probation. Subsidized by the federal government, a special support program helps 175 Latino ex-convicts per year. Fortune has expanded the program to non-Latinos meeting federal criteria. Latinos make up half of the Rikers prison population. They are confronted by cultural and linguistic barriers to their access to services. Through Fortune, inmates know that when they are released they have a means of finding medical care, housing, educational and psychological support, and training for clerical, chef, management, or security guard positions.

Workshops in correctional institutions are organized by Fortune in order to help violent juveniles change their attitudes toward violence.[27] In the Rikers workshops, Fortune mentors are both facilitators and role models for incarcerated teenagers because they experienced the same criminal careers. Their goal is to help the teenagers contain their anger and their frustrations without resorting to guns when they are in confrontational situations. More than 250 teenagers a year attend the workshops.

Some lessons can be drawn from the experience of the Fortune Society. First, it is an American way of stimulating civil society to work on itself to solve its problems. Second, it shows the importance of work performed by volunteers, a work which would, however, be impossible without federal, state, and city grants and private donations. In 1997, the Fortune budget included around $300,000 in private donations and $3 million in public grants. Third, despite the current emphasis on identity politics, it demonstrates the continuation of a universalistic approach, founded on values of social justice and solidarity toward the weakest elements of society – ex-convicts, youths at risk, those infected with HIV. Other organizations in New York share the same social justice values.

We are now in a position to consider the question: does New York offer some support to its most vulnerable and "suspicious" categories? The answer could be that, at least here, initiatives are taken, participating in what I call the local politics of pacification, through a mix of community work led by committed militants and public institutional grants. At a time when identity politics has taken on a larger role, universalist approaches and general efforts to benefit poor minorities are rarely recognized. On the contrary, some people lament the fact that so few of minorities' needs (for example, those of gay, HIV+ blacks) are taken into account by their communities. This may be the case. Instead of commonalities, claims for social justice seem more and more to divide into two types: claims for redistribution and claims based on the politic of racial, ethnic, age, and gender differences. Increasingly, claims tied to the recognition of identities tend to dominate, and current developments conspire to decentralize the politics of redistribution (Rorty, 1997). There is also a widespread decoupling of cultural from social politics and of the politics of difference from the politics of equality. Is this cleavage untractable? For Nancy Fraser (1995), justice today requires both redistribution and recognition and the two problematics can be integrated in a single comprehensive framework, accommodating both. If we move away from extremists' claims, we encounter hybrid forms combining features of the exploited with features of racially or sexually subordinated groups. People suffer injustices traceable to both economic and racial/ethnic, gender, and age spheres. They need to have both the injury of deprivation and the insults of misrecognition redressed. Neither can be reduced to the other. One set of actors' claims has dimensions affecting their other social identities and status. Fraser requires "perspectival dualism," that is, the imbrication of economy and culture with two conditions: a fair distribution of resources and equal respect/parity for all participants (1995).

The role of civil society

The involvement of residents of low-income neighborhoods in the fight against crime, drugs, and violence also explains why the latter have declined in New York (and other major cities). In spring 1999, hundreds of residents mobilized so that minorities in the neighborhoods where they lived would not pay a heavy price for the NYPD's triumphant "zero tolerance." The neighborhood indeed remains the major site for the enforcement of safety in the public space, for the respect of norms of commonality, for collective socialization of children, for grassroots initiatives, in brief, for the expression of a shared social capital, as Coleman (1988) expressed it.

I have already mentioned the protest demonstrations against police brutality in New York. All over the country, people have organized local projects to fight crime and exert some control over adolescents. Signs indicate that many neighborhoods are under video surveillance. In public housing areas in Brownsville, NY, the names of the tenant patrol captain and her team are posted on building doorways, indicating that the drug-dealers shoud keep away. Every single day, at 6 p.m., older African-American ladies set up tables on each floor of the building and keep watch. Frances X says that she has never been threatened, despite the very dangerous atmosphere of the neighborhood. Community police officers feel reinforced on their beat by the tenants' cooperation. This co-production of security measures, which might be interpreted negatively in France, belongs to the duties of citizenship which are part of the Anglo-, American, Swiss, and Scandinavian traditions.

The single parents, elderly, and adolescents whom I met at the James Welden Johnson public-housing projects in Harlem express a trust, a collective will, and a pragmatism leading to concrete preventative actions and self-help. Safety is indeed regarded as a right and an economic good. In the best of cases, it is a commonality organized around a shared micro-public space where conflicts are regulated.

The residents, all from minority ethnic groups, do not trust the police with their "go-get'em" mentality. Numerous examples are given to explain why and how they were framed by police officers, especially if they have some visibility in the community. As a consequence, they organize to co-produce their collective safety. Since the public housing police were merged with the NYPD and community policing was suppressed by Bratton, residents complain that they no longer know the officers on patrol, as they are replaced so often and move in "packs of six or eight." If they can handle the situation themselves, the women of the neighborhood say, they will not call in the racist and corrupt police. The same distrust exists towards firefighters, who are mostly white, and who are accused of robbing the poor whom they are supposed to help. This explains the tensions between the residents from low-income areas and the fire department, a phenomenon that extends beyond the USA. In general, relations between the police-as-enemy, the projects' tenants, and the management of public housing all over the county is mediocre.

The formal or semi-formal social control exerted through tenants' meetings in the housing projects represent a form of social organization of space, influencing individual trajectories as well as the social contexts regulating social interactions. They provoke or constrain violent interactions.

Public housing in New York is often held up as a success story.[28] The reformist influence of the trade unions which started building the first units for European immigrants and their children at the end of the 1930s can still be felt. Minorities, who moved in during the 1960s, imported their own cultural traditions. The housing authority also chose to interpret federal guidelines setting income caps in its own way. It thus avoided the over-concentration of families with heavy social handicaps that occurred in Newark, New Jersey, and Chicago.

The tenants whom I interviewed in Harlem did not feel abandoned by the authorities, as is often heard in the ghettoized areas of other large cities. They gave the names of the public officials that they would contact, if necessary. During demonstrations against police brutality, their political representatives were arrested, as were other political figures. Tenants mobilize on security issues. After twenty years of management passivity, they have organized their own patrols. They know the drug-dealers who operate in the buildings' lobbies and they tolerate them as long as they do not create public disorders. Those young "entrepreneurs" belong to the community. But, if they start turf wars and threaten public order, it becomes unbearable for the tenants and they have to be evicted.

Susan Saegert and her team conducted studies on the housing conditions of 6,000 of the poorest New Yorkers, in some of the most problematic areas of the Bronx in 1992 and in similar areas of Brooklyn in 1994. The findings are among the few researches showing that social capital in the co-production of safety can intervene even in very destructured housing projects, and improve collective life. Crime in public housing has rarely been studied: why would it be more criminogenic than private housing in similar neighborhoods? Public housing is not homogeneous, in size, building-type, age, or social composition; it has its own dynamics and crime rates vary widely. Saegert took an interest in the tenants of buildings foreclosed on by the city after the owners had evaded taxes, and in those in public housing managed by a central office or a decentralized system. Her conclusions are important: the more the tenants feel responsible for their housing and its environment, the more collective action becomes natural. Collective action and the sense of "community" attempt to respond to problems in general and to those of safety in particular. Unsurprisingly, the more the tenants are on welfare, the more the environment is problematic. The more they are employed with higher incomes, the more they get involved. Safety problems are perceived to be most acute in buildings sheltering former homeless people. Yet once the tenants feel a responsibility for their space, drug and alcohol problems vanish progressively (see figure 4.10). The level of education also has a positive impact. The more the residents are educated, the less serious the problems are perceived. Finally, demographic stability and smaller building size contribute to the success of collective action (Saegert, 1996; see figure 4.11).

Recently, the city of New York and the housing authority responded to pressure by tenants and made changes in the architectural designs of future housing projects, following Newman's (1973) theory of defensible spaces and crime-prevention: low-rise buildings have tended to replace high-rise projects whenever possible (with the support of the federal program HOPE VI, a scheme run by the US Department of Housing and Urban Development which organizes the demolition of large public housing projects and their replacement by low-rise buildings). But, there has been nothing to prove that the tenants will actually be better protected in low-rise

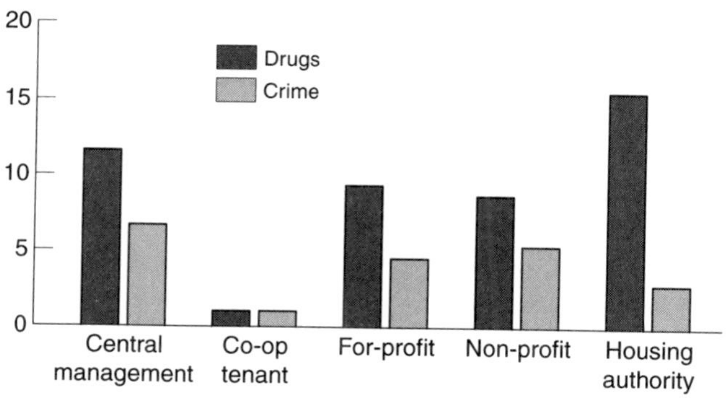

Figure 4.10 Levels of drug-abuse and crime in Bronx and Brooklyn public housing, 1992–1994 (%)
Source: Saegert, 1996

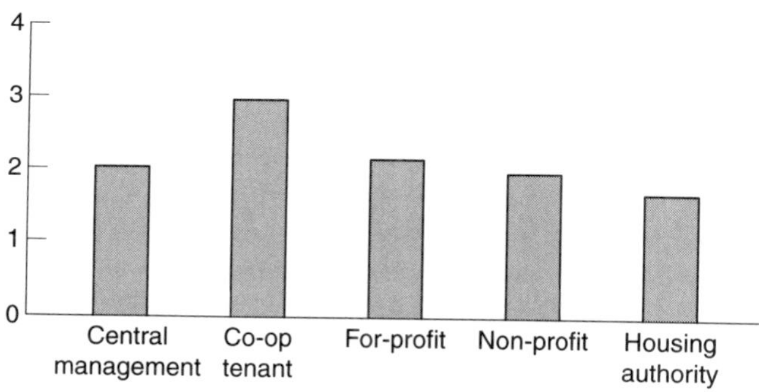

Figure 4.11 Levels of participation in community activities in Bronx and Brooklyn public housing, 1992–1994, on a scale of 1–4
Source: Saegert, 1996

buildings. Windows dominate the small squares where the elderly can get together and the children play – changes that meet Jane Jacobs's recommendations (Jacobs, 1961). Entrances are protected to prevent gangs from breaking in too easily. Alarm systems, lights, and gates have been reviewed with the architects in order to restore a sense of safety. Trained security guards have been hired. These improvements have been made progressively, and others, such as intercoms, video cameras, and electronic cards, usually follow. Is crime-prevention efficient in such projects? Many questions remain (see Fagan, 1996). Aren't adjacent areas – where the trouble-making relocates – as problematic as the buildings themselves? From where does the dynamic of violence start? How is social control measured and controlled in "interstitial" zones?

The "little brother syndrome" hypothesis

Finally, it has been argued that the participation of civil society in making public housing more secure may also come from inner-city youths themselves.[29] The socialization of children in the problematic neighborhoods of New York was greatly influenced by the extent of crack-related violence among their older brothers and sisters. Between 1986 and 1991, gang turf wars claimed thousands of casualties in ghettoized areas. In response, a considerable number of adolescents in these inner cities decided not to consume drugs or alcohol, or use guns. Were this a broad enough trend, we would shift to a lower point in the cycle of drug-abuse and violence, which would last at least as long as that generation's memory. This happened at the beginning of the twentieth century in the USA. There was an epidemic of cocaine abuse which claimed hundreds of victims. In response, the next generation practiced collective self-restraint.

Perhaps the fashion of carrying guns will thus disappear as fast as it came. The dozens of youths I talked to in spring 1999 confirmed this impression. Many had been given 6–7 years on probation. Unlike their Rikers counterparts, they did not carry guns that they considered as lethal. They had hopes and dreams like everyone else mixed with feelings of despair and an urge to leave a city that offered them so few opportunities. An awareness seems to take hold among the adolescents: not only does violence lead to prison and death, but it is no longer fashionable. It is still too early to tell. This behavioral transformation, the beginnings of which we may now see – witness the decline in gun-related emergencies in hospitals – would indicate a shift in value systems. That some of these youth might want to go to college and that colleges advertise their programs to inner-city students, as I recently saw in a deprived Bushwick high school, means something.

In brief, it is the co-production of safety by institutions – police forces, justice, prisons – combined with the dynamics of civil society – community organizations, self-help, innovative citizens' actions, etc. – which contributes to lower crime and violence in large American cities. But no one should forget the determining impact of economic and demographic forces, nor the larger federal and state policies interfering on those local dynamics.

Chicago

The initiatives undertaken by the city of Chicago will not be examined in as much detail. In Chicago, many of the tendencies observed in New York also prevail, including a hybrid policy that is tough but which has preventative aspects, shaped by the mayor and other public and private actors, by "resisters" inside the institutions determined to counteract the punitive rigor of the law-makers, and by citizens eager to fight against the threats of crime and delinquency. But Chicago also has its own profile in providing for pacification and space effect matters.

The global character of Chicago

Its population of 8 million easily places metropolitan Chicago among the world cities (Abu-Lughod, 1995: 176). Its cosmopolitan character is evident, with 11 percent of its residents foreign-born and 57 percent of its population classified as "minority." The population of Chicago (2,802,494 inhabitants) is 25 percent white, 39 percent black, 20 percent Latino, 4 percent Asian, and the rest falls under the category "other" in the 1990 census. From 1960 to 1980, the white population declined by more than 40 percent, whereas the black population grew by 40 percent to 812,000 residents (see table 4.1 for CSMA).

As for being a command center for transnational firms and the headquarters for large corporations (in manufacturing, advertising, finance, insurance, investment banking, and legal services), Chicago is second only to New York. Of the 500 largest corporations in the USA, 42 are located in Chicago (see table 4.2). Most of them have international operations and twice as many are located in the city proper than in the suburbs. Of the services to large corporations in the USA, 26 percent are in the Chicago area (Schwartz, 1982: 282).

Table 4.1 Percentage of Consolidated Statistical Metropolitan Area (CSMA) population that is black, Asian, Hispanic, * or American-Indian in New York and Chicago, 1990

Race/ethnic	*New York*	*Chicago*
Black (non Hispanic)	18.2	19.2
Hispanic	15.4	11.2
Asian	4.8	3.2
American Indian	0.3	0.2
Total minority	38.7	33.7

* A considerable majority of Hispanics define themselves as "other," thus confounding racial categories
Source: Abu-Lughod, 1995: 177

Table 4.2 Number of corporate headquarters in cities and suburbs of New York and Chicago, by type of firm, 1990–1991

Type of firm	*New York*	*Chicago*
Manufacturing (n = 200)	42	17
Advertising (n = 50)	34	5
Finance (n = 50)	20	6
Diversified services (n = 50)	8	3

Source: Abu-Lughod, 1995: 180

One important factor in Chicago's wealth is its domination of global trading in agricultural commodities through the Chicago Mercantile Exchange. In addition, the exchange's subsidiary, the International Monetary Market, effectively sets the rate of international currencies. Foreign direct investment in Illinois rose from $5.6 million in 1981 to $19.5 million in 1988, a 245 percent increase, greater even than that of New York (+198 percent). As Abu-Lughod remarked, "It is clear that even if we may have some reservations about which American urbanized regions are truly of world significance, foreign investors do not share our doubts. They single out New York, New Jersey, Illinois, and California as the preferred outlets for their investments" (1995: 183).

Chicago is not spared inequalities, as documented by William Julius Wilson's (1996) fine-grained quantitative and qualitative analysis of the transformation undergone by poor places and people in the last two decades. The polarization and segregation of the global city are observed in Chicago. It experienced a 61.5 percent increase in the number of ghetto census tracts (that is, where over 40 percent of residents are poor) from 1980 to 1990, even though the number of poor residing in those areas increased only slightly. In the ten neighborhoods that make up Chicago's Black Belt, the poverty rate increased almost 20 percent between 1970 and 1990 (from 32.5 to 50.4 percent), despite the fact that the overall black poverty rate for the city rose by only 7.5 percent during the same period (from 25.1 to 32.6 percent) (Wilson, 1996, 14–16; see figure 4.12). Looking at the map in figure 4.12, we can only be struck by the similarity with the map of delinquency drawn by Shaw and McKay for 1917–33 (figure 4.13), giving proof to their theory that some places are correlated with social problems, regardless of the populations living there.

Economic restructuring had a profound impact on the Chicago labor market: there were fewer well-paid blue-collar jobs requiring low skill levels and little education and the growth industries of the 1980s were

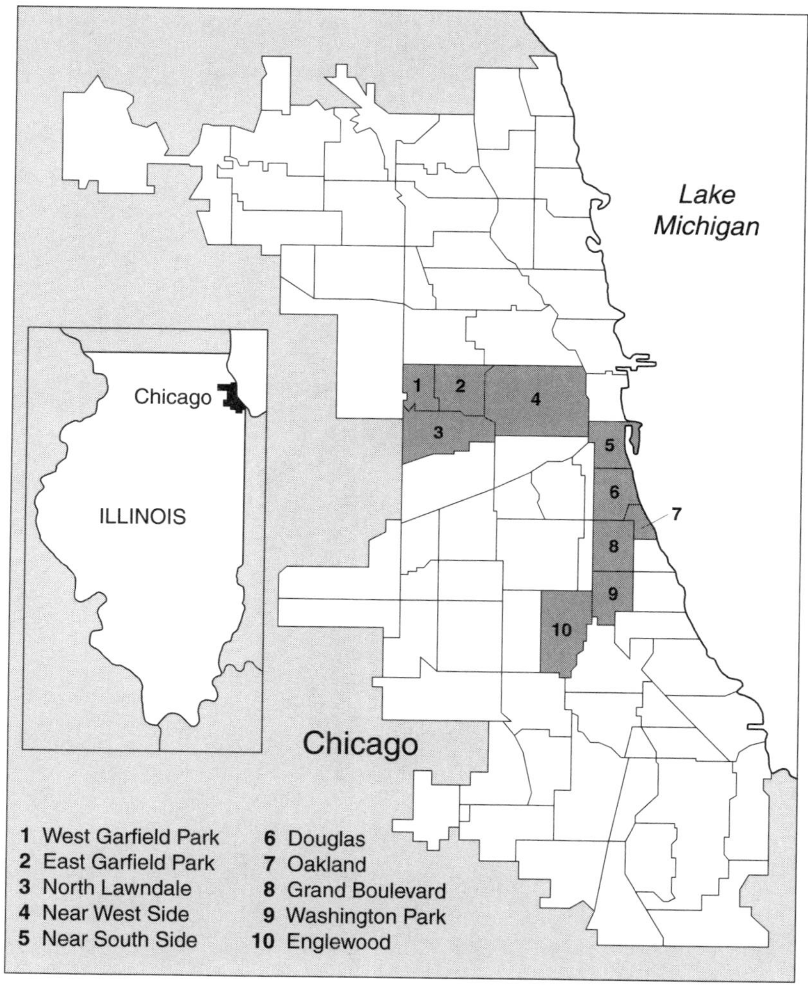

Figure 4.12 Community areas in Chicago's Black Belt
Source: W. J. Wilson, 1996: 13

characterized by low average wages, greater earning disparities, weaker unions, and a higher incidence of part-time workers. Between 1967 and 1987 economic restructuring cost Chicago 326,000 manufacturing jobs (24 percent compared with the 15 percent national average). Research on the way Chicago employers view prospective employees reveals a use of spatial designations distinguishing the inner city and the suburbs, both independent and interactive with notions of race and class (Kirschenman and

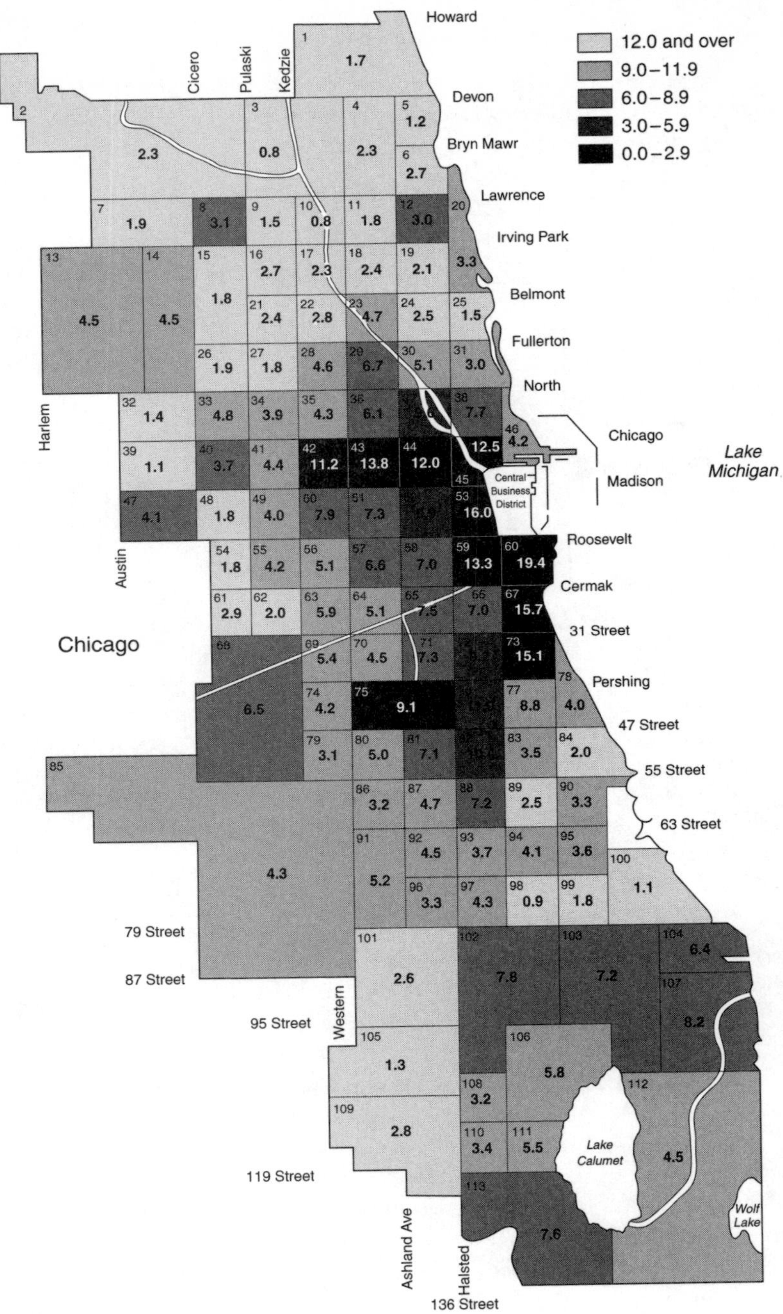

Figure 4.13 Rate of delinquents based on 8,141 male juvenile delinquents in the juvenile court, Chicago, 1917–1923
Source: Shaw and McKay, 1942: 62

Neckerman, 1991). In 1974, 57 percent of inner-city black fathers had jobs, as opposed to 31 percent in 1987. At the time, the black ghetto was still heterogeneous in terms of income diversity. The sharp decline in jobs is explained by the lack of public transportation that would allow workers to follow jobs, and their lack of education (due partly to poor-quality schooling in the inner city) required for high-skilled posts. For example, Wilson explains, in 1990 only one in three adults aged 16 and over in the twelve poorest Chicago community districts (South and West Sides) held a job in a typical week. In three additional areas, only 42 percent of the adult population was working during a typical week. Thus, in a population of 425,125 residents, only 37 percent of all the adults were employed in a typical week (W. J. Wilson, 1996).

Between 1985 and 1992, there was a sharp increase in the murder rate among men under 24; for males of 18 and younger, murder rates doubled. Drugs, weapons, gangs and violence seem to go hand in hand with joblessness, Wilson remarks. Inner-city black youths with limited prospects for stable and attractive employment are easily lured into drug-trafficking and find themselves involved in the violent behavior that accompanies it (Wilson, 1996: 21–2). In brief, despite general economic improvement over the last 15 years, the ghettoized poor and their concentrated, isolated communities have been hit the hardest. As remarked by Martin Goldsmith (see note 34), it should not be inferred, though, that a 100 percent black concentration in a Chicago neighborhood means that the neighborhood is a ghetto. There are many communities of this type which are middle or working class and crime is not significantly worse than in Caucasian or Hispanic communities. The black movement out of the ghetto has resulted in economic and spatial bifurcation resulting in an overconcentration of problems with contagious effects among teenagers in the ghetto.

The failure of public housing – particularly high-rise/high-density projects – had long been recognized in Chicago.[30] Its demolition was late in coming, if one remembers that the city of St Louis made the decision to tear down Pruitt-Igoe almost two decades ago. Under federal program HOPE VI, 116,000 public-housing units are targeted for demolition in the years to come.[31] Take for instance "The Hole," one of Robert Taylor Homes's most oppressive housing projects, which was demolished in July 1998 (Belluck, 1998). It was 99 percent black; 50 percent of the adults lived on less than $5,000 a year, and 96 percent were unemployed, as was the case amongst the rest of the residents in other Robert Taylor Homes. The Robert Taylor Homes were built between 1960 and 1962. "Initially comprised of 4,415 units of 28 identical sixteen-story buildings, the Taylor project has cast the shadow of the original Black Belt in concrete. Two miles long, barely a quarter of a mile wide, hemmed in by railroad tracks and the Dan Ryan Expressway, the development housed, when completed, some 27,000 residents (20,000 of them children), all poor and virtually all black" (Hirsch, 1983: 262–3). After the demolition of "The Hole," the city dispersed some of the families, relocating them, for instance, in "scattered site" apartments managed by the housing authority, but not part of any project or in the Department of Housing and Urban Development's Section 8 Certificate Program homes.[32] But how can members of one gang move into a building occupied by an enemy gang? How can families with criminal records or who pay rent only erratically be relocated? Housing officials try to help without stigmatizing the tenants. But it was a bit late. "Housing projects in Chicago are starkly identified with race and poverty. From the beginning (1959), city officials used them, along with the expressways being built at the same time, to contain the city's growing population of poor blacks. Of Chicago's 67,000 public housing families, 97 percent are black," one journalist writes, summing up in a few lines Chicago's history of racial segregation (Belluck, 1998). In 1969, Congress mandated that public-housing tenants pay a percentage of their income, which was already very low. The result was that working families moved out, while welfare families with handicaps remained concentrated and isolated on the same sites.

Recently, the city decided to reduce the size of its public-housing projects and to do away with "lifeless spaces, impediments to the optimal use of the metropolis" (Venkatesh, 1997: 35–40). Chicago's anti-crime policy in the late 1990s, addressing the consequences of hyper-segregration, consisted of razing or cutting off the top floors of the high-rise housing projects. Vandalism in housing projects cost the city $7.4 million between 1977 and 1986, with crime reaching astronomical proportions.[33]

Cabrini-Green is a housing complex with a bad reputation and, in 1996, 6,000 tenants, over 90 percent of whom were black. The project is located near a wealthy business center district, Lincoln Park, which is one of the country's highest income areas by zip code. Land speculation around the housing projects has been going on for years. But the process of developing market-rate housing there drew out considerable negotiation among the community, the City, and the developers. In 1996, the city and the Chicago Housing Authority, led by V. Lane, adopted a $315 million Redevelopment Initiative Plan for the Near North Corridor, where Cabrini Green is located. The developer selected for the first phase (there will be five) is Peter Holsten Realty – a development company with a reputation both for sensitivity to community interests and tenant management skills. The developer has teamed with a black development group for the first phase, 30 percent of which will be for public-housing residents, 20 percent partially subsidized, and the balance will go on to the open market. How will the social mixing evolve? Will residential stability be sustained?

The street educators whom I interviewed remain pessimistic (although their very function builds on this pessimism).[34] They admit that the new Operation Clean Sweep allows the management to kick out dangerous tenants, but the remaining gangs can make living very dangerous for the residents.

The new management selects new tenants, trains security guards, and installs computerized access cards; in other words, it develops a program of situational prevention. Armed guards are visible, patrolling the grounds. Visitors must be accompanied by residents to enter the buildings – most have to sign in at a 24-hour security office and are not allowed to stay overnight. Tough screening and lease-enforcement procedures designed to keep dangerous people out of public housing are enforced. "Public housing residents have the same rights as all other Americans to live in peace and free from fear, intimidation, and abuse. At a time when federal resources are scarce, public housing should be made available to only responsive individuals," Mayor Daley claimed, echoing French discourse. The difference comes with the follow-up: no one knows what happens to the tenants who are evicted. Do they live in the streets, thus perhaps becoming more dangerous for society? Do they move to smaller cities where they cause trouble? What happens to complex families with members who succeed and others whose criminal records result in eviction for the whole family? Does not dispersion have the effect of "vouchering out," of erasing the "social efficacy" of a community? Who defines "order" in a community?

Chicago has a long tradition of neighborhood activism and of innovative modes of organization, however, as witnessed by the number of community-based and community-development organizations. Even in places like "The Hole," parents form patrols to escort their children to school when the neighborhood becomes too dangerous. Three times a day, benevolent people – clergy, school staff, retired employees, former gang members, and other residents – wearing orange armbands watch over the children going back and forth (Braun, 1998). It is not enough to protect the children in the schools – safety corridors are needed outside so that children can attend school. Partnerships are at the core of the positive evolution of the city and some of the initiatives are likely to have an impact on the residents in terms of social justice. Chicago businesses have a tradition of getting involved in trying to address the city's problems. It is possible here too to evoke "the power of place."

As in New York, the crime rate has declined (by 12 percent since 1994), except for juveniles.[35] But again, one should be cautious with statistics. In 1991, a peak year, almost 200 juveniles were arrested for homicides, an increase of 127 percent since 1985; 623 murders were committed in six months due to the crack epidemic, resulting in 3,000 arrests (Wilkerson, 1991). It may prove either that more juveniles became violent

Table 4.3 Age of juveniles arrested for murder in Chicago, 1987–1992

Age	*Number*
17	313
16	68
15	134
14	197
11–13	22

Source: Scholastic Update, 2 November 1994

and/or that the police did a better job. Table 4.3 shows how, between 1987 and 1992, the age of juveniles arrested for murder in Chicago was distributed. In 1996, about 20,000 cases of juvenile delinquency were registered in Chicago; 46 percent of those juveniles arrested were between 14 and 15 years old (Reed et al., 1997). Chicago received national attention for horrendous acts performed by very young offenders. In the last twenty years, eight cases have involved children younger than 11, including the 1994 case of a 10- and 11-year-old indicted for dropping a 5-year-old from the 14th floor of a building. In another incident, during the summer of 1998, two 7- and 8-year-olds were accused of raping and killing an 11-year-old girl (they were eventually found innocent.) This epidemic of youth crime shows the difficulty in judging youths who are both authors of crimes on other poor and the victims of society's lack of support.

Strategies of local institutions

The Chicago police There are 13,000 police officers and 2,200 civilians in Chicago's police force. Under an Alternative Policing Strategy suppported by Mayor R. Daley, decentralized, community policing was adopted five years ago. For observers, this is a very interesting alternative to the zero-tolerance strategy of New York. According to Capitain Byrne in the 24th district,

> This reform had an enormous impact. The city is divided into 25 police districts, extending over two neighborhoods in most cases. Precinct commanders are encouraged to take initiatives and to develop their own programs. Why fill the prisons without solving any of the root problems? . . . Now that we have spotted the problems, we look at all the car license plates in crime areas. We help the population get mobilized: the residents, the store-owners,

> the community organizations, and our elected officials are pleased. Statistics on delinquency are declining and more victims come to file complaints at the police station. In this district, we have 9 patrols. We have focussed our efforts on the drug war and we have arrested a gang thanks to our computerized program, but we do not feel supported enough by the Feds nor by the justice system. We have two policemen in each of the 28 neighborhood schools, which are all equipped with metal detectors. We seize around 200 weapons here every school year. By our very presence, we act preventatively, but this is not always understood by our hierarchy. We can mount operations in cooperation. For example, 66 police officers recently gathered in front of a slumlord's house and, with our loudspeaker, we urged him to modify his behavior. We fight petty delinquency, which is what kills the quality of life. In our neighborhood, 50 languages are spoken. We have Latino policemen reaching out to the Latino community. This is difficult because of the undocumented residents.[36]

Richard Block, a sociologist, has created maps for the police.[37] In big cities, he says, public spaces can be dangerous spaces, but not all spaces are equally dangerous. Violence is more likely to occur in some places than in others. A "*place*" is a specific small area that reflects and affects the routine activities of the participants in the short run, and plays a role in the specific conflict at hand (Block, 1997). Taverns, liquor stores, and transit stops vary in their riskiness for violence according to many factors, including the characteristics of their surrounding space. Thus, a specific location provides the backdrop and the mechanism for violence.

Among other things, Block's maps may show that the location of burglaries over six months, of the drug markets, of public housing, and of the taverns and entrances to the elevated subway coincide (Block, 1997). At two o'clock in the morning, when the revelers are drunk, the thieves come to rob them and flee on the subway or through a low-surveillance parking lot. Such events echo the stories described by Clifford Shaw in the Chicago of the 1920s. If some places are hotter spots than others, by using the new computer technology, police can anticipate actions of the thieves, dealers, and gangs.

What is so innovative about Block's geo-coding information is the use to which it is put. It is being shared with community residents, such as those in the Thorndale neighborhood, so that they can understand their neighborhood better, control their fears, and suggest strategies to the police. Block admits that this co-production of safety between the residents and the police does not work so well with Latinos, who fear the police, or with insurgent poor minorities who have long hated the police, as residents of Grand Boulevard told me they did.

Like the NYPD, the Chicago Police Department, with 13,500 members, has a reputation for brutality – although only 2 percent of the police officers are the subjects of 20–25 percent of complaints – and for corruption.

The Chicago police have been reportedly committing serious abuse, including repeated acts of torture with racial undertones.[38] Between 1992 and 1997 1,657 lawsuits involving excessive brutality and false arrests cost the city more than $29 million and between 1991 and 1994 the cost was $16 million (Human Rights Watch, 1998: 171). The city ordinance in 1992 allowing the Chicago police to arrest gang members for loitering at street corners led to 45,000 arrests. In the spring 1999, to the great dismay of the mayor and the police, the Supreme Court struck down this decision as anticonstitutional (*Chicago* v. *Morales*), a proof of the existence of counter-powers in the city. It has been said that the maps used by the police are helpful for harassing the dealers and the gangs in what amounts to a continuous turf war. But it is a zero-sum game. When a tavern is shut down by the police in Granville for drug-dealing, another opens in Thorndale, as Block remarks (Block, 1997).

The Chicago Alliance for Neighborhood Safety (CANS) How was change initiated and the police reformed? According to Ani Russell of CANS, it began in 1981 with the federal program for crime prevention in cities.[39] The federal agency in charge of law-enforcement (LEAA) established nine demonstration projects throughout the country. Chicago was selected thanks to an umbrella organization which included the Citizen Information Service of the League of Women Voters, a non-partisan but politically well-connected network. From 1981 to 1984, Russell, a grassroots activist from the Alliance contacted and worked with ten Chicago neighborhoods to be selected for the demonstration project. Students involved in the Volunteers in Service to America (VISTA) program offered their help. Russell and other activists taught residents how to watch their neighborhoods, contact each other on the phone, and call 911 to establish regular contact with the police.

The mayor had always declared that he favored the mobilization of residents in the co-production of safety. After 1986, the expansion of gangs and carnage resulting from the crack epidemic generated turf wars and numerous victims among innocent residents, including children. The 911 service and the police themselves were overwhelmed. Until 1987, the residents' capacity for self-help had remained limited. But when the VISTA students decided to evaluate the efficiency of the 911 service, and of alternatives, it turned out that only 3–7 percent of the calls related to a real emergency. Besides, no one had an accurate definition of an emergency, except that if a homicide was taking place, a witness had to be there to tell about it. Opinion polls reflected a great discontent toward the 911 service and the Chicago police in general.

Police headquarters decided to install a system to allow them to better monitor whether emergency calls were serious. In addition, they tested two policies – one oriented toward problems (POP), the other toward places – and they realized that the latter was more apt in solving problems. But the reform that was needed to decentralize policing was to extend to the whole city so that race and class relations would not interfere with the new community police practices in the neighborhoods.

Several years of efforts were necessary to convince the Chicago mayors who succeeded Harold Washington (1983–7) that the police reform he had approved before his death would be beneficial for the city. In 1990, circumstances allowed CANS to improve its position. The city anticipated a violent summer. Chicagoans were more fearful than ever of gangs and the consequences of crack consumption. Few summer jobs for poor teenagers were offered. According to the president of the Chicago Chapter of Mothers Against Gangs, a support and advocacy group, "It [was] out of control. People [were] in agony. People [were] being held hostage in their own neighborhoods." In the South Side community of Englewood 71 people were killed and 40 percent of the homicides were related to drugs or gangs. The mayor was blamed (Wilkerson, 1991). The public supported CANS's crusade and 90 community-based organizations involved themselves in the surveillance of schools, public spaces, etc. But there were no funds. The MacArthur Foundation, located in Chicago, then organized a taskforce on community policing. The police were interested. The National Institute of Justice, until then reluctant because of the Chicago police's reputation for being racist, corrupt, and brutal, began changing its attitude. CANS mobilized the media to help initiate the change. A day after CANS's big press conference, the mayor gave the green light to police reform.

The success of CANS comes from its savoir faire, from its leadership capacities, and from its resilience (after serious troubles in 1989 which left it financially deprived, it did not quit). It also comes from its contacts. CANS had allies in the Clinton Administration, at the National Institute of Justice (which took care of the pedagogy of the reform via the media), and at the MacArthur Foundation (which provided Chicago police chiefs with eight weeks of training and two trips to Seattle and New York to observe community policing in the field). Finally, at City Hall, there were Democratic reformers who thought that all the neighborhoods were entitled to public safety services.

After the reform had been going on for a year, it was evaluated by private consultants and by Wesley Skogan, a professor of criminology at Northwestern University. The reports went along the same lines: the reform had a great impact in high-crime segregated neighborhoods (Skogan, 1996). It improved the quality of the relations between the residents and the police. Of all the respondents 85 percent thought that the police were helpful, and 80 percent rated them positively for expressing concern for people's problems and for dealing with people fairly. The police received high marks for politeness. Police training was enforced satisfactorily (for 200 patrols out of 279), while that of the residents done by community organizers helped by volunteers in 25 police districts was more dubious. Following the training, residents were able to solve 25 percent more safety problems themselves and they were more self-confident. In Cabrini Green, for instance, because a majority of the people considered themselves more as victims than criminals, the reform was welcomed.

But then tensions occurred, when the residents did not give City Hall enough credit. Funding from the city was halted, the number of community organizers reduced, and expectations were frustrated. "Then things started again, but it was never the same," Russell remarked. Nevertheless, the mayor claimed at an Address on City Management in the late 1990s that "community policing is one of the most important developments in our city in recent years. It has made police more responsive to people in neighborhoods, government agencies more responsible to community needs, and residents more aware of their duties to help build a safe neighborhood."

This experiment lends itself to comparison. Community policing does indeed reduce the problems of police abuse that were denounced in New York's minority neighborhoods. In numerous areas, the Chicago police have reassured the residents and work with them on their problems. An example of police–church alliance on the West Side of Chicago deserves attention. For years, the attitude of the police was to arrest local offenders and have them harshly punished. Meares and Kahan report a recent shift in drug-enforcement policy. Because drug markets flourish from a demand from outside the inner city, namely from white suburbs, "reverse stings" from the police target both the buyer and the dealer (Meares and Kahan, 1999). Instead of having just the dealers from the neighborhoods sanctioned, the police also pursue the buyers. This decision has restored a "working trust" with the residents. Zones of safety have been created, with juvenile curfews and the policing of minor offenses supported by the residents. The analysts of this experiment note that the leadership role of the local police commander, a long-time resident, contributes to its success. While I doubt that an alliance between church and police would be accepted in French neighborhoods of this type, what remains here is

> a new species of social capital that can be directed toward violence control: the police have access to new sources of information that can assist them in criminal investigations, and church leaders have been assured of greater police responsiveness to the crime affecting their congregants. Church leaders are now even playing an active role in recruiting and screening police

academy applicants from their congregations. (Meares and Kahan, 1999: 25)[40]

Will the police be able to act similarly in all the neighborhoods? "No-go" areas exist in the ghettos on the South and West sides where the police operate "sweeps" to reassure the public and obtain funds from the lawmakers. As in New York, racial antagonism then can be disastrous in the long run. But, on the whole, community organizations have cooperated with the reforms and have encouraged residents' attendance at the beat meetings. The mobilization has been more extensive in African-American neighborhoods than in white ones, the evaluation reports.

The justice system Like his counterpart in New York, Illinois Governor Jim Edgar supports a policy of "toughness" and the death penalty. Since his election in 1991, to date, five executions have taken place. An anti-death coalition, formed in 1976, argues, as in New York, that the death penalty is costly. It estimates that each execution costs $2.5 million, whereas life imprisonment is only $600,000 per person (Illinois Coalition Against the Death Penalty, 1998). But these figures are as arbitrary as those demonstrating that each incarceration saves society $430,000 a year, supposing that, if free, an offender would commit 187 crimes a year at a cost of $2,300 per victim (Donzinger, 1996). The Illinois Coalition Against the Death Penalty also claims that the process is discriminatory, since it is six times more likely to inflict the death penalty if the victim is white. Of those on death row in Illinois 67 percent are black.

Illinois's first state prison was built in Joliet in 1867. It still functions. The first "experimental" maxi-prison (in which the offenders remain in their cells 23 hours out of 24 with no contacts with the outside world) was built in Marion in 1972. Now this "model" has been reproduced in other states. As elsewhere, prison is a lucrative business. Illinois Correctional Industries is a private, centralized organization which trades the goods produced by 1,700 inmates. Manufactured and farm products, posters, clothes, glasses, and recycled tires are sold to private industry as well as to public authorities.

The juvenile court in Chicago's Cook County has experienced an increase in its caseloads, from 10,000 cases a year in 1987 to more than 60,000 in 1996 (Juvenile Court of Cook County Report). Youths aged 13 and over indicted for homicides, those 15 and older accused of rape, kidnapping, arson, assault, and other violent crimes, are judged as adults. Deterrence and public protection come before education. Any juvenile involved in dealing drugs in a school zone or a public building is automatically transferred to an adult court without an initial hearing. The transfer list includes more than twenty types of offenses, including

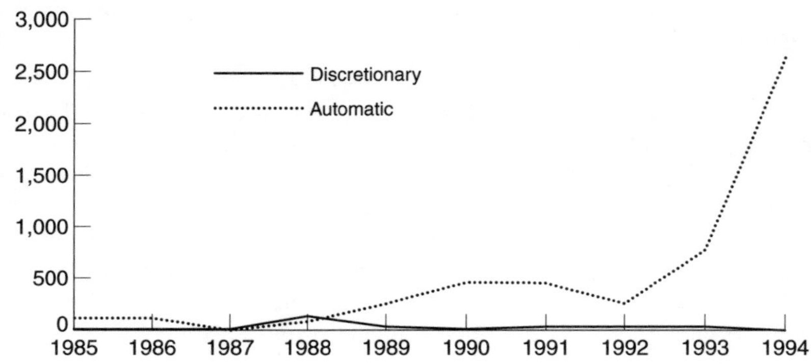

Figure 4.14 Automatic transfer of juveniles to adult courts in Cook County, Chicago, 1985–1994

Source: Statistical report from Cook Country juvenile courts, administrative office of Illinois courts and probation services

carrying a gun without a permit. Transfers have increased astronomically in the 1990s, in response to moral panics (see figure 4.14).

Like the New York judges that I interviewed, Judge Rogers in Chicago recognizes that "the judges' discretionary power in terms of waiving and sentencing has been strongly diminished by law-makers of the state of Illinois."[41] Judges are appointed by politicians, however, and cannot openly revolt, he says. He is skeptical about any decline in crime. On the contrary, he sees juvenile crime increasing due to drugs, the lack of drug treatment, teenage pregnancy, young mothers on welfare who are unable to properly raise their children, etc. The juvenile court works efficiently and social services attached to it do a good job, the judge admits. "It is the marriage of social welfare and justice," which is positive.

The juvenile court's interests are well represented in the State Assembly and the governor's office, and reforms are beginning to take shape. But judges, as elsewhere, are harassed by the conservative press, which molds public opinion by playing on its emotions and dragging justice through the mud. As chance has it, however, the judicial hierarchy supports rank-and-file "soft" judges.

The courts placed around 45,000 youths in locked facilities or under community supervision in 1996 (as compared to 8,000 in 1994, although there were many more in the 1920s when Clifford Shaw was observing them). As in New York, legal defense and numerous organizations defend juveniles' rights before the courts.[42]

Anti-drug policies The city of Chicago, with the help of the juvenile courts, has developed an important anti-drug program. Its aims are to develop:

- networks linking the drug treatment centers and juveniles in problematic neighborhoods;
- intervention strategies directed at the families of youth at risk;
- behavioral evaluations for school truancy and recidivism, performed even if the youths have only been in the program for a short time.

Counselors evaluate the situation and may decide to offer two weeks of treatment to first-time offenders and four–ten weeks to recidivists. Then, once released from treatment, the ex-offenders wear an electronic bracelet on their ankles for three months, during which time they are supposed to resume "normal" life. Nationally, in the 1990s, 75,000 people, a relatively small number, were being electronically monitored using those bracelets, that is, 3 percent of those on probation. This approach is not approved by a majority of the public, which feels safe only if the offender is incarcerated. Moreover, "to be at home with a good beer" does not seem punitive enough (Vogelstein, 1997). Rather than a coherent city policy, scattered, diverse, public, private, or mixed initiatives deal with problems on micro-territories.

Prevention programs at the city and state levels

These programs are diverse, yet not numerous enough. They may benefit the providers more than the clients in a number of cases. I offer a few examples.

Street Law Initiated by the State of Illinois, the Street Law program targets juvenile offenders' recidivism. The Judiciary Center for Families and Children (JCFC), directed by Bernardine Dohrn, used for the first time in 1994 an education-related law to extend education to incarcerated juveniles in Cook County.[43] The Street Law program recruits lawyers and law student volunteers to teach the juveniles about their rights and duties. It is unanimously recognized that the level of education is excellent and Street Law has been so successful that its scope has been extended into the neighborhoods. There, it aims at preventative goals, using specially trained "peers," including those who are bilingual, to intervene with youths at high risk. In 1996, 500 youths were thus contacted.

The program of prevention and of intervention against violence This initiative of the State of Illinois follows up on ex-incarcerated juvenile offenders and those on parole. The youths work with the probation officers on their emotions, on responses to provocation, and on situations leading to violence.[44] Videos made by the youths themselves lead to pragmatic discussions of alternative behaviors. Probation officers may decide, with the juvenile's consent, to transfer him/her to an adult who will act as a mentor. Mentors are generally volunteers supported by the Mentoring Network, which is chaired by a coordinator who establishes links between the professionals, the families, and the teenagers.

Teen courts Community justice courts operate in Chicago as they do in France, except that they require the intervention of a judge and of a jury composed of six–twelve adolescents. The indicted youths, from 14 to 18 years old, are selected by a probation committee and are non-violent offenders and non-recidivists. They defend themselves before probation officers speaking for the victims. The jury makes the decision and generally resorts to alternative and community-work sentencing. The youth's criminal file is cleared. About 150 community courts of this type are found in the country. An evaluation of 30 of them in Texas has shown that rates of recidivism were under 5 percent, compared with the 30–50 percent average, after juvenile courts' decisions (non-violent and non-recidivist offenders allowed access to community justice courts are strictly screened).

The Youth Net Program As in New York, numerous community organizations intervene in neighborhoods to fight drug-dealing. Operating in the north of Chicago, the Youth Net Program is funded by the city. It not only organizes after-school activities meant to prevent adolescents from being attracted to street life and drugs, but invites the families to attend school activities and offers linguistic courses and social services to Latino immigrant families. In this way, a sense of community may develop, allowing residents to involve themselves in efforts promoting public space safety.

A prevention program initiated by the University of Michigan Youths from 7 to 13 years old of African-American and Latino origin in the problematic neighborhoods of Chicago and of Aurora are supervised by a group of academics from this university, who intervene at three levels:

(1) In the classroom, their intervention looks like a conflict resolution class with peers. The classes relate to factors creating violent situations, to signals indicating coming violence, to motivations, the understanding of emotions, the legitimacy given or not to aggression, self-control, etc. The course lasts 40 weeks over two years and involves teachers from the local schools.
(2) The second level of intervention targets youths at high risk and works to build peer relations. Small groups meet 16 times a year and, hopefully, anti-crime leaders will emerge.

(3) The third level involves the families of young children. Groups meet every week. They work on attitudes and emotions and try to support families so that they can help their children resist the temptations of street life.

Initiatives undertaken by the residents themselves

In the problematic neighborhood of Edgewater, Ken Bucks is involved in a community board focussed on social prevention and delinquency. The neighborhood residents have decided to reconquer their public spaces and some of their buildings. One building has been squatted in by the Vicelords gang. The community police participated in the board's meetings and the residents asked the precinct commander to think over the problem. Then, on his advice, they went to a judge to require him to have the building sealed. "We needed to make life difficult for gangs, force them to move all the time and not wait for the police to become social workers," Bucks says.[45] The residents worked together on that specific problem and, being successful, decided to look at other issues together.

Beat meetings mix community police officers and residents, creating opportunities to offer suggestions to the police who then implement their strategies. Residents have accurate knowledge of what goes on in a neighborhood. At Edgewater, the poor residents, of mixed racial background, work together on the data provided by police researchers. They mobilize to intimidate the gangs and the dealers. They have obtained the closure of bars and clubs which attracted dealers and prostitutes, requiring the police to intervene. They admit that they sometimes have wrong ideas concerning some hot spots. The data may prove that they're incorrect. The police, for their part, acknowledge that the residents, including the youths cooperating in the beat meetings, offer very positive suggestions. But as will be seen, residents' involvement may become exclusive and reject categories (homeless, sex workers) entitled to live in the neighborhood.

Efforts undertaken in Robert Taylor Homes "Mentoring and Rites of passage" is a structure aiming at reducing violent behavior, injuries, and shootings in the most extensive ghetto of the USA, the Robert Taylor Homes. Youths from dislocated homes interface with adults who act as their tutors during the course of their adolescence. The "tutors" are trained for six months in order to counsel each of the 10–15 adolescents whom they meet with at least twice a week. Participants are encouraged to search for racial pride and work on their self-images, in addition to developing their verbal skills, communication, and decision-making. An anti-violence campaign reaches out to all the Robert Taylor Homes residents. But, it also sometimes happens that a gang takes control of the program, thus ending months of efforts. Such initiatives are indeed fragile.

Rebecca Stone has analyzed the famous program called "Beethoven" in this housing project:[46]

> It was initiated by a philanthropist, Irving Harris, convinced that the first years of life are crucial. He wanted primary schools to welcome children from poor homes in the best possible conditions. The Head Start Program welcomes only 4-year-old underprivileged children. Consequently, it was his hope that mothers, once they were pregnant, would receive the best care and advice at a support center and that a nursery would take care of infants' mental and physical development. But this program started in the worst conditions. After five years, at a very high cost, hardly twenty families had really benefitted from it. There were, as is often the case, communication problems between the very wealthy donators and these very deprived families. Who controled whom? The program got more and more costly and tensions increased. The program was duplicated – Child Development Program – although it was just beginning. It was a splendid vision but the practice did not meet the expectations.

The crucial role of two intermediary organizations

The MacArthur Foundation This Foundation is specifically involved in child and adolescent development in deprived neighborhoods and in juvenile courts in Chicago. Like the Vera Institute in New York, it subsidizes numerous research efforts in that field (e.g. research on recidivist young offenders and on youths about to become recidivists, on the promotion of institutional changes, on the restructuring and decentralization of juvenile community courts, and on multi-partnerships between schools, social centers, and courts). To examine the impact of poverty on urban problems, the Foundation tries to understand why teenagers fail and how change is initiated to remedy this situation in poor neighborhoods. A vast ethnographic study of 77 Chicago neighborhoods is currently under way.[47] In particuliar, seven age groups of youths of different ethnic, racial, and class origin are being observed. Three thousand people have been mobilized for the investigation and analysis. Video is used to record the physical aspects of the neighborhoods, the buildings, the graffiti, the children in the parks – an approach espoused by the School of Chicago 70 years ago. Over eight years, 11,000 individuals will be followed. The study tries to answer the following questions: Most communities are relatively stable and lawful – how do they remain so and why do others break down? What resources and "protective" factors enable some people to lead successful lives even in high-risk environments? Why do others become "career criminals"? What social control can be exerted on a given space? From the first results of this study, the idea of social efficacy, already mentioned, has emerged (Sampson, Raudenbush, and Felton, 1997: 918–24).

Chapin Hall Center at the University of Chicago The Chapin Hall Center at the University of Chicago, established in 1985, also focusses on policies, practices, and programs in the city's poor communities. Its goals are to initiate institutional change, such as Chicago's alternative policing policy, to improve the conditions of juvenile incarceration, and to assist problematic neighborhoods with the help of funding provided by the Chicago Community Trust (which is subsidized by corporations). It is a progressive, reformist vision that inspires the Center. For Joan Wynn, a research director, "money is often spent too late for the intervention on children and adolescents to be efficient."[48] Networks intervening with families with dependent children must be created while the children are very young. The Chicago Community Trust has spent about $30 million over ten years in eight Chicago neighborhoods and a deprived suburb. The length of its commitment is important. "Besides, we try to have the adults from these neighborhoods work together," she adds. "But there are always antagonisms linked to respective turfs, whether they are geographical or symbolical and the common vision fades away." Whatever the neighborhood, the issue of safety is a priority. Chapin Hall has the merit of bringing coherence to scattered initiatives already mentioned, evaluating them, and then making recommendations.

Other initiatives

Compared with the mayor of New York, who placed his political priority on safer streets, the mayor of Chicago made a less risky choice in opting for educational improvement in the public schools as both a preventative strategy and one answering the needs of the business community. With this choice, the mayor gathered the support of political, economic, and community interests which won over, in turn, the Illinois legislature. In the early 1990s, the Chicago school system was pilloried as the worst in the USA. In 1995, the Illinois State passed an Amendatory Law, replacing the previous Chicago Board of Education by a five-member Board directly appointed by the mayor. The new team, composed of experienced political, business, and academic leaders, implemented a tough, back-to-basics approach on the third largest public education system in the country (Johnson, 1997). Principals were removed and schools were put on probation. After failing tests, 4,200 students were given mandatory summer schools to attend and 400 "teen moms" received special training to better prepare their children for school. Conflicts were skillfully reduced with the unions through salary increases. Funding of $2 billion was provided to build new schools and kindergartens. In his effort to interact with a broad coalition of actors, Mayor Daley did not hesitate to put the emphasis on neighborhoods' positive influence.[49] In his 1998 address, the term was repeated more than 20 times. He insisted on the role played by parents as "attendance officers" acting against violence. Principals agree that parents have become more involved in their children's schools and schooling since the reform than ever before. According to *The Economist* (12 December 1998), "the people of Chicago seem to like their mayor, in part because of his passion for the city," in part because he wants to be remembered as "the education mayor." Of the two successful experiments that will be described below to illustrate the involvement of Chicago's civil society in the solution of its problems, the first, Youth Guidance, concerns the field of education.

Youth Guidance Youth Guidance is a non-profit, socially-oriented organization. Its goal is to help the young from low-income neighborhoods to become fully responsible and productive adults.[50] Although it had had seventy years of experience, it was only in 1969 that Youth Guidance settled in eight high schools and twenty-two primary schools in Chicago, which were attended by 7,000 children of various ethnic and racial backgrounds (67 percent African-American, 28 percent Latino, 3 percent white, 3 percent Asian). As Robert Taylor, president of Youth Guidance, emphasized:

> The approach via schools seemed better to handle children and teenagers so destructured that they would have been unable to go by themselves to a care service. In numerous neighborhoods, the school is the only institution that is left. The children are there, a link can be established. Problems may be contained if the young have a positive contact with the school, with an adult, or a center. It is this link that we try to strengthen. (Body-Gendrot, 1996)

According to Nancy Johnstone, director since 1973, seven principles guide the program: (a) to have employees in the field, four days a week; (b) to have a well-trained, inspired staff (Youth Guidance educators often choose to live in the projects to understand the stress of the families better); (c) to have a dual approach toward the individuals and their environment (some of the staff work with the children, others with the teachers and the parents); (d) to develop early intervention as soon as the child is in school and a problem occurs; (e) to get the parents involved early and to train them if necessary; (f) to follow up on students all through their school career; (g) to establish partnerships with other schools, business networks, and other organizations to help the older, skilled students get jobs. This multidimensional approach is based on the assumption that no actor can succeed alone – isolated services are overwhelmed by problems, while good partnerships meet with success.

The Roberto Clemente School is located at Humbolt Park, attended by 2,500 students, 50 percent of them African-American, 43 percent Mexican and Puerto Rican. The Cabrini Green project sends its children to this school. Problems in the surrounding neighborhood are very serious: high crime, gangs, drugs, teenage pregnancy, unemployment, high poverty. Many of Chicago's 120 different gangs come here to wage their wars, even inside the school. Some of the students have already been in jail, some are gang members and drug-dealers. Parents stand at the doors of school, acting as security guards. Community police officers offer midnight basketball in an area where no one else wants to linger after dark. Two officers work full time in the school, with one part-timer; they are helped by 50 volunteers. Parents with walkie-talkies escort the young from one class to the other, act as mentors, do fund-raising, develop their skills for the common interest. The school has been partitioned in four, to make its administration easier. Each unit has its own security infrastructure.

All kinds of techniques are employed to establish strong links with the young, as at the Washington Heights, New York school described earlier. The Graphic Arts through Education program has therapeutic virtues and helps improve the students' confidence. A number of programs have been inspired by a psychiatrist who observed that the children from inner cities did not learn much in school. This is why Roberto Clemente functions like a company, co-directed by the teachers and the parents. Not only are decisions jointly made, but parents are trained to become smart decision-makers. Parents are part of the solution.

It has been demonstrated that children who manage to remain three years in the school will probably graduate. For the most disturbed, alternative schools can provide a temporary or lasting solution. The Teen Parenting Center allows young mothers to complete high school – 118 such mothers attended the school in 1994, 72 two years later, following the spectacular success of a mentoring program involving older single parents interacting with teenage girls.

The Youth Guidance team, 30–40 percent of whom are non-tenured social worker students, receive grades from their universities after their training has been evaluated. The organization has an annual budget of $3,300,000 funded by the state, city, and monies from United Way and other large companies. It is hard work to convince the donors that the operation is worth supporting. Youth Guidance is staffed by 65 people, including 50 working full time in the schools; 35 volunteers, and about 30 students complete the team.

An evaluation of Youth Guidance's results, undertaken by the Milton Eisenhower Foundation, showed that only 11 percent of the guided students dropped out (as opposed to 19 percent of those not guided). Some of those in the program were among the better students. 61 percent of the Youth Guidance students had undergone at least two traumatic events in their life (e.g. their friends had been killed or raped) and half of them suffered post-traumatic symptoms (e.g. nervousness, lack of concentration, bulimia, etc.). In quite a large number of cases, Youth Guidance has succeeded in helping them to overcome their negative emotions and detrimental contacts. The young respondents admit that their relations with the organization were positive overall.

Bethel New Life In the West Side ghetto, Bethel New Life is an especially dynamic organization, founded by a Lutherian Church eager to work on problems in its neighborhood (Body-Gendrot, 1996). Not surprisingly, the social problems of the area, which has been taken over by drug-dealers, are astronomical: third-generation single-parent families, general disenfranchisement, surrender to misery. The population decreased from 60,000 to 24,000 residents in twenty years; small firms and trades left, leaving behind them polluted sites, which became public dumps as a result of inadequate public garbage collection.

Bethel, with a $7 million budget, employs 400 people in a neighborhood with 60 percent unemployment. The employees were formerly on welfare. Between 1970 and 1990, the number of unstable homes fell by 40 percent. Bethel renovated 1,000 of them with $65 million in investment. It opened a day-care center for 70 3–5-year-olds, cleaned the streets with volunteers, placed 4,000 people in job training, constructed a medical preventive center, and co-managed small stores. How? Bethel used a database to prove that banks had redlined this neighborhood. Referring to the 1977 Community Reinvestment Act, which forbids banks to discriminate against low-income areas, Bethel was able to convince the First Chicago Bank to invest in the West Side. After one year, the bank had not lost any money on its operations there. The Bethel activists asked it to convince two other banks to invest there also. Then the dynamics for change were in place. Bethel also took advantage of the area's empowerment zone status and the $2.5 million in loans available to new commercial and service companies creating jobs for the residents. It also helped employees become owners of their homes with bank loans that it guarantees. Ownership is thought to be the best means of stabilizing a neighborhood and of pushing residents to commit themselves to common goals. The modest goals relative to the deployed means are sometimes surprising: 25 homes here and 20 jobs there are at stake, but a cumulative effect is anticipated.[51]

Bethel fights the violence of these living conditions with a multidimensional approach. Every year, 700 youths are released from jail and have to be resocialized. Thanks to the support of a private firm, they are offered classes, which can help them be eligible for jobs in nearby suburbs, for instance, in recycling. This is a joke among local residents: they have been so often considered garbage by others that they can recycle garbage well and even get pride from it!

Bethel is very involved in anti-drug efforts and its activists lead door-to-door information campaigns. "It's a war, here," Marcia Turner, one such activist, exclaims. "I grew up here, but I do not recognize my neighborhood. What is the government doing? Why don't the media relate our efforts? . . . Every September, we chase the dealers, we occupy the sidewalks, night and day, we demonstrate our determination. OK, this is symbolic, but it is the sign that we fight back." But activists complain that the drug-treatment centers are so scarce and costly ($28,000 for 28 days) that they are unaffordable.

Lastly, for isolated and battered women as well as for abandoned elderly, Bethel has set up solutions with other partners. The organization expects nothing from elected officials who, as Marcia Turner says, "once they are elected stop being in service to the population. In problematic areas, the residents do not count, not even as a pressure group. Priority must be given to the reconquest of the streets by the marginals themselves." A militant, Marcia Turner is convinced that there will always be volunteers to pursue the community efforts. "Survive or die," she says. As for her, she has chosen to weave the threads of survival.

Such innovations should not mask the general trend: they seal breaches but do not solve the fundamental isolation of segregated neighborhoods plagued by crime and delinquency. At best, these initiatives prevent the American society from erupting. They perpetuate the idea that success is possible despite the worst odds. W. J. Wilson, who studied the Chicago neighborhoods in depth, has no illusions: the more the neighborhoods deteriorate, the more that safety problems are crucial, and the more difficult it is for the residents to lead collective action. When surveyed, such residents are 11–12 times more likely to say that assaults are a problem than people from wealthier neighborhoods; 80 percent of them complain about drug-dealing, and 63 percent about youths' behavior in their communities (W. J. Wilson, 1996).

Conclusion

It appears from the American cases that the decline of attacks on the quality of life in a diversity of neighborhoods and of cities is attributable in part to law-enforcement reforms,[52] in part to empowered residents, in part to other larger factors (demographic, economic) already mentioned. It can be said that sophisticated mechanisms of reconciliation are at work in safer cities.

The American solutions to street crime are brutal, however, and other Western countries should weigh the consequences before adopting the same strategies. Until recently, the question "how to punish?" distinguished a progressive policy from a conservative one. Currently, in most countries, an autonomization of the issue is observed. The debate no longer concerns "who should be punished?" "which crimes should be punished?" "from which collective values?" or "for which goal?" but "is this measure more appropriate than that other?" "should the incarceration age be lowered to twelve years old?" and "are boot camps efficient for juveniles?".

The massive incarceration policy chosen by the United States is likely to be short-sighted, as proven by recidivism data (unless one considers that life detention, the death penalty, "three strikes," etc. solve the problem of release). As noted by Abramsky, short-sightedness is also revealed by the neglect given to the phenomenon of increasing inmates' unprepared release from prisons.

> To use a worst-case scenario, some 660,000 will be released in 2000, some 887,000 in 2005, and about 1.2 million in 2010. Even factoring in lower release rates because of three-strikes laws and truth in sentencing, and even taking into account estimates that 60 percent of prisoners have been in prison before, there will still be somewhere around 3.5 million first-time releases between now and 2010, and America by then will still be releasing from half a million to a million people from its prisons each year (not to mention hundreds of thousands more from stints in jail). That is an awful lot of potential rage coming out of prison to haunt our future. (1999: 32–3)

It is as if American society had given up all hope of resocializing its offenders, externalizing them, casting aside their bodies in the hope that the problem of delinquency will be magically solved. From an ethical perspective, the criminalization of young minority males is disastrous for the civic culture of a nation.

Despite a decline, the problems of street crime are still measured in terms of homicides, school massacres, and riots. Certain cities – Los Angeles and Miami, among them – are said simply to exist between riots. Such urban dysfunctions reveal serious disenfranchisement, the result of spatial segregation, middle-class withdrawal and their secession from the public space. The laissez-faire approach is the prevailing response to such phenomena and when the state tries to intervene in such processes of social fracture, it causes an uproar.[53] I should add that ethno-racial cleavages discourage identifications and common mobilizations. Perhaps we could also suggest that a lot of the urban conflicts related to territorial coexistence which marked the end of the 1960s and the beginning of the 1970s were solved by mobility and the withdrawal of those who were able into protected enclaves. When three-quarters of white Americans are no longer living in the centers of cities, where a majority of minorities, immigrants, and poor people live, it reduces social conflicts.

An outsider is struck by the dichotomy of the majority view (us versus them) based on the belief that an individual must be able to ensure his/her protection and that of his/her kin. When solidarity takes place, and it does, it is within a smaller community based on an implicit social contract. The term "community", then, has a strong meaning. But for the unworthy and threatening poor, the carceral state is substituted for the welfare state. It is revealing that after the Watts riots in 1965, the Model Cities policy and all kinds of social measures were introduced by the federal government and the states. But, conversely, after the South Central riots of 1992, California's prison program expanded even more, as if a loss of faith in redistribution and regulation was prevailing.

On the positive side, one must recognize that once a policy is initiated, it is given the resources to succeed and, if it does, a very efficient marketing and communication campaign ensures that it is well publicized, something the French are very awkward at doing with their public policies. In response to urban violence in the French *banlieues*, the discourse is largely alarmist and defeatist, even though, each day, new positive innovations remain invisible to the public. Nobody talks about neighborhoods which are reborn, as is frequently done in the New World (cf. South Bronx).

Our two cases show that, to be successful, law-enforcement has to go hand in hand with preventative work: people welcome safer streets and elected officials can base their campaigns on restored peace. But their success may prove fragile, as shown by demonstrations in New York against police brutality.[54] Then, the political appeal of the politicization of crime backfires and grassroots resistance calls for less harshness. Several signals indicate that the general public may be changing: an Attorney-General in California campaigning on toughness was not elected; voters approved by referendum the medical uses of marijuana in seven states; grassroots branches of Families Against Mandatory Minimums opposing harsh sentences for minor offenses have sprung up all over the country (Gordon, 1999). In France, a rigid official discourse conceals widespread changes at work in cities. These will be analyzed in Chapter 5.

Notes

1 It cannot be denied that the redefinition of tax policies and transfer payments under the Reagan and Bush administrations brought about a redistribution of wealth, along with the restructuring of the economy.

2 These findings were corroborated by the Port Authority of New York and New Jersey's annual report. The average annual income of people in the 90th percentile of the income spectrum increased 26% to $80,000 in the last 17 years, while for those in the 10th percentile at the bottom, it fell 21%, to $15,000. The ability to benefit from the appreciation of companies' stocks explains in part such inequalities (*NYT*, 7 August 1998).

3 This information is provided by the Office of Management Analysis and Planning of the NYPD, 1997. I wish to thank Deputy Farrell and Luc Gwiazdzinski for providing material.

4 In 1995, 12 police officers in the Bronx were arrested for brutality, theft, and perjury. People died as a result of police violence during 15 documented episodes, and 30 people, two-thirds of them minorities, were injured or shot. In August 1997, a Haitian, Abner Louima, was violently brutalized by the police and the mayor understood a few months before re-election that he could not indefinitely support police behavior in all circumstances. Tensions between Haitians and African-Americans in New York might explain why demonstrations and protests remained weak. The tactic used by the mayor to defuse tension was to create immediately a taskforce with a $15 million budget to investigate the Louima case. The recommendations of the investigation were then ignored by the mayor. The Civilian Complaint Review Board (CCRB) was created in the 1960s. CCRB was revamped by Dinkins. All its members were civilians. Yet by 1993, after Giuliani's election, the CCRB became totally ineffective (Morales, 1997). It was underfunded and only had a $5 million budget to deal with hundreds of cases (interview with Dinkins, spring 1999).

5 In a recent Million Youth March sponsored by Khalid Muhammed, the mayor "abused the power of his office by turning a large section of Harlem into a police encampment . . . The cops came ready for war . . . with helicopters, horses, tractor-trailer trucks, buses, cars, vans and motor-cycles. Giuliani, by deploying his police as if all black people were a mortal threat, succeeded in intensifying the opposition among blacks toward him and his policy of overkill. The anti-Giuliani feeling among black people in New York City is overwhelming and growing," Bob Herbert from the *New York Times* reported (Herbert, 1998).

6 These figures come from the Correctional Association of New York.

7 When I interviewed Capitain McCann, who supervises 100 men and women in Brooklyn, he only wanted to talk about the prevention methods that he had developed: sports programs for youth, from the South Bronx five evenings a week, anti-gang and anti-drug programs lasting 17 weeks in the schools and reaching 20,000 students, summer programs for 100 kids supported by public and private funds. These programs are quite similar to the French ones. What was probably different was the tone of McCann, who was convinced that the mayor was "one of us."

8 Interview by the author, spring 1997, with Judge Fried.

9 Interviews with Judge Newt Bingham and Judge Bernard Fried, State Supreme Court of New York, by the author, on separate occasions, spring 1997.

10 Known as the "subway vigilante," Goetz was made into a hero by the media after he shot four black teenagers who tried to extort money from him in the subway. While self-defense has stronger roots (linked to the frontier tradition) in the USA than in France, Lilian Rubin has shown that, in New York, opinion was divided on the matter (45% approved of Goetz's actions; 51% of men approved, while 46% disapproved). Yet the media chose to ignore the complexity of public opinion.

11 Live observations by the author during criminal court hearings.

12 Yet only 12% of the nation's drug-users were black at the time, according to official statistics (Harris, 1995).

13 Old boats are used for penitentiary purposes: the *Bibby Venture* shelters 380 inmates and the *Bibby Resolution* 280. Drug-treatment is performed within such prisons.

14 Interviews, spring 1997.

15 Interview by the author, spring 1997. M. Jacobsen is currently Professor at John Jay College.

16 The city is the "creature" of the state and is constantly begging for more resources. In 1997, the mayor suggested eliminating the jobs of 500 prison guards as well as the methadone programs. However, prison has become a drug-treatment center for minorities who would not be able to afford health care if they were free. If drug-treatment policies are eliminated by the mayor, eager to balance his budget, more minorities will be hurt. The NYC Corrections Bureau itself has been threatened with elimination.

17 Interview with the author, spring 1997.

18 I wish to thank anthropologist John Devine for organizing field visits for me to the inner-city schools that he has observed for over 10 years.

19 "The tradition of the oppressed teaches us that the 'state of emergency' in which we live is not the exception but the rule," Walter Benjamin wrote, quoted by Devine, 1996: 26.

20 A similar visit confirming those perceptions took place in a high school of Bushwick in the spring of 1998 and I wish to thank Professor Peter Lucas for sharing his insights with me.

21 I wish to thank Joy Dryfoos for her help in organizing this field visit and for the insights she shared with me.

22 When I visited IS 218, the students put on an Alvin Ailey-style dance performance. In fact, the show has become so successful that other schools invite the group, which helps to motivate other students.

23 It costs $800,000 a year to finance the resource center, the health center, the multi-service center, and all kinds of activities at IS 218. Public and private funds support the school (Dryfoos, 1994; 1998). Half of the students benefit from Medicaid. The Board of Education subsidizes the maintenance, security, and insurance, which are very costly due to the working hours and the reputation of the area.

24 Interview with Paul Steele, former research director, March 1997.

25 Interview with Joanne Page, executive director of Fortune Society, April 1997 and fall 1998.

26 The American system of law-enforcement is shockingly repressive, no one will

deny it, but it is not the Gulag. Receiving damages while being incarcerated is something which could not be possible in an emerging country like Brazil where repression is exerted with impunity (in 1995, 1,100 civilians were killed by the police, according to interviews I conducted in São Paulo in 1997). Summary executions in small local prisons are frequent (Pinheiro, 1998). Even from the US death row, Moumia Abu Jamal could address the whole planet via Internet.

27 Two mentors from Fortune, Stanley and Wyllis, were able to reconstruct their lives after taking exams in prison. Then they registered at a community college, as J. Vasquez had done, and received a training to become community educators.

28 There are about 15,000 public-housing projects in the USA, representing 1.3 million units managed locally by 3,225 public-housing authorities. Officially, there are 3,003,000 tenants. But in reality there are probably twice as many (Popkin, 1996: 361–78). Official statistics on public housing, which was cut back during the Reagan Administration, provide little information. Fewer than five hundred apartments are contained in 90% of the projects, and half of the tenants officially live alone. But the few empirical studies done on public housing mostly concern the crime-ridden exceptions: the very large structures with over one thousand units that account for 1% of all public housing.

29 This idea was developed by Jeff Fagan, the director of the research center on violence at Columbia University.

30 50,000 units distributed among 1,500 buildings are officially occupied by 5% of the population, that is, 150,000 people.

31 In Chicago, housing projects contain eleven of the nation's fifteen poorest census tracts. All 28 buildings of the Robert Taylor Homes, which accommodate 11,000 people, will be demolished by 2015, the city says. Stretching over two miles, the buildings shelter 0.5% of the Chicago population. Until recently, on average, 11% of the murders, 9% of the rapes, and 10% of assaults in Chicago occurred there, according to police reports.

32 Under a Federal court order in the 1970s and 1980s, the city moved 7,100 families out of segregated public housing and into rent-subsidized apartments. The tenants were screened and counseled, which explains the success of the operation.

33 Interviews with Martin Goldsmith, director of the Edgewater Council, 1996 and 1997.

34 Interviews with Carmolita Curry, Catherine Collins, Michael Onofrio, early intervention program, Cabrini Green. See also McRoberts and Linn Allen, 1998.

35 Chicago sociologist R. Block is convinced that when the economy is booming (the unemployment rate was around 5% in Chicago in 1998), thefts and larcenies decrease because employers will hire any employable person and provide some hope of a future in poor neighborhoods.

36 Interview with the author, spring 1997.

37 Interview with the author, spring 1997.

38 These tensions between blacks and the police of Chicago date back to the beginning of the century, as witnessed by the riots of 1919. But it was

the police brutality towards protesters during the Democratic National Convention of 1968, the famous order "Shoot to kill" given by the Police Chief and the police violence against Martin Luther King which is widely remembered in the black community.

39 Interview with the author, spring 1997.

40 My thanks to R. Sampson for sharing this information with me.

41 Interview with the author at the Cook County juvenile court, spring 1997.

42 During one of the hearings that I attended, however, a judge found no other solution but to send a Latino homeless youth back to jail, as no one, the state services included, would take care of him.

43 Interview with the author, spring 1997. Dohrn, a former activist from the Black Panthers, illustrates how in some cases the system allows a pluralism of voices to be expressed by co-opting dissident voices for its own benefit. Conversely, the dissidents see an opportunity by working with the system to push for progressive causes.

44 Interview with a probation officer, Lieutenant Lorigio, also a professor at Loyola University.

45 Interview with the author and field visits, spring 1997.

46 Interview with the author, spring 1997.

47 Based in Chicago and directed by F. Earls from the Harvard School of Public Health with A. Reiss from Yale University, it brings together scientists from a wide range of disciplines from across the country.

48 Interview with the author, spring 1997.

49 Meeting with the mayor, spring 1996. Daley is convinced that companies will lack manpower, due to the boom in the American economy, and that schools and vocational training have their part to play in addressing the coming needs.

50 Interviews in 1996, 1997, and 1998 by the author, supported by field work.

51 The severity of the problem appears when, for instance, in a nearby area on the West Side, 66,000 residents have access to only one bank and one supermarket but to 66 betting parlors and a hundred liquor stores.

52 In Boston, police and probation officers walk together through the streets with success. No street death has been registered in recent months at the time of writing.

53 I am alluding here to the law signed by President Clinton in October 1998, requiring income (consequently racial) integration in each public-housing building of the nation.

54 The 22 October Coalition to Stop Police Brutality organizes marches nationwide and documents claims of police brutality.

5
Managing Polarization: Paris, Marseilles, and Lyons

In contrast to the national perspective, French cities express a wide range of perceptions concerning urban threats. Institutions and local actors do not share the same discourse, or the same perceptions of events, and in fact draw notably different conclusions from them. In particular, actors from the public service hardly speak with a single voice, thus confirming how diverse the views on the phenomenon of urban violence are. More broadly, local antagonisms reflect the increasing segmentation of society and the loss of common references. As the case studies below will show, social polarization by income is reflected in visible segregative practices differentiating the neighborhoods. Public or pseudo-public space then becomes the object of appropriation by different types of actors who claim it as their own.

How do French cities maintain their social cohesion? Neither zero tolerance, nor incarceration have been political options. For a long time, mayors emphasized social prevention. Has it been efficient? Is this choice now questioned? What kinds of social control do large French cities exert?

To start with, it is important at this stage to point out the large contextual differences between American and French large cities. It can only be repeated that the size and nature of street disorders are not the same. If Americans have reason to be alarmed by homicide rates and racial divisions, it is not to murders and to the consequences of hypersegregation that "urban violence" in France generally refers. Polls give a very different view of the fear of violence, when addressing the local sphere. In a Sofres national poll of November 1998, when asked whether the situation of violence was "extremely" or "very worrying" in the locality where they lived, only 22 percent of the French respondents answered positively. As for the mayors of large cities, only 28 percent agreed. 43 percent of them thought that the situation was stable in their city and 80 percent of them denounced the media for their negative role (distorting events, reinforcing the public's fears).

A second major difference concerns the intervention of the state in the management of French cities. On the one hand, policies of redistribution exert forms of regulation. On the other, the local authorities who are in the front row when questioned by their constituencies on the issue of safety do not have much leverage. If French mayors have gained power with the decentralization laws passed at the begining of the 1980s, in particular over land use, they have little control over public safety (the Préfet is in charge),[1] in criminal justice decisions (the prosecutor and the judges may decide never to speak to the mayor) or the choice of residents in public housing. In 1997, the state council declared unconstitutional local ordinances forbidding panhandling and imposing curfews that had been initiated by a few mayors with diverse political affiliations. For his part, the French Minister of the Interior has attempted to forbid municipal police to bear arms.

Yet, in contrast to large American cities, political representation is tighter in France: there are 36,000 mayors for 60 million French and through multiple office-holding[2] and a back-and-forth movement between local and national issues, local authorities may be influential and initiate political changes.

Each city observed in this chapter experiences specific forms of disorder and finds remedies which are better understood in the context of its history, its socio-economic profile, its demography, the composition of its elites, and its own resources. Sites were selected, according to the following criteria: the size of the cities, specific urban disorders, and the media's coverage of events. However, this selection has been voluntarily distorted for two reasons. First, in the case of Paris, it was a better choice, it seemed to me, to deal with a nearby suburb like St-Denis which is experiencing all kinds of symptoms of crisis and violence and taking positive action; affluent Paris is not representative of the urban threats which worry the public. Second, I thought that it would be just as important for the research to understand why a city like Marseilles has not erupted so far, despite the accumulation of all the symptoms of crisis, and why a wealthy metropolis like Lyons has.

St-Denis in the Paris Region

St-Denis is located in a polarized and segregated region. It has experienced all kinds of disorders, the most spectacular of which occurred in the public schools and buses. The mayor of St-Denis has opted for spectacular actions of reconciliation, some of which were taken during the Soccer World Cup, a "global" event in a local space.

A polarized and segregated region

St-Denis is regarded as a *banlieue* of Paris and can be easily reached by subway from the center. As described in the Introduction, since the middle of the nineteenth century, the term *banlieue* has defined a portion of space between the city and more distant and loosely urbanized zones and bound into a compact whole,[3] but recently, the term has taken on very negative connotations. St-Denis is located in the Île-de-France region (11 million residents, 19 percent of the French population, 8 *départements*)[4] which partly subsidizes the locality. Another source of financial control and support comes from the *département* of Seine St-Denis.

The Île-de-France region is the most demographically and economically successful in France. It has gained 3.5 million residents since 1960, yet lost more than 500,000 residents since 1990 (*Le Monde*, 7 July 1999). In 1990, three-quarters of nationals and two-thirds of immigrants living there were employed in the service sector, a shift which compensated for the losses in manufacturing and construction jobs. It should be remembered that when France lost 1.3 million industrial jobs between 1975 and 1990 (68 percent of which were unskilled or semi-skilled), 400,000 (27 percent) of them were located in the Paris region (Beckouche, 1994). The region is home to 18.8 percent of the population of France, 23 percent of the country's jobs in 1990, 29 percent of the GNP, and 83 percent of the headquarters of the 200 largest firms. Growth in recent years in Île-de-France (more than 2.5 percent) favored nationals as well as foreigners.

The theory of social fragmentation refers to the processes of segmentation within the labor force. It is intertwined with the process of spatial polarization (Rhein, 1998: 430; see figures 5.1 and 5.2). Age and gender interfere in these processes. For instance, youths of foreign origin are particularly vulnerable: three out of ten are unemployed (five out of ten North Africans under 25 are jobless). If youth and national origins are considered jointly, more than one North African out of four under the age of 25 is unemployed and the general rate of foreign unemployment is twice the national rate (except for Europeans settled in France). When it comes to gender, in 1990 female workers accounted for 47 percent of the metropolitan workforce (Rhein, 1998: 432). Foreign women, especially the unskilled, experienced an average 35 percent rate of unemployment (again, except for EEC natives). Cultural and religious factors provide some explanation (INSEE, 1994a: 83).

New unstable spaces of economic activity, some of them at the margins of the legitimate economy or marking a transition between legal and illegal occupations, are generated by social transformations occurring in the Paris region. At a time of harsh competition and due to insufficient dissuasive mechanisms, it appears that the underground economy has expanded and

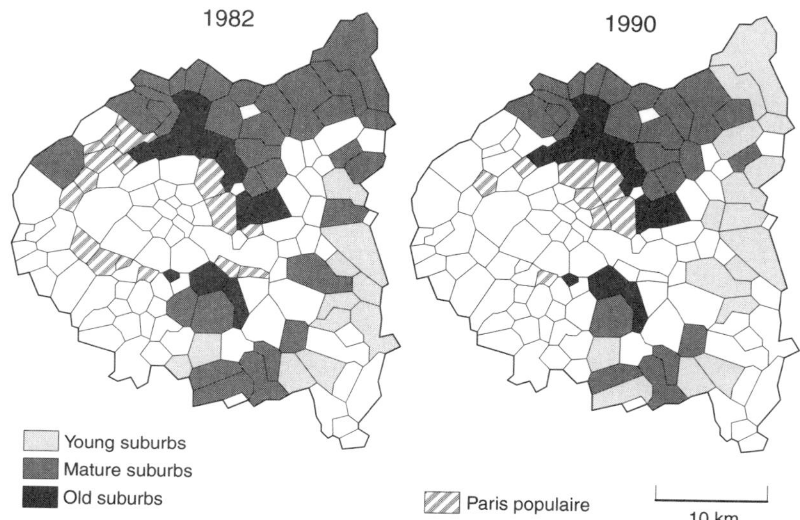

Figure 5.1 Working-class areas in Paris CMSA, 1982 and 1990
Source: Rhein, 1998: 441
(By permission of Taylor and Francis, PO Box 25, Abingdon, Oxfordshire OX14 3UE)

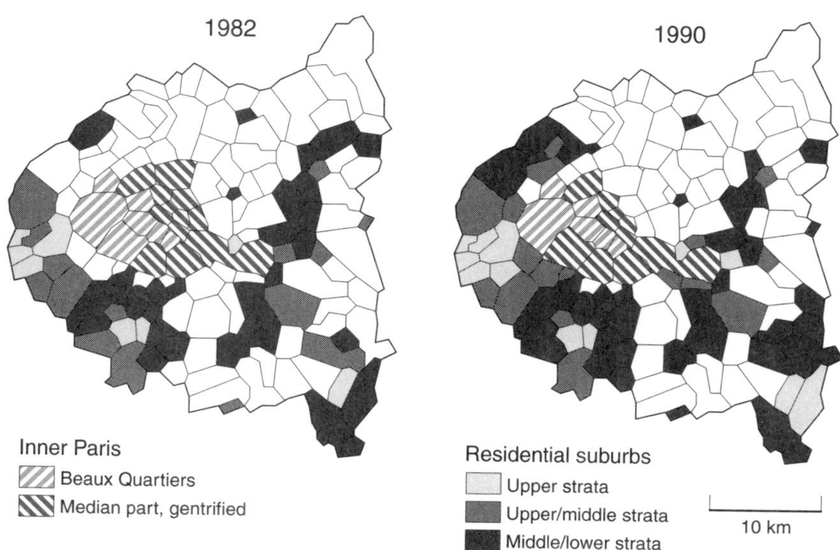

Figure 5.2 Residential areas in Paris CMSA, 1982 and 1990
Source: Rhein, 1998: 442
(By permission of Taylor and Francis, PO Box 25, Abingdon, Oxfordshire OX14 3UE)

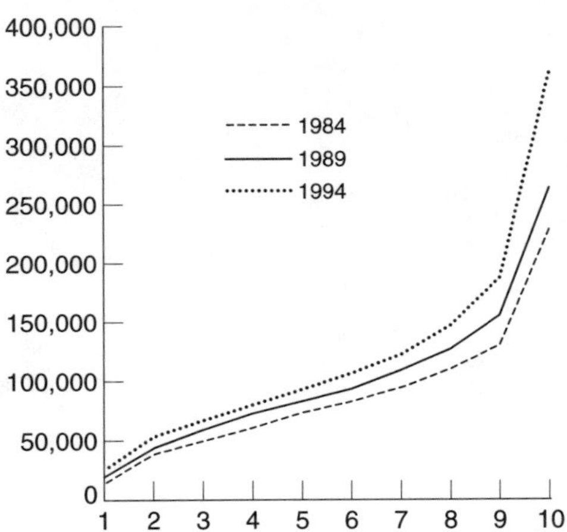

Figure 5.3 Total income of households in Île-de-France, Paris, by unit of consumption, 1984–1994 (French francs)
Source: Préteceille, 1999: 5

diversified, a sign of adjustment to the new requirements in the production of goods and services in the Parisian context.

Préteceille (1999) has carefully compared the evolution of economic inequalities over three income surveys from INSEE (Institut national de la statistique et des études économiques) in 1984, 1989, and 1994.[5] His observations show that the situation in the Île-de-France region is strikingly different from that in New York. First of all, he notices that the distribution remained almost constant over twenty years: there was no increase in the poverty of those on the lowest incomes. On the contrary, by the late 1990s, due to "income socialization" – i.e. unemployment schemes, family allowances, and pensions from the social welfare system – this group's average income improved by 55 percent in relative value. There was no "hollowing-out" of the middle, either. The total average income improved, especially between 1989 and 1994 (when it climbed by 38 percent in absolute value). This situation is more pronounced in Île-de-France than in the rest of the country. The growth of inequalities is due to the progress of the highest income group: business owners' incomes doubled, which is not the case outside of the region (see figure 5.3). This stratification confirms the hypothesis of the global city on one point, Préteceille remarks, but not on others. First, the growth of inequalities comes from those at the very top. Second, the wealthiest are not "yuppies" in the financial sector, nor those forming in global cities what Sassen (1991) calls "the central commands of the world economy," but business-owners, craftsmen, storeowners and employers. This phenomenon implies that the change comes from the redistribution of capital and salaried work rather than from the restructuring of occupations and skills.

The spatial polarization of Paris and its regions

Spatial polarization refers to the evolution of residential segregation. On the one hand, more than half of France's wealthiest socio-economic categories are concentrated in just a few arrondissements and in the west and south belts of Paris. A relatively large number of employed foreigners also live in the same areas. On the other hand, the concentration of poorer and immigrant populations in the old degraded areas of the periphery of Paris is not a new phenomenon. What is new, however, is the overconcentration of "recent" immigrants with numerous children in certain areas and their difficult coexistence with the French working classes. Seven sectors of Greater Paris have more than 25,000 foreigners each (in decreasing number: the 18th arrondissement, Aulnay-sous-Bois, the 20th, 19th, 16th arrondissements, St-Denis, 15th arrondissement). But, unlike the situation in the USA, no neighborhood can claim a majority of minorities or just one non-white race. At work, because the public sector is so large, it is likely that middle-class French of French ancestry are in close contact with poorer immigrants, as teachers, police officers, post-office workers, welfare administrators, etc. This social and cultural interplay prevents a "casbahization" or "Little Vietnam" situation from developing in certain territories, although signs of ethnic identity are numerous enough in some neighborhoods in the Paris region that they appear not unlike some parts of Brooklyn, NY, or Pilsen in Chicago.

With the price of real estate ever rising at the center, an important fraction of the immigrant population has decided to move to the periphery of Paris where housing is cheaper. In the Paris region, 14 percent of foreign immigrants live in public housing, beside French citizens of both immigrant and French origins. The public-housing projects are concentrated in the localities strongly affected by contemporary deindustrialization.[6]

Public housing represents, however, upward mobility in comparison with the slums of the inner city where so many immigrants lived after World War Two. But, as in London or New York, the public-housing issue reveals modes of ethnic segmentation – although the term "ghetto" would be irrelevant here – triage phenomena, and institutional discrimination in the modes of housing distribution so that local standing or some form of legitimacy is required in order to be housed in the better public housing.

The malaise of the periphery localities is revealed in figure 5.4. The map shows the distance between 82 problematic – and very diverse – neighborhoods and their localities' centers. In these neighborhoods,

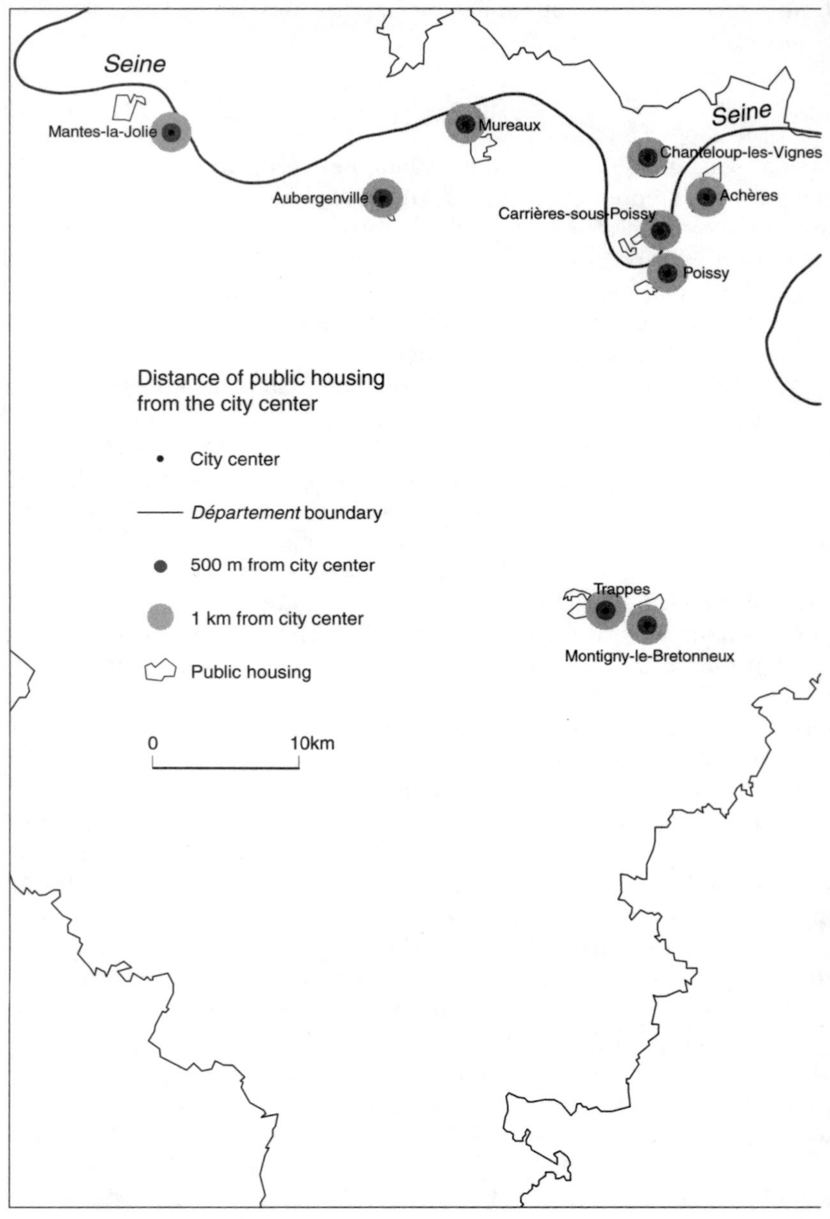

Figure 5.4 The 82 problematic neighborhoods of Greater Paris
Source: Institut d'aménagement et d'urbanisme de l'Île-de-France, 1999

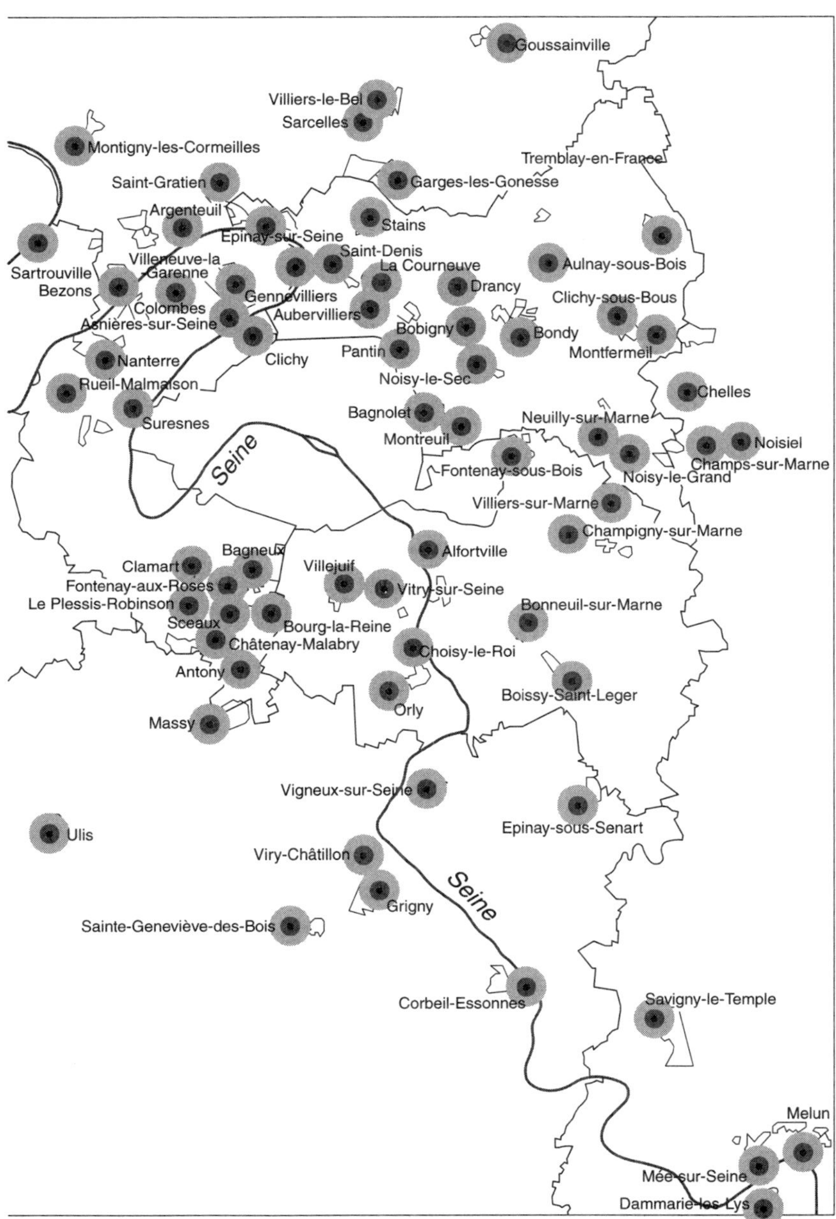

Figure 5.4 Continued

residents have to travel one or two miles, crossing railway lines or a highway, to reach a movie theater or a commercial mall. Not including Paris, 10 percent of the regional population (840,000 people) live in such areas, which are hit by unemployment and gather around 25 percent more youths, blue-collar workers, and immigrants than other localities. The main function of these neighborhoods, with a high level of deteriorating public housing, is precisely to shelter the poorest populations of the region. Unlike the American cases already described, 70 percent of these neighborhoods benefit from national and regional urban programs.

St-Denis, a city of 86,000 residents (in 1999), is incorporated in a scheme of globalization marking the evolution of Greater Paris. It is located within the *département* of Seine St-Denis (1,382,000 inhabitants) which encompasses an international airport, two universities, prestigious research centers, corporations, and which houses the most famous eleventh-century basilica, the necropolis of the kings of France. But can these assets on one scale counterbalance liabilities on the other? High unemployment, poverty, crime, school drop-outs, a concentration of new immigrants, disastrous architecture – all the social difficulties and complex problems of French society seem to concentrate in the problematic neighborhoods of Seine St-Denis.

In 1990 14 percent of St-Denis residents were jobless (23 percent of those under 25), but since then, their number has kept growing, while the safety net has tightened: out of 8,226 unemployed people in 1997, only 4,715 received unemployment benefits and 2,021 (in 1995) welfare (*revenu minimum d'insertion*).[7] In 1990 58.7 percent of the St-Denis population was foreign, a decline from 1975 (when it was 64.6 percent), 21.1 percent was non-European; 36 percent was under the age of 25, and 13.2 percent was over 60, according to INSEE (1990). Half of the population (51.5 percent) lived in public housing.

The analysis of the active population over eight years reveals a decline in manufacturing jobs (36 percent of the occupations in 1982 as opposed to 27.8 percent in 1990) and a growth in service jobs (63.7 percent vs. 72.2 percent). The active population is distributed as follows: 4 percent are employers, craftspeople, or store-owners, 7.6 percent have intellectual and "professional" occupations, 15.8 percent are executives and have intermediary occupations, 33.2 percent are employees, and 39 percent are blue-collar workers.

This pattern can be traced back to the 1950s, as an outcome of growing industrialization and differentiation in the labor and housing markets, and, more precisely, of the public-housing sub-market (Castellan et al., 1992). This old working-class suburb[8] in the first ring, north of Paris, is part of the red belt (*banlieue rouge*) and has been run by communist mayors since the 1930s.[9] Numerous immigrant blue-collar families in large housing projects live side by side with elderly retirees of French ancestry in detached housing. "In the 1970s, the supply of public housing was considerably delayed. [Communist mayors] are keen on public housing, from where their votes are likely to come. If the working class wanted better housing, it had to move out," Mayor Braouezec told me. Upward social mobility was seen as an electoral threat by communist mayors. Misery was then deliberately concentrated and corporations were the capitalist enemy. Welfare dependency was regarded as normal and, unlike others, the *département* made no attempt to require recipients to engage in workfare (8 percent of welfare recipients are signed on in workfare as opposed to 53 percent nationally) (Fenoglio and Herzberg, 1998).

Disorders in public schools and buses

Urban violence in the *département* of St-Denis amounts to "200 offenses and crimes a day," according to Préfet Duport (*Urbanisme*, January 1996: 69), largely committed by young men between 16 and 25, but also by 13–15-year-old marginalized youths, with no horizons (*repères*) and no future. Juvenile delinquents are responsible for half of the crimes in the *département*, according to prosecutor Rosenczeig, the president of the juvenile court in the nearby locality of Bobigny. Families are destructured – polygamous families, single-parent families, and jobless families experiencing more difficulties than others at transmitting a model of authority to their children – and they are not visible enough in the public space, when it comes to controlling petty crimes.

The city of St-Denis is represented in the media via two or three housing projects (*cités*) which are problematic: Francs-Moisins, Salvador Allende, and Les Cosmonautes. There, some young (French) males of immigrant origin are said to be responsible for high rates of school failure, unemployment, delinquency, and drug-dealing. Although these projects are physically less isolated than others, outsiders may have the feeling that they are penetrating a fortress when at the outskirts of a housing project – young men are on the lookout and do not make them feel welcome.[10] The police have not attempted to patrol these *cités* for years, so as to avoid clashes. Many companies' service and delivery people have refused to climb the Francs-Moisins staircases after a truck was entirely emptied within a few minutes of arrival in 1991. Medical services avoid going there at night. But two institutions in particular have become more or less dysfunctional: public schools and public buses.

In the public schools of St-Denis, such as the middle schools Elsa Triolet or Garcia Lorca (or nearby ones like Louise Michel in Clichy-sous-Bois and Jean Vigo at Epinay-sur-Seine), teachers require police intervention more and more often. Teachers' protests against violence culminated in two long strikes in 1998. Some fights starting in the *cités* with knives, baseball bats, and box-cutters may continue within the schools. Most of the victims are the youths themselves. But principals and teachers have also been assaulted by armed youths who, they say, do not belong to their school. Last year in St-Denis, Mayor Braouezec reports, a youth from another *cité* broke into a school of St-Denis in Francs-Moisins and ransacked it.[11] His rage was fueled by an old Romeo and Juliet-like story between two opposing *cités*. The mayor convinced the young man to turn himself in to the police. But then the media distorted the story, making St-Denis sound like Fort Apache. Political manipulators (anarchists, the far left) were said to have intensified the tensions. But, as noticed by the mayor, "when a demonstration finds such wide support, it reveals a problem. We all underestimated the extent of despair."[12]

The public schools of St-Denis face deeper challenges related to social integration than most, and some teachers feel incapable of ensuring social peace in their classes. They regard themselves as "the last bastion of the République still standing, up there alone with the post-office" (Debarbieux, 1998). Denouncing violence is a way of demanding more resources to do their work in decent conditions. In the spring of 1998, a report by Rector Fortier described the situation of inequality in the schools of St-Denis compared with schools in neighboring *départements*: Seine St-Denis counts the largest number of schools in metropolitan France, with more than 1,000 students in each; more than a quarter of the younger students have transferred out of Seine St-Denis; one-third of the teachers are transferred each year (half of the rest have requested transfer); 3,000 violent incidents were reported in the schools, including 138 armed assaults. Despite territorial affirmative-action programs initiated by the state (Zones d'éducation prioritaire, or ZEP), its withdrawal in the past 15 years has caused damage to the schools of the area: one student out of four drops out of the system and only 12 percent go to college.

The local authorities of the *département* also bear some responsibility. To avoid the ghettoization effect, they have avoided having the local schools classified under ZEP (which would mean more resources, smaller workloads for the teachers, and fewer students per class). Only 160 schools out of 800 and 20 middle schools out of 110 are in the ZEP scheme in Seine St-Denis (Gurrey, 1998). Communist mayors also refused to bear their part of a financial burden that they think belongs to the state, while the National Education Ministry, for its part, was claiming insufficient resources to restore equality in the poorer districts.[13] Real distress and politicization have created explosive local conditions in the schools, on which the mayor of St-Denis and his team have little impact.[14]

With the public buses, the situation is also alarming and here, also, the violence actually camouflages institutional dysfunctions. On this issue, the mayor is rather powerless, though he is vulnerable to public opinion about the problems. Violent incidents in the public buses of the *département* increased by 30 percent in 1997. In the region, almost 2,400 crimes and incidents were reported by RATP, the Parisian transport network, including 925 assaults on agents and 200 on safety brigades (Body-Gendrot et al., 1998). Public buses are used as school buses three times a day, and this is when problems occur. Take bus 615B: it drives 10 miles through tough localities such as Bobigny, Bondy, Aulnay, and Villepinte – that is, through places hit by unemployment and welfare dependency, with high rates of minorities and young children (see figure 5.5). The bus is used by 12,500 students each month, picking up youths from two *cités*. At lunch time, one group returns home from a middle school of 950 students for a 45-minute break, while the other group commutes between another *cité* and another middle school, just as big. Here also, an old feud pits the two *cités* against each other, a situation that worsened when turf conflicts and drugs-related problems piled up on one another. When I attended the end of a school day at Villepinte, it seemed obvious to me that crowds of hyper-excited minority adolescents were all trying at once to get on the same two buses, cheating to evade the fare, and trying to settle scores – all the ingredients needed to start a conflict. As remarked by a security guard in charge of safety on the bus, violence is a means of expression, a compensation after hours of having to sit still, something normal for kids who had learned to communicate only through violence. Dialogue can defuse tensions, he said. Serious incidents may, however, contradict his statement. In recent months in Seine St-Denis, a bus driver was injured for reporting the presence of three youths with pitbull terriers to his headquarters, another was taken hostage so that stolen motorbikes could be transported on his bus, while a third, sprayed with tear gas, had to drive endlessly while local gangs had a battle on the bus and threw one kid off it while it was moving. Such incidents are infrequent, but they create real tensions for the bus drivers.

Sometimes, they can be explained by the need for revenge against a specifically hostile driver who, in turn, defends his manhood ("we are not here to unbuckle our pants"), but sometimes not. The youths are often heard saying that "they own the bus," since "it is on their turf," that "all the driver has to do is drive" while they lacerate the seats, break the windows, pull the alarm system, and jump out of the windows. Needless to say, the far right is popular among bus drivers. Yet not all the bus drivers are victimized and their repeated strikes in response to urban violence mostly reflect their general malaise at being on the lowest institutional rung, feeling disrespected and pressured by new logics of profitability. Transit administrators claim that even trivial incidents are criminalized: a few snowballs hitting a bus are interpreted by some drivers as aggression and they stop working.

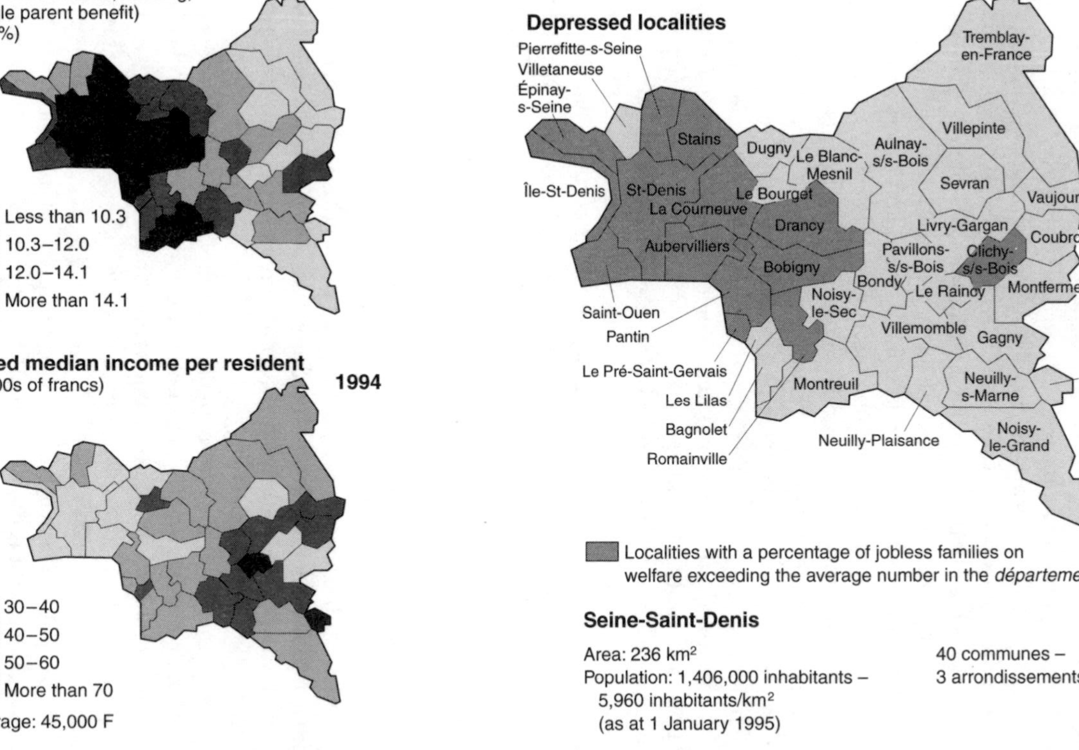

Figure 5.5 Disadvantaged localities in Seine St-Denis, Paris, 1994–1997

Source: Caisse d'allocations familiales de Seine St-Denis

The decision of a local criminal court of justice in Bobigny to speed up the process of judging offenders, to require reparations or community work, and to summon and fine the parents has been met with praise. It is said that only 8 percent of those arrested for the first time will re-offend. But because of a lack of staff, 400 decisions related to hard-core young offenders are not being enforced in Seine St-Denis. In 1999 there were only five institutions able to house the serious offenders. Partnerships between local public transporters and the schools have yielded good results, however. Other partnerships, between the transportation management and the police, are organized every two weeks. More drivers are offered anti-stress and conflict resolution training. It seems that here, as in the schools, youth violence is an excuse used by drivers to express a deeper malaise towards their institution. As a response, the RATP has tried to replace the concepts of victimization/protection with those of respect/quality and it has launched a large campaign on this theme to boost the drivers' morale.

Specific actions of reconciliation taken by the mayor

What is a mayor to do to combat such negative trends? Braouezec, who is also a Parliament member and a "questor" in charge of the administration at the National Assembly, says that he aims at finding actions which will "federate" energies and reinforce the cohesiveness of the population, while avoiding those stigmatizing certain areas, such as the policy of enterprise zones. For him, violence reveals a societal crisis which also reaches rural areas. The primary causes of death among French youths are suicide and car accidents. "As long as our society is so unfair, violence will prevail," he says. But local preventative actions may restore the social links which are of so much concern to the French mayors. For instance, Braouezec organizes regular block festivals that allow interactions between multicultural populations. The Préfet also believes in cultural events where youths, adults, and the local authorities discuss and share views. The mayor initiated the campaign "Hello, neighbor," during which twelve thematic meetings took place, with residents invited to express their views and talk to one another. Every six weeks, he meets with residents, at the doors of the schools, in the markets, near the post office, etc. When he realized that graffiti flourished even on the walls of his own building, he simply thought, as did the police chief of The Hague in the Netherlands, that "adolescents need spaces of transgression." He kept his cool and offered to hire them to create murals. He is not alarmed by crime and violence because he is confident that his prevention actions can restore hope in the long term.

Others are at work in St-Denis on these acts of reconciliation. I will describe how two original experiments are aiming to reinforce the concept of citizenship.

Auto-école at St-Denis A few years ago, an *auto-école* opened in St-Denis, run along the same lines as the Central Park East Secondary School in Harlem run for ten years by Deborah Meier. *Auto-écoles* – "self-schools" – welcome kids who have dropped out of the system. Using ad hoc methods and no rigid curriculum, they attempt to restore the self-confidence of the students using positive reinforcement. They stimulate their motivation and give them a sense of working together. For the first time in their lives, maybe, adults listen to them, give them special attention and encourage them. Although staff work long hours, they like what they are doing, they do not have to work along strict guidelines, and they form close-knit partnerships. The one in St-Denis was a sub-branch of a larger middle school, Garcia Lorca, and each year it welcomed thirty students who had dropped out of the system. At the end of each year, the school was able to put a majority of the students back on track. How was this possible? Innovative methods were put to work at the school, and the teachers – all volunteers – acting as mentors were devoted to the experiment, through which they gained a lot of responsibility and satisfaction. The students also felt respected and participated in the decisions made by the school. The city offered space to the school, and the experiment received a lot of attention from the media. As the principal became famous, others became jealous. In a centralized system, innovation is not easily tolerated, especially by the teachers' unions. While in the Harlem school, Meier had been asked by NYC to work on the broader pedagogical reform of secondary schools, the principal at St-Denis was transferred and the experiment was terminated.[15] Despite this apparent failure – the principal experienced the fate of NYPD chief Bratton, for the same reasons – the *auto-école* became a symbol of what could be done as an alternative to a rigid system, if there was a political will to do so. A similar experiment has been tried in Marseilles.

Nicolas Frize at the Lafontaine hospital Another innovation has been carried out by composer Nicolas Frize at the hospital of St-Denis. The Lafontaine hospital is like a city within a city: 2,000 people visit, work, or suffer there every day. By playing and composing music for those people in the hospital who were interested, Frize was attempting to restore citizenship – that is, equality and links between people who usually did not talk to each other.[16] During a concert or a musical break, a famous surgeon would sit next to a hospital cook or to a patient and share with them music and emotions. Frize has created his own firm in St-Denis, Les Musiques de la

Boulangère, and at the request of the Minister of Culture, he is currently recording the sounds of a city as they are at the end of the twentieth century. He stores the tapes in a vault as a sort of time-capsule for people in future centuries to discover. In the hospital, he played to the staff sounds he had registered but which were too familiar for them to notice. The experience, conducted at the request of the hospital management, lasted six months, with mixed results in terms of cultural and social transformation. In my view, it failed at transforming power relations in the hospital; but it changed the view many staff had of the place (some of them would return to the hospital to "do" music after a day's work) as well as alleviating the pain patients – especially children and their parents – felt when there. Again, it was a symbiotic experience with lingering symbolic effects.

Gobalization reconsidered: the Soccer World Cup

The mayor of St-Denis, who has a reputation for being progressive, has been actively involved in the promotion of his city's image. He has taken part in minor activities such as workshops which try to resurrect and record the history of the community (with public writers capturing the words of those who cannot write) and in larger ones like the Soccer World Cup, which was played in the new stadium of St-Denis in July 1998.

The mayor saw the World Cup and the new stadium as an opportunity to tell the kids from his city that they too could be "winners".[17] The World Cup phenomenon has been called a "historical ecstasy" by philosopher Edgar Morin. It was indeed a feat to transform an event which only addressed amateurs' passions at the start into a national paroxysm which would make St-Denis famous forever. In the course of this epic venture, St-Denis lost its downtrodden working-class image to become an ephemeral world village. A gigantic wave of happiness submerged a usually pessimistic country and depressed suburbanites. A big "Us" seized youths, adults, immigrants, nationals, men, and women into a state of fusion, identity-sharing, poetry, and non-military patriotism. There was no agression, contempt, or ethnocentrism accompanying defeats or victories during the games and brother/sisterhood prevailed. The integration of brown, black, and white players into one plural action came out visibly – "we won because we were French," the players said – and marked the triumph of a multicultural team trained by a hitherto insignificant figure on whom the media had looked with contempt for months. A non-aggressive globalizing event linked the planet to the tiny, now fully integrated territory of St-Denis. The mayor had invited adolescents from the rest of the world to come and play soccer, in parallel with the big world players. The latter participated benevolently in these games. In the subsequent months,

St-Denis became the twin city of Porto Allegre in Brazil, where stunning experiences of local democracy have also taken place. During the World Cup, youths from St-Denis, from Francs-Moisins, and other problematic projects, were involved in public security: no single incident marred the event.

Marseilles: A Deceptive Calm

To date, Marseilles has been spared riots. This does not mean that it will always be. But according to the police chief, several serious incidents could have degenerated into urban violence. Since they did not, it is interesting to understand why.

(1) In February 1995, a young French of Comorian origin, Ibrahim Ali, was murdered by poster-stickers from the far right (recently, the far right got 28 percent of the vote and came first in 10 of the city's 14 districts). Immediately, the authorities created a crisis squad, provided a rapid response to an angry population, and launched preventative schemes. This reaction reminds us of the reaction of the city of New York during the LA riots. There, too, a crisis squad was activated and networks that had been created months before were operated 24 hours a day for four days to prevent disorders in problematic neighborhoods.

(2) In the Mortada case, a young motorcyclist was chased by a police squad. He killed himself and immediately the rumor spread that the squad was responsible for his death. Instantly, police and judicial authorities issued a coherent and strong public statement that was convincing enough to calm public opinion. The attorney-general gave a speech painting the police as innocent. He met at length with the most exasperated community organizations and appeased tensions.

(3) Four young people, including a minor, were taken to the police station after they threw Molotov cocktails at specialized police units (CRS) during disorders erupting in the working class area of Belzunce. Over three hundred youths gathered there intent on looting. All night long, a delegation of them met with the police chief and a priest acting as mediator. The tension faded away.

(4) In September 1997, N. Bourgat, the son of a well-known local physician was stabbed to death by a 15-year-old youth of Algerian origin. Emotions heightened and demonstrations attracting thousands of residents took place all over the city, including one under the auspices of the far right. The Préfet took the risk of allowing such a demonstration. Riots did not erupt.

In these four cases, it seems that the immediacy of the responses, time given to listen to grievances, the cohesiveness of the police and judicial elites, and the mediations provided by civil society prevented the diffusion of disorders.

A difficult institutional and economic context

There are many shortcomings, however, in Marseilles, especially inside the police structure.[18] Middle-ranking police officers are insufficiently trained and the rank and file, whenever they try to innovate in a tough context, are not encouraged or supported by their superiors. Moreover, the attorney-general remarks, managerialism is imposed everywhere, requiring statistics and quick-fix solutions at the expense of obscure investigative work. An understaffed court and the absence of a Maison de Justice leave an impression of institutional inefficiency.[19] The mayors of arrondissements are not allowed to play a prominent role and the mayor of Marseilles is torn between claiming responsibility himself at election time (local elections take place in 2001) and remaining passive as long as there are no major crises. For a few years, the mayor of Marseilles, J. C. Gaudin, was at the same time the French Minister of the City (1993–6).

Economically, Marseilles is less a global than a depressed city. It has lost 10,000 residents every year since 1985 (Peraldi, 1996). The second or third largest city of France, depending on whether one considers the city or the CMSA, it is currently home to fewer than 800,000 residents, including 100,000 people of Algerian origin. Marseilles lost more than 400,000 manufacturing jobs between 1975 and 1990, yet it did not make up for that loss in service jobs. Growth came only from the public sector, both national and local (13,500 employees work for the city). Of the population – 350,000 people – 44 percent receive their income from public support: 110,000 of them are below the poverty line ($19,000 for a family of four); 20 percent are jobless; 130,000 are on welfare. Immigrant families are hit the hardest: 62 percent of immigrant single-parent families are in the lowest economic decile, 48 percent if only one parent is foreign. For those under 30, the unemployment rate often reaches 60 percent, and their fathers are also unemployed.[20] But Marseilles is also a dual city, with 55,000 executives living in nice homes along the beaches, and a development of European significance thanks to its exceptional location. The large public program EuroMéditerranée aims at revitalizing an area located between the port and the railway station. It will include an enlargement of the stadium, an operation which already started in advance of the preliminary round games played there during the World Cup in July 1998.

The north of Marseilles (see figure 5.6), whose potential for violence raises fears and where the underground economy is especially active, dis-

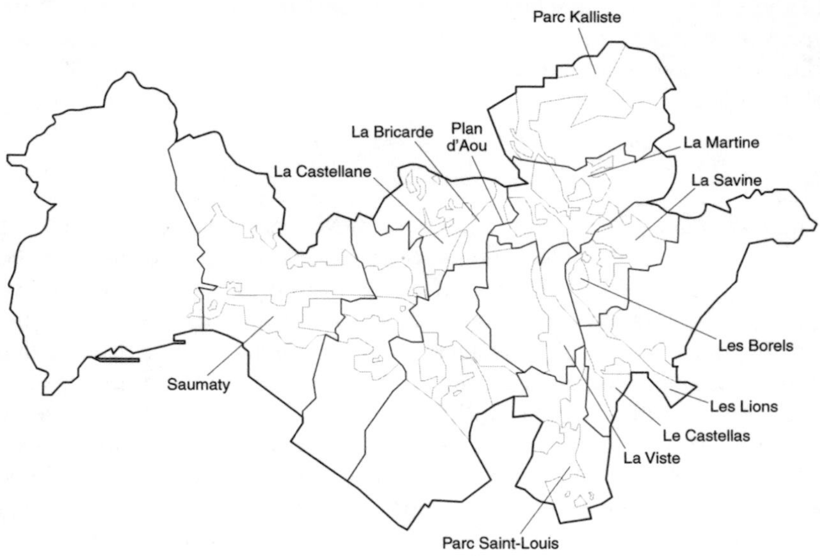

Figure 5.6 Density of population in the northern neighborhoods of Marseilles
Source: IGN, 1990; INSEE, 1997

plays the most indicators of distress (Vichery, 1997: 223). Half of the population lives in colossal public-housing projects which are twice as numerous there as in the rest of the city. Older private-housing stock is also visible. Seven residents out of ten have been in the same neighborhood for more than eight years. This stability generates a "space effect" and an efficacy, which can be both positive for the insiders and negative for outsiders. If workers were not so concentrated in the one area, some people say, the unemployment rate would be halved (from 55 to 22 percent); if there was not such a preponderance of young people there, the gap would be reduced by 10 percent, and if immigrant families were not overconcentrated there, the differential would be even less. Indeed, the handicap is not the location itself (the landscape is splendid, one of the best overlooking Marseilles), but rather, as a sociologist remarked, it is in being near "those people" – the workers, the youth, the immigrants – the sign of one's downward mobility (Bordreuil, 1997: 232–3).

Mansuy and Marpsat, two researchers at INSEE, have analyzed the social space of Marseilles and they confirm that it is a mosaic of 24 types of neighborhoods representative of all the different kinds found at the national level (1994). The core of the city is itself diverse: working-class areas like Belsunce or the 14th, 15th and 16th arrondissements and more affluent zones in the 1st and 13th arrondissements. Foreign households, including affluent store- or business-owners, make up 28 percent of the core of the city in the 1st arrondissement, living next to elderly or to single people. While Algerians run the stores, Sephardic Jews are in finance, Tunisians open new markets with Lybia, and the Senegalese are into retail. Between 1980 and 1988, before visas became necessary for Algerians to come to France, 400 stores had a turnover of 3 billion francs due to the presence of 30,000 transnational commuters from Maghreb and from nearby cities every weekend (Tarrius, 1992). Thanks to its past and its geography, the city is multipositioned. On the one hand, it is made up of a juxtaposition of small historical villages, gradually absorbed by urban growth, yet retaining their identity (see the section on La Lorette below). But, the pattern of the Marseilles region is also polycentric, a puzzle of diversified economic zones, each locality playing its own cards to get hooked to the world economic flux.[21]

The exodus of affluent populations to the southern part of the city has been accelerated by the construction of public housing in the northern neighborhoods, thus marking a trend of spatial segregation between working and middle classes. Social polarization has occurred less inside the city, which is heavily Mediterranean with richer and poorer families, than between Marseilles and nearby Aix-en-Provence, where the "old wealth" and the professional bourgeois have chosen to live. Although Marseilles is a city of immigrants, its northern neighborhoods are not "ghettos." Public transportation connects the projects to the center. Public and social services are distributed in the neighborhoods (but never in enough quantity or quality to meet the needs, according to the residents). Everywhere, French nationality prevails alongside a multicultural mix – North African, Jewish, Armenian, Italian, Comorian, gypsy, Asian – the city having been molded by multiple migratory flux. "But the last ones in have always tried to lock the door behind them," a local politician says. Families from Maghreb are frequently the oldest in the neighborhoods, having settled there more than 25 years ago. This factor of stability is an important asset in countering violence.

Policies of prevention

A community policing unit works with youths at risk, but not at times of crisis. It has contacts with gang leaders, as well as with those attempting to avoid delinquency. It plays an important role in violence-ridden schools and it even runs the social centers in some dilapidated sub-neighborhoods, working with undocumented residents, ex-convicts, elderly, newcomers, etc. Composed of ten police officers, this flagrantly understaffed unit (in comparison, there are 250 police officers in the special anti-crime unit in Marseilles) needs to have its image and resources reassessed within the police institution. It is well integrated and accepted in the neighborhoods where it operates.

According to the Préfet, the work performed by community organizations is at the heart of the city's identity: "connections, networks, mediation: it is a genuine culture in Marseilles." Community organizations, such as that run by the women of Shebba, youth organizations, and organized residents from the three northern arrondissements, are involved in all kinds of mediation with children, adolescents, and drug addicts. Immigrant women acting as official mediators (*femmes-relais*) intervene in the schools as they do in other cities of the country. They listen to students of immigrant origin who feel devalued because the image they have of their parents is so negative (some undocumented parents, for instance, give up their role as parents and expect the state to act as educator for their child). They also listen to teachers and act as go-betweens. They offer literacy classes to immigrant women in order to get them out of their homes, to generate exchanges of information about the larger world, including how to regulate domestic tensions. These older immigrant women have a long-standing savoir faire and they form the foundation of community life in Marseilles. They are at the heart of community networks, can mobilize human resources, and generate support from institutions and from civil society. The legal right of foreign immigrants to associate after 1981 gave rise to a very dense social fabric, with community organizations being supported by public funds. The militant community-based organizations have attempted to compensate for public-sector dysfunctionality and the scarcity of services. Social centers, houses for youths, health centers, social work agencies, and schools are frequently the only public institutions in neighborhoods where public and private actors never invested resources.

The local anchor is also defined by a global interface. Special connections with other regions of the world are sometimes more significant to the residents than the national reference. While some of the women I met in Marseilles had marched throughout the country to demand equal rights at the beginning of the 1980s, one of them had been to Istanbul for the summit of Habitat II. She had become part of a world network linking residents, professionals, and institutions. Others had gone to the Rio and Dakar summits and exchanges "from the bottom" about urban violence had woven cross-national connections.

Angry residents

"For youths living in a context of violence, is burning a car an act of violence? No, it is a mode of expression. It would be dangerous if they were unable to express themselves thus," one resident says.

For twenty-five years, youths in Marseilles have set fire to cars, frequently in the wake of rodeos with stolen cars. The phenomenon is nothing new. Yet the residents I talked to seem less eager to blame the young offenders than the local institutions and the "arrogant" police in particular. Mediators recently launched a survey on the way the residents from the northern neighborhoods resent institutional "violence."

A lot of rumors inflame opinions. Who diffuses them? Who lies? For what purpose? Can youths build identities for themselves only through car-burning, conflicts with the police, or through the wildest stereotypes? Do continuous ID checks and the abuse of police powers that the youths experience reinforce the disrespect they have for institutions in general? Are such checks demanded by other fractions of the community residents? Tensions separate groups from one another in the housing projects.

Yet, if the *cités* of Marseilles explode one day, it will be for legitimate claims to which the city never paid attention. Not only did monstrous architectural housing projects in the 1960s at the fringe of the city's core marginalize 350,000 residents, but a lack of adequate and cheap public transportation, especially at night, reinforces their exclusion. Express highways cut the buildings from the stores and pose permanent danger for young children. Social segregation fosters a sense of physical and mental exclusion. Years of planning projects responding to the problems get nowhere; urban renewal only takes place before election time, it is said. Angry residents ask: "How can we find public services adjusted to our needs? How can an extremely diverse population, thrown here at random, find its own commonalities and mobilize? Should residents accept more underground economic activity and drug-dealing in exchange for a sense of sharing and internal rules?"

The residents resent the indifference of local authorities and the lack of public facilities. Young mothers from the *cités* had to wait forever to get playgrounds and benches. The children of one *cité*, Cité des Flamands, were not accepted in the nearby school when the families moved in, as if no one had anticipated their settling in the area. The lack of speed bumps on the roads crossing the *cités* has contributed to the deaths of several children and adolescents. One day, the mothers had to make a chain with their bodies to contain the rage of the young people. "If we had not been here, there would have been fire," they recall. The disrespect of local authorities for the residents' demands reinforces a rancor which has crystallized over the years. "But we no longer have the energy to talk about it," some residents sigh. Sometimes, however, mobilization gets results, when it is accompanied by the support of the media and the right connections: two police officers who had abused adolescents were indicted, for example, as were others who had tortured an immigrant worker. Residents plead for the right to speak out themselves about their concerns, rather than have civil servants talk for them and without them.

The economic revitalization and challenge of the Grand Littoral

The challenge of the Grand Littoral illustrates the difficulties French society has in recognizing multiculturalism and promoting the equal rights of groups discriminated against because of the color of their skin, their culture, or the place where they live. It also reveals the efficiency of civil society's mobilizations in certain circumstances when institutions are ready to negotiate to promote their own goals.

In November 1996, a large, beautiful commercial mall opened at the edge of the northern neighborhoods on the hills of Marseilles. A first plan, initiated in 1985, had failed: leftist elites opposed the idea, saying that building a temple of consumption there would provoke the poorer residents. The second plan, in 1991, pushed by socialist mayor Vigouroux, emphasized the necessity of revamping the image of these northern *cités*, of "reconciling" them with the rest of the city by creating jobs for the residents (in three *cités*, La Castellane, La Bricarde, and Plan d'Aou, 47 percent of the residents are officially jobless). These *cités* are feared by other Marseilles inhabitants and it is said that even gangs from the southern and peripheral areas do not venture there. Exclusion is more socially than ethnically based. Bringing patrons from the whole region to this area therefore represented a large commercial risk. The developer of the Grand Littoral, the Thema corporation, was hoping that the exceptional landscape, at the top of Marseilles with a view over the sea, would attract at least two hundred store-owners. It did. The project has been a success.

One should concede that the mall entrance does not face the *cités* and that no pedestrian path links the two entities.[22] It should also be reported that numerous problems slowed down the construction – intimidating sit-ins, continuous thefts on the working site, engines set on fire – as the residents feared that they would be passed over when jobs would be distributed. They thought that they had only violence to put on the negotiating table. The case of La Lorette, as analyzed by Hénu (1998), illustrates these tensions.

The residents' mobilization at La Lorette After immigrants had left Kabylie, a region of Algeria, to work for a tile factory in Marseilles, they built their own homes, near the neighborhood of St-André in northern Marseilles.

They grew small gardens and became almost self-sufficient. But unemployment hit the fathers and slowed down their children's social integration. The youths were caught between two cultures, while feeling marginalized by the larger society. Their marginality would sometimes lead to delinquency and to a stigmatized image.

In 1990, the state and the city decided to remove slums, including La Lorette. There was disagreement between the different generations living there about the decision. While the oldest were attached to their homes and did not want to be evicted, the young aspired to rehousing in better homes and to social mobility. The decision to create the commercial mall, Le Grand Littoral, was accompanied by plans to build a public-housing project at La Lorette. The developer, the city, and the public-housing managers got involved in mediating sessions with the residents. In the end, 74 families wanted to remain on the site and they were offered a choice of 91 detached houses that would be built there.

In October 1995, however, 13 families had to be evicted by the police. What had happened? The slum removal and the building work had hit numerous obstacles and the residents had become hostile and radicalized. Tensions between the social workers acting as mediators and the militant young residents became acute. The young people could not see their home base disappear smoothly and, as a matter of principle, they resented institutions. They were reluctant to cooperate. Everything therefore had to be negotiated sharply: the maintenance fees in the new housing project, the amount of space which would be devoted to collective premises, jobs going to the residents in the new mall, and the indemnifications for the old homes' destruction. The residents – like the Brazilian mojabitos – had indeed settled there without asking and they had no property titles. The factory had allowed them to use the land which now belonged to the Thema corporation. Official estimates of the old homes' values raised mutual suspicions. Rivalries divided the community, the fathers and the sons, the neighbors and the mediators. This is when the building operations were slowed down.

The building contractors responded by offering short-term employment contracts to the residents, provoking jealousy in nearby communities also plagued by joblessness. It was high time that reconciliatory measures were taken by the authorities. Labor inspectors and decision-makers at the *département* level and the Préfet adjunct for the *politique de la ville* began to work together on the residents' economic integration into the commercial mall operation.

Territorial affirmative action at the Grand Littoral Extravagant promises had been made by political demagogues speaking to the residents and the media: 5,000 jobs – 10 times more than necessary – would be created, they had said. This context made more difficult the negotiations with groups of radical residents, mostly men between 20 and 35 years old. Yet, after months of meetings, a consensus emerged on the definition of stakes, and opinions from both sides influenced each other.

As they had nothing to lose, the militant residents gained a lot. Hypermarket Continent agreed to give priority to jobless residents from the nearby arrondissements when it came to hiring. A national foundation acted as an interface, selecting and training 90 residents, 58 of whom were hired. Then training sessions for the Continent executives were organized to facilitate their understanding of the cultures of the local residents with whom they would work. The hiring process lasted a year and a half. It took into account the local culture. Recruiting agents were sent to see how other supermarkets operated in the Marseilles region. The residents' training was carefully constructed in order to help them use their skills if they moved out. The cost of the whole scheme was partly subsidized by the European Social Fund.

The information sessions on the hiring process, the jobs, the working hours, and the requirements were attended by 8,500 residents. The sessions took place at city hall and at the mall. In all, 4,000 job applications were filled out, 3,000 were sent to the local branch of the national unemployment agency, and 1,500 to Continent. Workers were selected from the applicants by twelve national professionals (the hiring process involved tests of their abilities and letters of motivation, but did not take into account school diplomas). Half of the applicants passed the tests. Anyone who failed was personally informed by phone in order to avoid violent collective reactions. There were 450 people placed on priority lists for hiring, on condition that they also met the following qualifications: (a) they lived close to the Continent hypermarket, (b) they were residents of the northern neighborhoods, and (c) they were inhabitants of Marseilles. At each step of the hiring process, control, screening, and acceptance of candidacies were performed by the local branch of the national unemployment agency. The emphasis on hiring disadvantaged ethnic residents was deliberate. Indeed, 55 percent of labor contracts approved by Continent went to the residents of the 15th and 16th arrondissements, with a weighting distribution observed among *cités*; 75 percent went to the residents of the northern neighborhoods and 83 percent to those of Marseilles. The same goals, although with smaller quotas, were followed by the other stores in the mall.

Can this successful territorial affirmative action, an action of social integration and crime-prevention, be reproduced elsewhere? Regulation worked and there was a reciprocity between the residents and the employers, the Préfet remarks. The whole operation remains fragile, however, and everyone concerned must be on the lookout. According to the commercial manager of the mall, the operation did not aim at profitability and thus will remain limited; its goal was buying social peace. "Where two employees are needed, here we hire five of them because the residents are so unskilled," he said. The security is all taken care of by the residents of the *cités*, La Bricarde, Castellane, Plan d'Aou, and La Lorette. "It is a job appropriation. They are at home here. They do not trust us because we are not from the same community. We have no leverage and it is most difficult to fire an incompetent employee and yet we need the most skilled employees. Blackmail and intimidation are used by the residents' spokespersons to get hold of the jobs," the director says. While complaints about reverse discrimination are being heard, a highly respected social mediator has been hired. He goes back and forth between the *cités*' social centers and the mall management.

So far, no incidents have marred the operation. The mall is used by 45,000 people on Saturdays and 30,000 each weekday. There have been no reports of graffiti or stolen cars. It has become an element of pride for the residents and it partly revamped the image of the northern neighborhoods.

The choice of the Grand Littoral can be analyzed in the context of equal opportunity in French corporations. According to three specialists, no one firm stands out as a guide to good practices that can be used to fight racism (de Rudder et al., 1997). Here and there, there are isolated and discreet actions, like at the national firm Electricité de France in Seine St-Denis, at the car factory Renault in Flins, or at La Redoute, a mail-order firm in Roubaix, not to mention the new jobs for youths recruited in the *cités* by various public partners. But the solidarity approach is social and territorial and ethnic discrimination is never denounced. The taboo of discrimination against those who are French but of immigrant origin is never lifted. For instance, the laws punishing those who prevent French people of Arab origin from getting housing and jobs and from entering discothèques are not enforced. There is no explicit legal protection for collective discrimination against minorities (the term does not even exist) and statistics relating to this issue are censored. The laws only take into account individuals. This lack of protection and the related inability of groups discriminated against to redress the harm done is all the more damaging in the context of growing inequalities, rabidly conservative discourse, and the global ethnicization of social relations which identifies immigrants and their children with the new dangerous classes (de Rudder et al., 1997). The 1997 report of the state council on the principle of equality discussed affirmative action and concluded that it was inappropriate for French society, which has to remain one and indivisible and without minorities. To compensate and redress admitted inequalities, though, the report justified territorial affirmative action, which is less divisive, it said, than action based on race and gender. Territorial affirmative action provides justification for the creation of ZEP, enterprise zones, the hiring of security guards in charge of security in public transportation, and public youth jobs for the disadvantaged neighborhoods.

In the case of the hiring at the Continent hypermarket, the residents imposed a "local preference," as a response to the "national preference" claimed by the far right. (The four *cités* run by far right mayors are all near Marseilles.) What was compensatory action at the start was translated into a desire by the employers to hire specific residents, after the succession of demonstrations. Territorial criteria and flexible belonging in terms of identity processes allowed this operation to succeed. The territory thus defined acquired a positive image when it came to getting a job. The compromise of the Marseilles hypermarket allowed a game of

> ethnic hide-and-seek, a genuine manner of talking about ethnicity without actually mentioning the word and of reintroducing the mobilization of communities within an institutional frame . . . The use of the socio-geographical criterion "northern neighborhood" made it possible to fight racial discrimination without talking about it, without setting ethnic boundaries by formal labelling. (Poiret, 1999)

One should not be blinded however. Providing 400 jobs for 8,000 unemployed residents does not solve the problems of the *cités*. It should be noticed that city hall was not involved in the whole process. Its parochialism and policy of clientelism have prevented Marseilles so far from developing global or European large-scale operations.

In conclusion, Marseilles is economically and politically a Mediterranean city, but not yet a global city. Culturally, because of its openness to the world for over 2,500 years, life in the city is spared acute ethnic conflicts. All of the young in Marseilles like to loaf about in the Old Port. The calm of Marseilles is frequently explained by three conflict-solving factors: "dialogue, sun, and soccer." Every youth identifies with Marseilles and "does not feel like breaking what belongs to them."

The underground economy certainly plays an important role in the non-explosion of the northern neighborhoods (Tarrius, 1992). Informal young salespeople are seen going from door to door in the housing projects in the evening to sell goods out of suitcases. Networks and codes of honor perpetuate Marseilles's historical savoir faire (Peraldi, 1999). As explained by Tarrius, echoing Braudel:

> men impose their success on modes of circulation, institute networks and patterns where, not so long ago, they all were the objects of a mobilization alienating them from their kin and from others . . . It is from the fertility of crossed agreements that links have been woven between generations, links

> which were revealed, officialized and made explicit. Local authorities' consciousness of the wealth coming from this situation is very recent . . . The transnational individual unknots the city managers' blocked amnesia. Social boundaries and networks experience a mature vision and a broadened scope that they impose more and more on Marseilles, thus bringing it a recognition of its history and of the skills the city has as a site for mixing, absorbing and producing. (Tarrius, 1995: 185–90)

In-depth work by community organizations and networks of former residents emphasize a preventative approach. In parallel, the *politique de la ville* brings recognition to the youths' efforts by providing social workers and other mediators' voices. There is also the University of the citizen (not a real university but a forum funded with public funds from the Ministry of the City), which is directed by a former street educator, Jon Ros, with other street educators, community organization leaders, institutional representatives, and concerned individual residents. Their basic goal is very modest: to provide contacts and linkages between residents beset with their everyday problems and anonymous bureaucracies. Based on the principle that residents who pay taxes have a legitimacy as "we, the people," the team at the head of the University of the citizen attempts to make institutions in Marseilles and its region more accountable. It finds the experts' names (judges, policemen, administrators) – not an easy task, with bureaucratic procedures, decentralization, and pluralistic systems of decision – and it organizes meetings with the residents who then travel out of their neighborhood to express their grievances, an important process. At the same time, a colossal effort is made to help the voiceless residents express themselves, to train them to speak out in public spaces that are not their familiar community area, to talk to one another, to register on voting lists, to become committed and involved in changes. The plays put on by the theater-forum, which is part of the University, follow the residents' own stories and are written with them. They are based on role exchanges and audience participation during the play, thus de-dramatizing daily life problems. Are these efforts to promote dialogue and grassroots dynamics enough to counterbalance the weight of negative forces? Strong proposals are needed to maintain a fragile peace.

Turf Battles in Lyons

Lyons and Marseilles are rivals for the title of second largest city in France. In 1999 Lyons, with 445,000 residents, is actually often viewed as the largest provincial metropolitan area, because, if one includes the Greater Lyons area (also known as COURLY (Communauté urbaine de la région lyonnaise), its population jumps to 1.3 million residents, out of 5.6 million

Table 5.1 Evolution of the population of the city and of the CSMA in Greater Lyons, 1936–1990

	1936	*1968*	*1982*	*1990*
Lyons City	570,600	524,600	413,095	415,000
COURLY CSMA	829,800	1,048,800	1,106,000	1,134,600
INSEE Greater Lyons	883,600	1,130,700	1,220,844	1,214,900

Source: Bonneville, 1997: 18

in the Rhône-Alpes region (see table 5.1) (*Le Monde*, 7 July 1999; Bonneville, 1997: 1). But the city lacks the economic, demographic, and international dynamism of Milan, Barcelona, or Frankfurt and, in 1988, it ranked only twentieth out of European cities (based on demographic and economic criteria) (Brunet, 1989). Lyons's laggard evolution is due to heavy centralization and territorial administration, which penalize all metropolitan areas apart from Paris. Lyons has seven times fewer residents than Paris, and while Greater Paris is home to 53 percent of executive positions, Lyons holds only 4 percent of them (Bonneville, 1997: 35). In the high-tech sector, Lyons represents a fortieth of Paris's activity. The same is true for auditing, consulting, insurance, and law firms: between 75 and 84 percent of them are concentrated in Paris (Carrez, 1991). Lyons has 40 banks, as opposed to 278 in Paris (Froment and Karlin, 1993: 69).

Greater Lyons is made up of 55 localities. Its interface with the larger region accelerates its potential international expansion. The promotion of technological and superior tertiary occupations support the global strategy of the region. The assets of Lyons are numerous because of its location between Paris and the Mediterranean and its proximity to Switzerland, Italy, and Germany. The city is linked daily to national and European sites by 48 high-speed train connections, European freeways, and by an international airport (4.5 million commuters in 1995). International firms have headquarters in Lyons and anticipate its potential as a "Eurocité." The IMF and G7 meetings (in 1996) helped confirm that impression. A cancer research foundation, operating under the auspices of the World Health Organization, moved to Lyons in 1987, followed by Interpol in 1988, Euronews in 1993, and the Aspen Institute in Europe (Bonneville, 1997: 42). Research resources are developed at elite higher education schools such as École normale supérieure, Framatome, nuclear engineering, Urban Transportation, etc. New office construction increased 2.5 times as fast as in Marseilles between 1989 and 1994. Yet, looking at several indicators such as household income, real estate, diplomas per capita, Lyons is not so different from other French provincial cities. As underlined by Braudel,

"The tragedy of this city is that it is only at the international level that it finds its order and the conditions of its fulfillment; it depends on very wide-ranging logics. It needs the implicit support [*complicité*] of outside forces. The fairies in favor of [Lyons] are foreign" (1986).

Economically, Lyons's profile is similar to that of larger cities: a decline in manufacturing jobs (minus 16,000 in the 1980s) and a growth in service employment, especially in corporate services (plus 20,000) heavily concentrated in a few arrondissements. At the beginning of the 1990s 25,000 jobs were cut, hitting Lyons (a loss of 13,000 in two years) and the eastern periphery (minus 5.2 percent) (Bonneville, 1997: 105–7). Lyons's eastern industrial periphery is still home to 100,000 jobs. On the whole, the active population of the Lyons region is distributed as follows: 26.4 percent blue-collar workers, 28.4 percent white-collar workers, 23.1 percent intermediary occupations, 15.1 percent "professionals", 6.4 percent craftspeople and store-owners, and 0.2 percent farmers (INSEE, 1994).

The recomposition of space in Greater Lyons

Contrasts mark the demographic and spatial evolution of the area. The changes are similar to those experienced by large French cities since the 1970s: suburban sprawl marks the differentiation of space functions and the dualization of urban organization. Segmentation results from economic polarization and spatial segregation processes, some of them leading to urban violence.

While the core of the city has demographically stabilized, the first ring (Bron, Villeurbanne, Vaulx-en-Velin) is now losing population (it has lost 44,000 residents since 1962), while the urbanized region expands further. Housing policy reflects these mutations. The voluntary policy of concentrating public housing in the eastern and north-eastern parts of the first circle created colossal social tensions between the core and the periphery and between the east and the west of Greater Lyons. This national policy was enforced in the 1960s by the mayor of Lyons, Maurice Pradel. Claiming that he would eradicate slums from the core, he had the poorer families deported to the periphery, where communist and socialist mayors welcomed them as political constituents. Between 1950 and 1975, 60 percent of the housing built in a few arrondissements of Lyons and in the fifteen localities of the first circle was public housing (Bonneville, 1997: 103; see figure 5.7). The land value reflects these inequalities. Pradel's successors invested in the center and beautiful renewal works were launched. Land there quadrupled in value during the 1980s. In 1992, Lyons was the most expensive provincial city for new housing, with a 78 percent increase over ten years, while the eastern localities' housing depreciated in value.

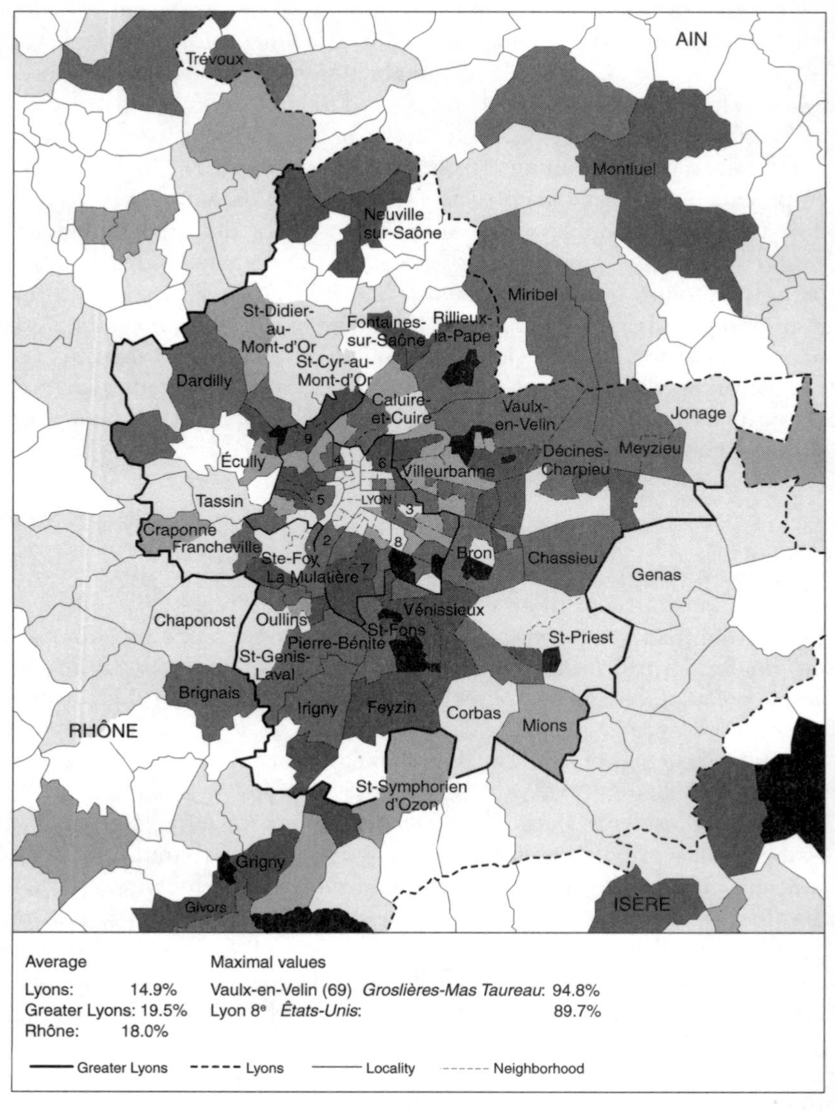

Figure 5.7 Public housing (HLM) in Greater Lyons, 1990
Source: INSEE, Atlas du Grand Lyon, 1994: 26

Demographically, the apparent dualization of Greater Lyons is visible: older populations and smaller households are overconcentrated in Lyons and in the western parts of the first ring, while those younger than 20 years old from large families tend to live in low-income localities: Bron, Vénissieux, Vaulx-en-Velin, Rilleux (see figure 5.8). The 91,000 immigrants who live there represent 10.5 percent of the population of Greater Lyons and 9.3 percent of that of Lyons. European immigrants arrived in the 1920s and when they became French, they merged into the mainstream. The North-African population makes up 52 percent of the foreign population (29,000 Algerians) and the Portuguese, 13.7 percent. After 1970, many of the recent immigrants moved out of Lyons to the periphery into public or older housing, although some arrondissements with older or public housing still hold between 13 and 18 percent of foreign immigrants. The 20,000 most marginalized households are single, single-parent, and immigrant, 45 percent of them in the city itself (Bonneville, 1997: 130).

The situation, however, never approaches that of American cities, and the image of a wealthy core with a poor eastern periphery or north/south-type divides would be wrong. Most neighborhoods have a complex ethnic mix. It is accurate that at one extreme, Vaulx-en-Velin's population is 23 percent immigrant and Vénissieux's is 18.6 percent, with concentrations of over 33 percent in specific sites, but in no area are minorities in the majority. The trend, however, is toward more ethnic concentration due to the marginalization of poorer immigrant families, to their inhabiting the most dilapidated public housing, and to their desire to live together.

Ethnic and income differentiation are therefore marked. While, on average, 40 percent of the Greater Lyons households do not pay any income tax, in Vaulx-en-Velin and Vénissieux, the proportion is between 50 percent and 60 percent. These are also localities with a high rate of unemployment, reaching over 60 percent in some zones of public housing (Bonneville, 1997: 129). These localities have become poorer while some neighborhoods in the wealthiest western localities and some arrondissements of the city have become richer, but again, income polarization is also apparent within Lyons itself and within the poorer localities.

Violent incidents at La Part-Dieu

This context helps one understand the disorders which took place on 31 January 1998 in the commercial mall at La Part-Dieu at the core of the city. They echoed other violent incidents which occurred two years earlier in the center of the city, Place Bellecour, and some other less visible disorders. Both events happened on the last day of Ramadan, usually marked by the feast of Aïd-el-Seghir. In both cases, groups of youths demonstrated

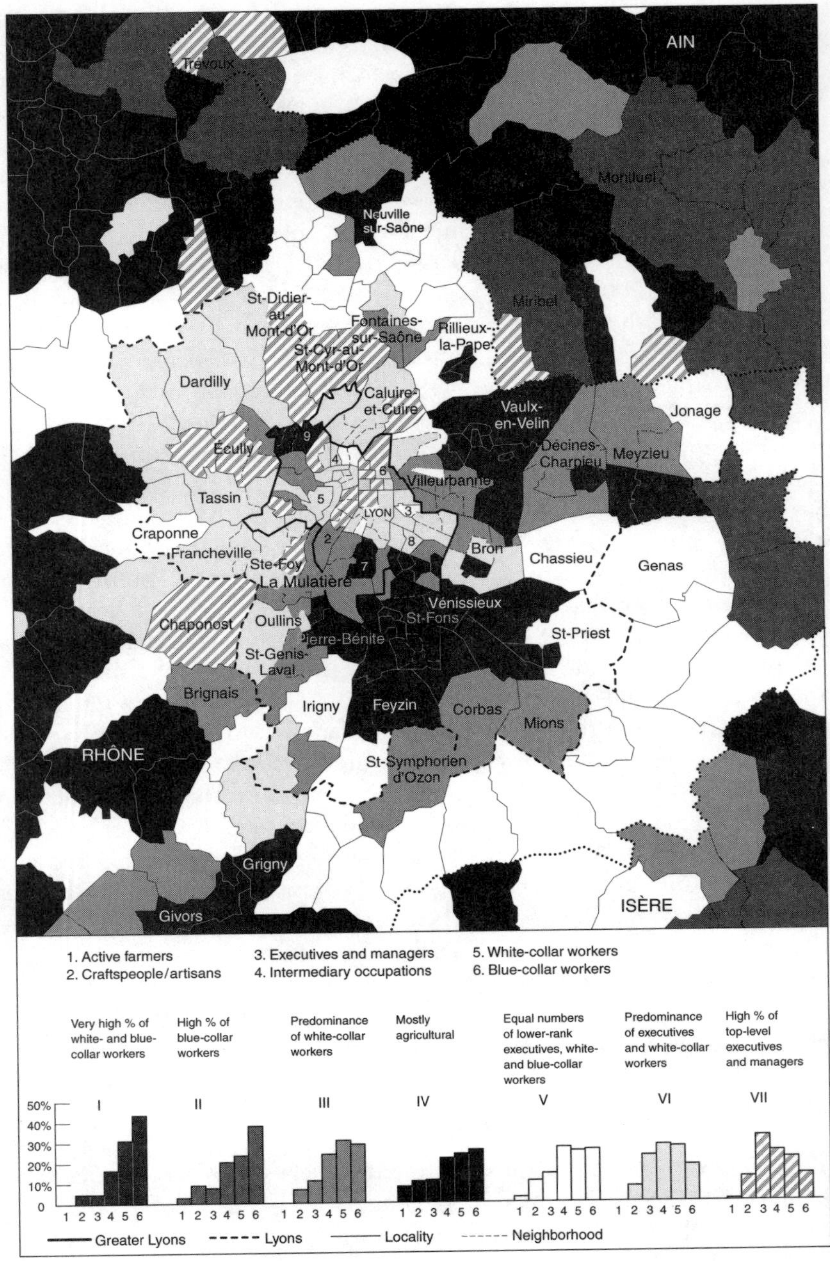

Figure 5.8 Distribution of social categories in Greater Lyons, 1990
Source: INSEE, Atlas du Grand Lyon, 1994: 20

noisily in the public space, throwing firecrackers, breaking shop-windows, and provoking the anger of store-owners. These provocative gestures may seem to be a response to the segregation that these youths experience. It can be interpreted as the youths' forceful appropriation of a part of the wealthy city from which they feel excluded. How did Lyons respond to this violence?

A diversity of perceptions While the media rolled all the events into one, presenting them in stark terms, my own investigation revealed other nuances.[23] According to the national newspaper *Libération*, the commercial mall in the posh 6th arrondissement of Lyons had been annexed by the *cités* (the housing projects from the east of Lyons). The Préfet agreed: the young people were using this private space, which they perceived as public, as if it was their own. "They goof off, break windows and attack security guards when they have nothing else to do." Every day, idle school drop-outs or jobless youths were hanging about in the mall. Truancy was not punished; the private management did not want bad publicity and recent local ordinances aimed at punishing truancy by involving the police had been declared illegal by the state council. Inaction prevailed.

Anticipating disorders, the Préfet had twenty special teams of anti-riot squads – CRS – assigned around the mall on the last day of Ramadan. He did not imagine, in part because of cold weather, that so many youths would mobilize at La Part-Dieu. For his part, he estimated that 100–200 hyper-excited youths were present by the end of that Saturday. The police raised the number to 1,000, while the mall's manager suggested a figure of 10,000. "It was a frightful invasion, as in the film *The Birds* by Hitchcock," one observer exclaimed. "Wild hordes" was how a police officer described it.

The mayor of a nearby city has offered another interpretation: he saw a plot from the far right or from Muslim fundamentalists. The word had been passed around in advance: "Everybody at 6 p.m. at La Part-Dieu," he says. Very young kids were driven by car to the mall. How could police authorities not have anticipated the events, he asks?

Dramatic voices insist on the electric atmosphere that prevailed that afternoon in the mall where between 80,000 and 100,000 people shop each Saturday. According to the mall manager, adolescents were moving in groups of 15, coming in and out of the stores rapidly, throwing firecrackers and stones, provoking the store-owners. "They came to intimidate and to hit where it is soft." The disorders started at the McDonald's restaurant when noisy teenagers refused to leave. The owner called the police and the security guards. Events then accelerated rapidly. Fire hoses were unhooked by adolescents, and water hydrants opened. Windows were broken and half a dozen stores were looted. The police intervened quickly, the mall and the subway exits were closed, buses detoured, and officers posted at the mall's gates.

How could this small "mayhem à la française" have occurred? La Part-Dieu is an older commercial mall (unlike Le Grand Littoral in Marseilles). "It was designed for peaceful customers," the manager remarks. There is no internal security chart and not enough video cameras registering what is going on. The mall is so big that an incident may occur in one part while everyone else is unaware of it. "We are powerless," he says:

> If we intervene to chase the youths, we are perceived as racists. We are not allowed even to check IDs. We could use security backup on certain days; the police could act as a deterrent. We also need better cameras and a strong will on the part of the prosecutors so that the store owners' complaints will not be ignored. . . . We cannot tolerate needing 80 special police officers so that store-owners can do their work as usual, or, simply because some people have a religious feast, we need to close the commercial mall that day. It goes against the freedom of trade and the freedom to operate . . . Will the mayor of Lyons, the mayor of my sector, my PM, the President of the region hear my concern?

But not everyone shares the mall manager's anger. In the minds of benevolent observers, there were just small groups of fifteen youths that day in the mall. They had come from the periphery to celebrate the end of Ramadan. They had dressed properly and their parents had given them some money to go to the mall. The shopkeepers were not ready to welcome such clients, who were looking for cheap jeans, sneakers, or records. The presence of security guards with dogs in the mall's dim lights could be interpreted as a provocation by the young people. They were, indeed, made to feel that they were not welcome. The lack of understanding and the fear between the two worlds were reciprocal and flagrant. The disorders were provoked by panic. Most of the mall's patrons have never been to the eastern localities and will never go. "Even if the statistics of juvenile delinquency are small, the fear generated by those youths is beyond words," an observer remarked.

The local police authority's discourse

In Lyons, the perception is that delinquency targets a few bourgeois areas – the 3rd, the 6th, and the peninsula areas – and is attributed to the youths from the eastern periphery – Vénissieux, Bron, Vaulx-en-Velin, Rilleux, St-Fons. For the Lyons police, the urban violence, in contrast, expresses a new type of offense committed collectively by smaller or larger groups in the poorer communities rather than the bourgeois areas. The manifestations of this delinquency include banging into cars in automobiles with reinforced fenders, burning private cars (around three a day – 1,000 in all in 1998), assaulting police officers, firefighters, bus and ambulance drivers, throwing stones at them, vandalizing, looting, and rioting. For the police chief, these forms of urban violence have become a sort of rite that has grown faster than juvenile delinquency during the 1980s. The authors of urban violence are 15 years old on average and are too young or too unskilled to get some of the 900 jobs on offer to young people by the *département.* Such jobs are more likely to go to the 22-year-olds who are considered more stable.

Moral panic has seized the Lyons population. Some residents say that they feel exiled in their own city, that they are scared to go to some sites such as the main street, rue de la République. Since it became a pedestrian street, groups of youths have permanently been hanging out there, according to the police. The local population has abandoned the space and the police dare not intervene: these are not illegal gatherings (as they would be in Amsterdam).

On the other hand, the residents' complaints appear legitimate. They demand that authorities take a tougher stand when young people create disorder and noise in public spaces. But the local police are reluctant to get tough. They know that should a youth be hurt as a result of their intervention, the population would change sides rapidly and the media and intellectuals would launch inflammatory rhetoric against them. When cars are burnt, for instance, the police prefer to let insurance companies solve the victims' problem rather than arrest the authors and have the youths retaliate. "A mix of social, police, and justice responses needs to be deployed," the Préfet says. "For too long, institutions have denied the existence of urban violence, even though hundreds of incidents were reported both by the local schools and the police." In the last three years, juvenile delinquency has reached a climax. "The mayors know the ten to twenty youths who are troublemakers in their community and who should be put away for some time . . . But do we have the appropriate structures to contain them?" the Préfet asked. Who is willing to be in charge at both the national and local level?

Several local security contracts, signed recently between the state and low-income communities, are aimed at finding adequate resources. But education and justice are distant partners. The *politique de la ville,* the Préfet remarks, frequently offered jobs to youths from the housing projects who were quasi-delinquents. The consequences were disastrous. Appearances are treacherous. Students with good grades may be gang leaders outside. Out of 700 youths trained in Vaulx-en-Velin via such social/urban measures, 100 can make the situation unbearable for others and yet their actions are not criminal enough to be punished by law.

The Préfet has made some proposals:

- The most problematic youths should be expelled from their communities for some time. This decision is costly and citizens should be warned of the cost of resocializing them for the better.
- Each transgression should be punished appropriately. The punishment should have an educational content and should appear to be fair.
- The size of the public schools should be reduced to make them more manageable.
- The schools should be used as sites for enforcing judicial decisions. When released, juvenile delinquents should attend cultural classes or practice sports until 8 p.m. (midnight basketball programs in the USA have similar aims).
- Prosecutors should be given adjuncts to participate in local governance and have their administrative tasks alleviated.

Urban violence in the Lyons region: a summary

These events in Lyons did not occur in a vacuum and to understand how a city is dealing with a problem, history can help. The Lyons region has been the theater for numerous urban disorders, not to mention the famous socio-political riots of the nineteenth century. In 1981, the Minguettes' rodeos in Vénissieux alarmed the country and the media gave them a wide coverage. Riots shook the city in 1983: cars were burnt, tensions erupted between youths and the police. Then they faded away. In the meantime, youths of immigrant origin calling themselves Beurs (Arabs) from the eastern localities began marches against racism. These marches, which were a form of political statement, attracted a lot of attention. They were falsely stigmatized by the far right – expressing ethnic differences was un-French – since the goal of the marches was meant to be universal (Jazouli, 1992). Then dissents occurred amongst the marchers themselves and a new top-down organization, SOS-Racism, fostered by the socialist party, absorbed the grassroots movement. In 1990, riots erupted once more at Vaulx-en-Velin. Two hundred cars were burned in ten nights, public buildings were vandalized, and firefighters and police were assaulted by irritated youths convinced that the police had caused the death of a youth on a motorbike. Such incidents became a reference for the current police chief of Lyons every time other events occurred: "The media were on the site before we were, it was premeditated." When President Mitterrand unveiled the national policy for cities in Lyons, in 1990 he claimed that the extended city would be a social lab for the future (cf. Chapter 3). Riots were all the more amplified by the media. Then a series of isolated incidents took place. A public library, a gymnasium, and other buildings perceived as symbols of society by the *banlieue* youths were vandalized. Between 1992 and 1997, at least eight buildings were the victims of arson, and 5,000 residents moved out of Vaulx-en-Velin.

In 1995, another case linked to terrorism focussed national attention on the Lyons region and on its youths of immigrant origin. In July, during a kind of man-hunt meant to catch a young French-Algerian named Khelkal from Vaulx-en-Velin, Khelkal was shot by the police, an incident that was caught live by TV cameras. Youths retaliated by attacking the police in Vaulx-en-Velin.

Since 1995, youth violence in these peripheral communities has targeted the "other" – the firefighter, the power utility agent, the postal deliverer, the janitor, the educator – as well as themselves. The incidence of suicide and self-destructive actions has increased, as has membership in fundamentalist organizations. Because the local police are not followers of the broken window theory, fear of street violence increases in the vulnerable population and disenfranchised adults are pitted against young people. Although racism is never discussed, the dominant image of an inner-city youth is of someone poor, provocative, and of immigrant origin. Storekeepers organize for self-defense and weapons circulate in the housing projects, although this is against the law. During the elections of 1995, the politicization of the theme of law and order revealed conflicts of interests and the far right made large gains in the region. Currently, the same thing is predicted for the local elections due to take place in 2001.[24]

Are the police in control? The police of Lyons have three functions: deterrence, repression, and, to a very limited extent, prevention. Out of 100 police officers, 20 are actively engaged in the field.[25] They are confronted by more and more aggressive youths and drug-dealers. They recently had to exchange their police cars, which were too visible and constantly crashed into, for ordinary cars.[26] Police from the judiciary section need to catch delinquents in action and bring evidence to the prosecutor, but, too often, the presence of a hostile media discourages them from acting, they say. During the Part-Dieu episode, just fifteen arrests took place and only seven youths were put on probation.

How long will the city of Lyons be able simply to resort to its police and relegate explosive problems to the margins? Will it avert the boomerang effect?

Policies of reconciliation For the mayors of the low-income localities at the periphery, the issue of safety has reached dramatic proportions. How to redress heavy structural handicaps? How to reduce high rates of unemployment in former working-class neighborhoods that affect Algerian immigrant groups more than others? How to fight against school dropouts, drug-addiction, domestic violence, and more or less overt discriminations? How to change the residents' collective perception of second-rate citizenship?

According to the mayor of Vaulx-en-Velin, urban violence is a very complex phenomenon caused by a small group of actors in the presence of 100 passive observers perhaps, who watch the turmoil, their hands in their pockets, as they would watch TV: "The degree of violence is related to the victim's degree of delinquency, in cases of police or judicial abusive actions." A lot of incidents confirm this statement: when a notorious juvenile delinquent is injured or killed, a will of visible revenge seizes his or her friends. Then the context weighs in; it is just as important as the feeling of vengeance.

Poverty characterizes Vaulx-en-Velin, a city of 45,000 with one-third of its residents under the age of 20. According to INSEE data from 1995, 60 percent of the households are on welfare and 25 percent have less than half of the income level that defines the poverty line. Of youths under 25, 26 percent are jobless. The deterioration of living standards is visible in health statistics, the mayor says. Of children in elementary school, 70 percent need dental care. But observations are not followed by action. The community's difficulties are also illustrated by problems encountered by the local mission, which enforces the national policy for cities and aims at training youths for jobs. Out of 2,200 youths with whom the local mission had worked, one-tenth are in a state of extreme distress and suffer from psychological and emotional disorders, due to deprivation. "The parents and the global society make children feel guilty when they do not meet their standards of success," the mayor says:

> School is experienced as a place of knowledge consumption where the tester is hostile. The Khelkal case showed that perceived discrimination in high school alienated this adolescent. Our society treats youth with violence. Offering them unemployment or precariousness as a future inhibits their dreams. Even those who get jobs via city hall suffer from a lack of recognition. Because they live in specific neighborhoods, these youths will not get jobs or will be denied access to other places.

The mayor says that he wrote to 6,000 companies pleading for the youths from his area. To illustrate the discrimination that these young people experience, he talked about Luna Park, which was to be created nearby. Some fundraisers proposed a site inaccessible by public transportation "in order to get the families and not the youths." "We need to be both very humane and very strict," he adds. He has created a crisis unit where the main actors can work together – the Préfet, the local prosecutor, the police chief. But efficiency has unanticipated effects. Because the current police chief managed to lower the rates of crime in Vaulx-en-Velin, six police officers were transferred, replaced by adjuncts who still need to learn their job.

The mayor emphasizes individual skills over institutional will in the local governance partnerships. He appreciates the quality of his relationship with the current prosecutor, characterized by implicit agreements and flexibility. But the lack of probation officers and of educators in his city may come both from a national lack of commitment or from a reluctance of these educators to be submitted to the patronage of a strong mayor. As in St-Denis, the mayor aims at reinforcing social cohesiveness. He devotes much attention to the pre-adolescents and, like other reformist communist mayors, he opposes the suppression of family benefits to delinquent parents. Addressing the issue of parents, the mayor thinks that sanctions should have an educative value: "If we make them feel irresponsible, the burden will increase for the children. We should avoid penalizing the whole family and sanction only the trouble-maker." Indeed, dysfunctional parents should be supported by public services, he says. He offers several suggestions. First, the locality should help the poorer families take a vacation with their children instead of sending the children alone to fresh-air camps. But this would be very expensive. Second, instead of subsidized meals in schools, children should receive the same subsidized meals at home (in many families, at least one parent is unemployed). If a juvenile delinquent has to be punished, prison is not the solution, the mayor observes. Specific educative and disciplinary camps are enough when kids need to be set apart from their delinquent friends. Victims also need to know what happens to their cases.

Besides institutions, the civil society is involved. Active grassroots organizations repair the social fabric, often thanks to the strong presence of militant women. Proselytizing religious organizations in Vaulx-en-Velin and in nearby localities may appear to some as a threat to universal values. The close relationship of some Roman Catholic priests to fundamentalist leaders has been widely debated locally and nationally in the general context of Islam in France (Kepel, 1989; Césari, 1997).

According to the mayor of Vénissieux, André Gérin, the national society is responsible for collective life, prevention, and commonalities and government should accept that they must pay the price of social peace. Then, initiatives should be taken locally. After the Vénissieux public library had been burned by angry youths, the mayor hired a black Muslim social worker and asked him to become a boxing and karate coach for the unemployed youths of the community. Good results followed. The mayor also required four city counselors to develop better relationships between the city and the youths: they were asked to make contacts, listen to them, give them recognition and respect.

Gérin thinks that too much fuss is generally made about incidents involving the burning of cars. One out of two have been stolen, frequently in Switzerland. People then try to cheat their insurance companies, and spouses retaliate against one another. Young people have little to do with these crimes. For the last ten years, the mayor has developed a good relationship with the police chief ("an intelligent understanding" man, the mayor says) but not with the criminal justice system, which, he says, is too slow. In this regard, he is similar to many mayors around the country who have little to say about who in the police and justice systems is sent by the Parisian decision-makers to work with them and who will be abruptly transferred.[27]

Yet the mayor who has to form a team with these civil servants of passage is also the first authority to be held accountable by the public when disorders occur. If he or she appears to minimize events, the far right will politicize the issue and inflame discontent. The mayor therefore needs to take (at least discursive) action. After a fight between two groups of youths in the subway had caused the death of a respected young man from the community, the mayor of Vénissieux organized night activities near two subway stations at weekends. He mobilized community police officers, bus agents, newstand agents and public employees to combat the fear of street crime.

With social links in mind, mayors have to deal with all the grassroots organizations in their locality, including the fundamentalists "who should be treated without complacency" in order to defend the principle of secularization. The mayor of Vénissieux observed that youths should be

> helped to grow, encouraged to vote and to express their views, and their incorporation into the job market should be supported. When they find jobs, they get so much pride out of it that they need not resort to violence. Then they feel integrated and when they better understand other people, they are also better understood . . . Preventative actions are crucial: they teach the youth how to structure themselves and how to negotiate.

Negotiations imply, indeed, that youths are perceived as critical actors in the city. They need better political representation (Begag and Delorme, 1994). Offering them access to (legal) "voices" and not just to the job market would reduce their need to "exit" and strengthen their loyalty.

In brief, my inquiries in the Lyons region have revealed the difficulties experienced by polarized and segregated urban cities when they have to manage identity and turf appropriation conflicts as well as their resources. How unique is Lyons in that respect? Is there a space effect related to this city's social history?

In the past, Lyons has been perceived as both a secretive, cold, conservative place and an entrepreneurial and hard-working society. Recent economic and social evolutions, in the context of waves of immigration, have altered these features without erasing them (Bonneville, 1997: 137). With a history marked by industrialization and working-class revolts, the city has retained a tradition of strong social antagonisms and of struggles. The lack of compromises has delayed newcomers' social integration. On the contrary, an era of broad mutations and forceful challenges has hardened the positions of the established and of the outsiders. Face-to-face conflicts have been replaced by atomized and identity/territorial conflicts.

New forms of local governance attempt, however, to confront these challenges. After the long domination of Lyons's mayors and their ignorance of the low-income localities' problems, the mayor, former Prime Minister Raymond Barre, has offered periphery mayors the position of vice-presidents in the Greater Lyons structure. That the reformist communist and socialist mayors of Vaulx-en-Velin, Vénissieux, Villeurbanne, Rilleux, St-Priest, and Bron should sit together at the end of the century at the same table represents an important innovation and a rupture with the past. A search for consensus is imposed by budgetary constraints, by the need for the Lyons region to expand internationally, and by the problems of social exclusion which limit its ambition. Local governance is substituted for a model under which institutions with precise but limited skills managed a territory: in the new model, more and more diversified actors intervene on larger functional territories, no longer bound by administrative definitions. Systems or networks of actors thus aim at producing coherent policies via cooperation, negotiation, and contracts. Private actors – local, national and international companies – progressively intervene in these policies and become responsible for public action. In Lyons – as in Marseilles – the city management obtains its recognition henceforth less from the state elites or from institutions than from local initiatives. A consensus favors constructive processes with numerous actors tackling important issues (Lefèvre, 1995; Lorrain, 1995).

Fighting segregation and favoring the integration/recognition of relegated youths – of immigrant origin or not – is one of these issues. Are police actions and the development of private security and law-enforcement technology the solutions? Is a harmonious coexistence of different populations in the same city impossible? The criminalization of low-income, jobless, immigrant youths is a strong trend that the Greater Lyons mayors' preventative efforts, supported by the resources of the national socio-urban policies and by the actions of the grassroots organizations, may counter. They all call for the reinforcing and the re-explanation of citizenship.

* * *

Several assumptions have been tested in the case studies that I have presented. It has been shown that at a time of economic globalization and despite colossal constraints, cities are the spheres where social innovation takes place and where a general local interest is pursued. Cities – New York, Chicago, St-Denis, Marseilles – can reduce crime. They are laboratories for the transformation of institutions, professionals, and their decisions. New York and Chicago demonstrated their capacity to transform an institution like the police, and even though some of the New York policemen, unchecked and untrained, went too far, subsequent measures taken after demonstrations by citizens proved once again that the city must negotiate, accommodate, and compromise. In St-Denis, Marseilles, and Lyons institutional partnerships were created, supported by grassroots organizations, to work on a crime-prevention approach. As citizens forcefully express expectations about their right to protection, they precipitate institutional change. Cities are sites of articulation between formal institutions which have to be reformed and civil society which has to mobilize to exert pressures on institutions. Moreover, it has been shown that thin regulations define relations between the center and the periphery. In brief, the local arrangements and the social engineering taking place at the local level explain why urban societies in Western countries remain generally pacified despite destructuring and "de-nationalizing" trends and occasional social eruptions. Local governance and empowerment remain, nevertheless, ambiguous notions. The reasons behind the decreases in crime are multivariable or multifactorial.

The interest of the case studies was to raise a cross-national perspective. Having reached this far, I suggest that some convergences cannot be denied. For instance, in response to the weakness or indifference of central governments, cities have shown that they could display both imagination and action to prevent the threats of disorder (making the World Cup happen in St-Denis was such an example, as well as developing symbiotic actions in the schools of Chicago, enforcing respect for the quality of life in New York, and launching a socially integrated commercial mall in the northern neighborhoods of Marseilles). Innovative partnerships and multilevel commitments link cities' actors with other scales' units of decision-making (transnational, national, regional, local). In both France and the USA, city authorities are confronted by negative and destructuring forces: unemployment, family disruptions, normative economic changes, the media destroying months of multileveled efforts.

National differences remain very important, nevertheless. They include, among others:

- the size and nature of the problems;
- racial divisions and urban segregation in the USA vs. culturally atomized troubled localities in France;
- the gun culture of the USA and the strength of local autonomy;

- the scarcity of public services, especially in ghettoized areas in the USA;
- federalism vs. centralism;
- state intervention in the French economy and the reluctance to accept a Third Way;
- the advantages or disadvantages of a national police and judicial systems insulated from local political pressures in France;
- the capacity of mayors to terminate local programs in the USA and to launch new ones, as opposed to corporatist reactions and the weight of intermediate hierarchies discouraging change designed at the top in France;
- the relative powerlessness of French mayors as opposed to the accountability of major USA cities' mayors on the issue of law and order, but their relative capacity to influence national policies via multiple office-holding and political connections;
- the disproportionate influence of the business community in the USA on urban decision-making;
- the role of intellectuals and of a mediatized culture of opposition in France, and the political effectiveness of lobbying decision-makers;
- the weakness of civil society in France in terms of "self-help instincts" as opposed to the relative strength of civil society in the USA, although, as will be seen in the Conclusion, an evolution is occurring in France.

Notes

1 The Préfet, the national representative at the local level, is in charge of the police. The District Attorney can refuse to have any informal interaction with the mayor and the school principals. He will retain data regarding new undocumented families' children in the local schools and the housing projects, which is positive in terms of civil liberties yet questionable for the mayor in charge of social peace in his constituency.

2 *Cumul des mandats* allows mayors to be also Parliament members or members of the government or regional councilors, thus softening the rigidity of centralization. It explains the success of local elections.

3 See Introduction, n. 9.

4 The expression referring to the Paris region has no scientific basis; slightly larger than Île-de-France, it implies that the impact and attraction of Paris is felt on even more distant areas, as far as 80 kms away (in the case of 240,000 commuters, for instance, coming from the most distant reaches).

5 These national surveys are administered to 10,000 households, including about 2,000 in Île-de-France. The first decile includes the lowest incomes and the tenth the highest.

6 This type of housing is a far more diverse stock than in any other European

country, given the many agencies – public as well as private, local, regional, and national – in charge, particularly in Greater Paris (Rhein, 1998: 440). The management of public-housing projects is rarely handled by a single authority and 13 or 14 different managers (*bailleurs*) may intervene in the same locality. The allocation of housing units depends on (a) the Public Housing management, (b) the Préfet for "distressed families," (c) the firms participating in the financing and the construction and (d) a municipal board. This pluralism results in a lack of accountability.

7 I want to thank B. Madelin from Profession Banlieue at St-Denis who transmitted these official data to me.

8 The urbanization of La Plaine St-Denis began at the end of the nineteenth century when an industrial and logistic pole developed in the north-east suburbs of Paris. The northern railway network expanded then, as well as the warehousing and storage of coal and goods (Rhein, 1986: 55).

9 In the former Seine *département*, 48 out of 80 communes had a communist local government in 1959 and, except for 5, they remained communist after the election of 1989 (Rhein, 1986: 55).

10 I was told by an educator in 1998 that were I to come to Francs-Moisins, I should come before 10 a.m. to avoid trouble.

11 Field work in St-Denis is based on interviews with the mayor of St-Denis, former Préfet of Seine St-Denis, J. P. Duport, police precinct commanders, school authorities, students, artists, educators, and members from Profession Banlieue, 1997–8.

12 It is true, one teacher said, that we could write a 10-volume *Les Misérables* based on the children's social situations that we come across.

13 Over the last ten years, Seine St-Denis claims to have invested $100 million a year on average in the middle schools, 90% of which came from taxes and only 10% from the national government (Gurrey, 1998).

14 The mayor, P. Braouezec is a former schoolteacher. He remembers that in his class at Francs-Moisins, he had sixteen students: fourteen Portuguese, a Spaniard, and an Algerian.

15 Many acts of violence were perpetrated against the school and the staff. The principal had gunshots fired at her windows, received death threats, and had her car vandalized.

16 Nicolas Frize had conducted the same experiments in factories, in the schools of St-Denis, and in prisons. I accompanied him to the maximum security penitentiary of St-Maur where not only did he initiate inmate volunteers to sophisticated techniques in the sound industry, providing well-paid jobs for them, but he invited them to create pieces of music which they performed in the prison. Finally, he was allowed by the prison administration to invite representatives from the civil society to informal discussions with inmates and in thematic round tables in the prison.

17 When the highway cutting St-Denis in half was covered, a carnival was organized with people from the community working for weeks to decorate the floats. The city had a garden made with magnificent lighting on the regained space. The mayor supported endlessly the residents who, for 30 years, had demanded that their city be one again.

18 The field work took place during the spring of 1998.

19 In the Houses of Justice, all the mediation programs are based on a contract on which all parties agree, before discussing the context of the conflict and the subsequent relations that the parties will have. The direct contact between various contenders anticipates new modes of social regulation. Where there was distrust toward formal institutions, there is confidence in the process of discussion and deliberation.

20 The figures come from the family benefit center.

21 A resident will say "I am from Belsunce" before saying "I am from Marseilles," which reflects the fragmentation of territorial belonging in a context of economic restructuring and its expression in new polarized urban patterns.

22 The official reason was that, built on a quarry, the mall needed to have its foundations supported by the hill with the parking lots on the pilings.

23 Interviews were conducted with the Préfet, the police chief, the housing department, the mayors from eastern localities, their administration, the director of the commercial mall, social workers, teachers, progressive militants, and families.

24 At the end of 1998, the President of the Rhône-Alpes region, Charles Millon, was forced to resign because of public outrage caused by his association with the far right.

25 Other police officers are in training, or on vacation, or on sick leave, or resting.

26 Buses are frequently the targets of youth violence. Surveillance systems allow drivers to warn the police. Police and drivers have formed partnerships.

27 When I was heard by the French Association of Mayors, all the mayors complained about the delays and lack of communication from magistrates (fall 1998).

Conclusion
The Social Control of Cities?

I began this research with several questions. From my (limited) empirical work, what insights have emerged? The questions I started with were as follows:

1. Does globalization have an impact on the disorders experienced by cities? How do national contexts buffer cities, if at all? The future of policing and of crime and violence control lies in the interconnectedness of global change and local experience and through pressures towards social and system integration. Social systems are stretched across what Giddens (1984) calls "time-space distantiation" by the "disembedding" nature of the global economy. Yet this phenomenon connects with a sense of local contextuality (Crawford, 1997: 203), all the more reinforced by threats of social disintegration.

In my research, I have demonstrated that the market does not favor social cohesiveness but generates tensions: it reinforces economic polarization and inequalities in cities, the recomposition of space unveils power conflicts among major actors, and hundreds of thousands of marginalized people and their children may use their "voice" as a threat to express their claims. Even if a few global cities like New York and Chicago take advantage of globalized processes – by increasing their wealth, their export sectors, their foreign investments, their capacity for technological innovation, their job creation in the service economy, their potential for extensive networking all over the planet – here, as elsewhere, processes of segregation and problems of social disintegration do not disappear, despite the good health of the local economy. Even in the midst of an economic boom, inequalities persist and deepen, racial tensions abound, and public services and infrastructures remain neglected (Fainstein, 1999). The obvious fact is that the whole system based on global exchanges of capital and goods is fragile, as illustrated by the 1998 economic crises in Asia, Brazil, and Russia.

The destruction of neighborhoods and the development of localized crime problems are caused as much by individual abandonment as by economic disinvestment and local political decisions. The range of local political decisions varies from destruction of public housing, the termination of rent-control, evictions, and the withdrawal of municipal services, to tax-break incentives, mortgage assistance, and zoning restrictions pushing for the spatial replacement of poor populations by reliable taxpayers, either companies or private individuals.

Two processes are at work. If recession occurs in post-industrial societies, institutional abdication could be disastrous in terms of protection for disinvested zones detached from the world economy (except for crime purposes – see the first scenario I painted at the end of Chapter 2). Unwilling to initiate the kind of redistributive social action capable of incorporating the dispossessed and unable to countenance criminality, the US government and society turn to anticipatory acts of surveillance, repression, and exclusion (see my second scenario painted at the end of Chapter 1) (Scheingold, 1991: 187–92). Such acts construct circumstantial reasons for exercising control over that part of civil society they feel is particularly threatening (the non-white poor, undocumented immigrants, youths at risk, the homeless, etc.). The sovereign function of state policing, then, is not diminished by the abundance of global deregulations. In the field of crime control, Crawford remarks, many of the "hollowing out" processes are less developed than in other spheres of social life. There is a trend toward increased central control, as seen in Chapter 1, even at a time when budgets are being tightened. But the trend has to be seen in a context of centralizing and decentralizing tendencies, of blurring of spaces, of displacement and restructuring of powers in process (Crawford, 1997: 220). What constitutes the legitimate responsibilities of individuals, collectivities and the state? Where is the rearticulation of new socio-political relationships played out and contested?

Crime is a case in point to evaluate the context of globalization weighing on cities. The increasingly diversified and interconnected global criminal industry operates within cities' low-income neighborhoods. In certain neighborhoods in global cities where unemployment is high, poverty deep, and social regulation limited, part of the population is enmeshed in the workings of the world economy via drug-dealing, arson benefiting uptown promoters, etc. (cf. my second scenario). Political blindness and the security apparatus of organized crime, the network of law-enforcement agents, judges and politicians who are on the payroll reveal the institutions' internal weaknesses. At the end of the millennium, Castells remarks, the global criminal economy represents a formidable challenge for states and cities (Castells, 1998: 180).

What amount of depacification is then tolerable? It would be erroneous to think that the degree of civilization being such (cf. Norbert Elias) and the pacification of morals so anchored that decivilization can be tolerated in circumscribed zones. It would be a short-sighted view to ignore the power of the powerless. The question of Barrington Moore (1978) concerning injustice is still valid: instead of "why do men rebel?" we should wonder why they do not rebel more often. The margins and the center are interdependent. The margins of society can hit the center at its core, as the Lyons case showed. Under the effects of the imposition of a new centrality, and once there is nothing to negotiate, the marginalized, whether they live in the inner cities of the first world or elsewhere, are not powerless: they can always negate what is a precious good for the others: social peace. The transgressions of the "haves" (excessive accumulation, deregulation, chaos, non-redistribution, etc.) must therefore remain limited for their own sake.

2. Which national and urban conditions allow certain types of local policies to develop, if any? Here, I will examine institutional arrangements.

In the USA, the relationship between states and localities and the federal government has changed. As the crime issue has shown forcefully, the federal government uses financing for measures to exert leverage at subnational levels. Conversely, states and localities also bargain, as they need the grants and the guidelines to defuse multiple internal conflicts. Unlike in France, where assistance is still mostly distributed according to universal principles, American states select cities to receive aid on a competitive basis. Studying local governance and the resources used by cities to fight crime and violence, we move from multiple discourses and laws passed at the summit of the state or of the states in the American case on the one hand, and, on the other, to processes which operate at the bottom, empirically, between a plurality of actors.

While national approaches differ, cities in both countries appear as the locus of social control. At the request of US mayors, traditional institutions, police, the judicial system, and corrections are committed to redress the delinquency which poisons the collective life of cities. In New York, the mayor used both the prevailing conservative mood of the country and statistics on crime to enforce a "zero tolerance policy," while the mayor of Chicago used an array of measures – the decentralization of the police, an emphasis on education, and the destruction of crime-ridden housing projects supported by HUD – to restore safety in the streets. The mayors of both cities have national stature and aim at upgrading the image of their cities for business and tourism reasons. Other actors' practices also count. Criminal justice professionals attempt to thwart the harshness of the laws and re-establish discretionary practices. Judges, prosecutors, and defense attorneys are eager to express their reluctance about federal mandatory minimum laws, federal sentencing guidelines, "three strikes" laws, etc. Not only can they not accept having their resources stretched to breaking point by the enforcement of laws based on moral panics, but they refuse to have their discretion restricted. Consequently, they attempt to bend the harshest provisions of these laws to reinstitutionalize traditional procedures and to rectify disparities and inequities caused by the laws (Feeley and Kamin, 1997: 136) which prevent them from using mitigating circumstances and their own judgment to alleviate the sentences. An example is offered by a Supreme Court decision of June 1999 which reversed a judicial measure against gang members loitering passed in Chicago two years earlier. The majority argument was that gang members' constitutional rights – including freedom of movement – had to be protected. Police officers have also come to distance themselves from policy directives from above and to develop pragmatic views, knowing that as they rescue one street, they may be losing another somewhere else. Local governments then adapt law-and-order policies to their specific circumstances.

The blurring of boundaries between the global and the local reinforces the role of urban mayors; despite alarmist discourses, the mayors of French and European large cities have "to slow down the clock",[1] which is not always the case in the New World. In the Paris, Marseilles, and Lyons regions, mayors have taken advantage of national programs. The Soccer World Cup and the consolidation of Europe, for example, give leverage to mayors from the geographic periphery. The concept of multiple office-holding is also an asset for the mayors of large cities: the current mayor of Marseilles was the Minister of the City until 1997, that of Lyons was Prime Minister, that of St-Denis holds an official position in Parliament. These mayors all play important functions in their respective political parties, which are centralized in France, and can lobby the government and address the media via these channels.

3. How important is the role of mayors? At first, it would seem that the mayors of large American cities are no longer as important a political force as they were in the 1960s. What is the weight of the Mayors' Association? If a city were to experience a fiscal crisis nowadays, would its mayor be told to "drop dead", as the mayor of New York, A. Beame, was instructed by President Ford when the 1975 fiscal crisis revealed the bankruptcy of the city? For reasons just mentioned, the mayors of large US cities seem to be more politically isolated than their French counterparts. City councilors are also more distant from their constituents because the constituencies are so much larger. There are 36,000 mayors in France for more than 60 million people, and about as many in the USA for a population of 260 million.[2] While French mayors see themselves as "guardians of the locality," supervising everything and a priori hostile to external intrusions, the American mayor is frequently one actor among others and the business community is particularly influential. There are several mayoral policy styles: among others, the caretaker and policy mediator, the policy innovator and champion, the opportunistic policy-broker (Ross and Levine, 1996: 126). At the same time, big city mayoralty

has been defined as an office of massive frustrations: limited formal powers and staff, a shortage of money, conflicting demands and expectations, wrenching personal strains (Caraley, 1977: 208–22). In that respect, the American mayoralty is not very different from the French one. In the USA, minority mayors do run large cities. But they rarely come from the inner cities. Even if their identity does not always provide them with clout to solve the inner cities' problems (Judd and Swanstrum, 1998; Body-Gendrot, 1998a, 103–8), they are more at ease when it comes to appointing minority police chiefs with instructions to ease tensions in the minority ghettos. In France, mayors and police chiefs of North African origin simply do not exist.[3]

Despite a general feeling of powerlessness due to the deterioration of the social situation in poor neighborhoods, to bureaucratic inertia, to top administrators' "clipping their wings," and to financial constraints, the mayors of large French cities have often used their cities as laboratories for daily experimentation on social control (see Chapter 5). For most French, the mayor is the most legitimate and probably the most efficacious "last bastion of democracy." Yet while only 42 percent of the French would grant them more power (especially for the control of crime/violence), 73 percent of the mayors wish they had more resources to fight delinquency in the vicinity (CSA poll, *Le Monde*, 18 November 1997). A typical letter addressed by a resident of a run-down locality to the mayor would say something like: "I am making you aware of these facts, so that city hall can take preventative measures against this form of meaningless delinquency against people." For this resident, the mayor is the decision-maker, yet we have seen that the mayor depends on other levels for law and order. Mayors complain bitterly about their lack of clout over the national magistrates (prosecutors, judges) who frequently refuse to have contact with them, lest their judgments be influenced. The mayor, as an officer of the judiciary police, is in charge of order. Mayors are divided over the opportunity to strengthen the municipal police accountable to them or to rely on the national police. There are 12,500 municipal police officers (an increase of 121 percent since 1984), compared with 113,000 national police officers (+4 percent). Each mayor develops his own strategy. Some of them require police stations to be open 24 hours a day with surveillance cameras and radios, but they do not arm their municipal police officers, the decision of Amiens's mayor, for instance. Some point to Strasbourg and claim that the municipal police have been armed since the nineteenth century and should not change. Others, as in St-Denis, rely only on the national police, but also have social mediators for community safety in order not to mix the functions.

A brief typology may clarify the various "styles" of mayors on these questions.

Exclusionary politics

In the USA, some mayors have developed enough autonomy and twisted institutional arrangements in order to pursue exclusionary politics. The size of the city concerned makes a difference. In smaller localities, mayors feel more narrowly accountable to voters and "respect" the local culture. Not only do a lot of them, in the southern and western parts of the country, support the National Rifle Association and its policies, but they would not disavow in public an armed homeowner who shot a stranger trespassing on his property. Repression, surveillance, and cuts to the budget for "unworthy" poor frequently seem the easiest way to win the local business community's and conservative voters' approval. Playing on moral panics is a way of policing the city.

In France, as in the USA, a symbolic policy which harshly treats immigrants and street youths has been frequently used as a way of cementing the "other half" while playing on the ethnicization of social relations. Such mayors strictly apply the laws. In such localities, ethnic antagonisms are fueled by local bureaucracies, including the management of public housing. City Hall's refusal to register immigrants' and old-stock French children in the same day-care centers, schools, leisure camps, etc. contributes to the enforcement of spatial segregation and to the NIMBY syndrome. Courts, human rights networks, and civic organizations have to be seized by progressive activists to protect the civil rights of "suspicious categories". French cities, run by the far right or its equivalent, skillfully "market" public safety in their weekly newsletters in the same way that gated communities in the USA celebrate privatization and the reconquest of space.

Progressive politics

Local US culture, as in Burlington, Vt., or Santa Monica, Ca., or Boston, Ma., may be the most important variable here. But a progressive mayor is also likely to be elected, thanks to divisions within coalitions, as was the case with Harold Washington in Chicago. Progressive politics means redistributive policies and support to liberal grassroots organizations. Few American cities fall under this profile, but smaller neighborhoods do and this is not due to the influence of the national level. Such progressive episodes however do not and cannot last. The ideology of redistribution is not consensual, tolerance toward the ways the "othered" organize to survive cannot be stretched for too long, maximum feasible participation leads to maximum feasible misunderstanding. Strict gun control is not popular and in some states it is unconstitutional. In matters of redistribution, jealousy occurs at the margins of the stratification between the law-abiding, "almost poor" taxpayers and the very poor benefiting from public or private "largesses" and being perceived as unfairly favored. Race, ethnicity, and age increase the tensions. As Fainstein aptly remarks:

> recollections of persecution of one group by another or feelings of group superiority based on color, nationality or religion will not go away simply because of economic equality . . . An economic programme with redistribution as its central goal . . . alienates substantial portions of the stable working-class, who see their security in homeownership and in separating themselves from the social strata beneath them. (Fainstein, 1997: 23–5)

The same phenomenon exists in France, all the more because the institutional arrangements of national socio-urban policy in cities run by leftists are thought to support "unworthy" families, especially illegitimate immigrants with numerous children, seen as draining the benefits of the welfare state. It has been constantly difficult for progressive mayors in Vaulx-en-Velin, Vénissieux, Grenoble, Mulhouse, Dreux, Epinay, Les Mureaux, St-Herblain, and Amiens to undertake a genuine redistribution in the inner city for people who cannot or do not want to vote. Yet such mayors and their teams develop strategies to reduce the risk of disorders with various partners in public housing, public transportation, and public education. They use their knowledge of the community by providing support for vulnerable parents, city jobs for their children, education and training for drop-outs. They develop all kinds of mediation, as shown by the campaign "Hello, neighbor" in St-Denis, which involved artists or the University of the citizen in Marseilles or the street festivals on weekends in Vénissieux. Those mayors provide practical support to victims, promoting insurance and compensation schemes. Night "correspondants" circulate throughout the neighborhoods to signal incidents to the police or to meet the parents of children on the streets. These mayors are at the forefront of unifying actions and regulation among public employees on safety matters. They also defend the construction of mosques, although they know that the cost for a mosque in their city is likely to mean a 30 percent increase in votes for the far right in the next elections.

As in the USA, they cannot overcome the jealousy of those at the edge of the poverty line, whose votes count. The problems of delinquency, drugs, and terrorism, amplified by the media, catch up with their redistributive measures, the war is never won, and the mayors are frequently overwhelmed. This style of mayor cannot last: inner-city residents themselves are never satisfied, and when they are, they move out. Then other poor categories move in, new problems arise, and the mayor is called inefficient. But these mayors' democratic efforts are symbolically important and they become references in the urban imagination.

Status quo politics

According to circumstances and pressures, mayors will have local measures of prevention passed, benefiting their majorities (education, training, jobs, community policing, public housing renewal, and all kinds of partnership) and they will resort to repressive measures when they are facing other pressures, for example after violent episodes generating majorities' discontent (curfew ordinances, cuts in family allowances for delinquent parents, deployment of armed municipal police, video cameras in public spaces, etc.). If they can avoid doing so, they will. Most French mayors – 60 percent – do not wish to ban panhandling or impose midnight curfews on youth under 12 years old, in a conciliatory move.

The national mood may nevertheless influence local public opinion, exerting pressure on the mayor and the city council via vocal lobbies (e.g. store-owners, tenants' organizations) for more social control and more policing. Local episodes become salient issues: a gathering of gypsies in the locality, vandalism of public buildings, violence in local schools. Suppression is rarely demanded on its own, but is dictated by events, the media, the presence of competitors. The difference in status quo politics between France and the USA is related to national resources which mayors can use for local problems. The national police can be reinforced by special squads before or during incidents; mayors can tap into diversified networks for help and if public–public partnerships have a life of their own in France, the mayor is their coordinator.

This typology is only an indication of various styles of governance. Instead of having cities or mayors which embody one type or the other, what occurs most often is that mayors and their staff resort to the three patterns according to circumstances and issues.

Take the city of New York. Mayor Giuliani and his team were able to capitalize on their previous experiences in matters of public safety and use a powerful metaphor (of broken windows) to convince the public that New York was being more innovative than other large cities, that the mayor had responded to his constituency's demands for a safer city. The zero tolerance approach was supposed to pacify the city. Supporting aggressive policing, surveillance, and excusing police abuses, the mayor has appeared as clearly authoritarian and intolerant. For a while, he benefited from all kinds of favorable demographic, economic, and cultural circumstances. The political disorganization of minorities, the first victims of police brutality, and the ambivalence of majorities allowed him to forcefully market his innovations. At the same time, the mayor, compared with leaders in California, for instance, was using a liberal rhetoric praising immigrants, "the worthy poor," and pro-choice attitudes. His "liberalism" here has been in keeping with that of his predecessors. Yet he will also have worked to maintain the status quo: if, in interventions with street vendors, he encountered public opposition, he would promptly back out. I call it the status quo because this line of action is inspired by circumstances. A measure, if proved unpopular, can be changed within weeks. The aftermath of the Diallo and Louima cases (see Chapter 4) will show if the mayor's popularity has been severely damaged or if he can rebounce rapidly.

Pierre Bédié in Mantes-la-Jolie in suburban Paris is one of the young mayors among many who, in contrast, practice a mixed mayoral style. Mantes-la-Jolie, population 45,099, is made up of an older village and an urban zone, Le Val Fourré, which, with 25,638 residents, is where all the symptoms of urban crisis are concentrated.[4] Two successive riots within a fortnight in 1991 made this neighborhood famous. Pierre Bédié's predecessor, social-democrat Paul Picard (1977–95), supported by militants from community organizations, used Mantes-la-Jolie at the beginning of the 1980s as a showcase for social-democratic urban reform. A mosque was built, despite controversy. All the national measures designed for social prevention were tested at Le Val Fourré. But, as said before, progressive experimentation and massive redistributive policies could not last. The problems were endless. As a result of new liberal national policies on home-ownership, the lower-middle classes moved out of the high-rise projects, poorer residents stayed, while a "second" generation of immigrants moved in. The neighborhood's reputation got worse. Paul Picard was accused of playing the national card too much and deserting the grassroots. People in Le Val Fourré did not vote and those in the old village felt that they had not gained at all from the mayor's choices. After riots in 1992, four high-rise projects were razed in Le Val Fourré and new local leaders of immigrant origin used the media for their claims. Picard lost the elections in 1995 and conventional wisdom had it that a lesson had been learned: it was no use piling up resources on one pathological neighborhood, the whole city had to be taken into account by urban measures (now, the appropriate scale is thought to be the metropolitan area). The new mayor, a political centrist, was in favor of curfews and of the penalization of delinquent parents. He has supported repressive police strategies to reinforce law and order at Le Val Fourré. He has forcefully demanded a reform of local taxation (launched in the fall of 1998). A lack of resources prevents such small peripheral communities from taking more initiatives and, on this point, Bédié has practiced status quo politics, launching ad hoc actions whenever the circumstances demand them. On the progressive side, he has supported territorialized affirmative action policies and measures of redistribution for his poor neighborhoods in order to maintain social peace. Making use of the 1996 entreprise zone law, he has been able to attract 60 small businesses and to convince them to locate in Le Val Fourré, bringing 200 jobs to a neighborhood with 3,000 jobless. It more or less "pacified" the area, he says. One wage-earner brings the whole family back into the mainstream and gives it hope. A contract with a private company, the Sogea, allows "tough" boys from Le Val Fourré, frequently with criminal records, to get construction jobs. The firm has negotiated with the residents' council, which screens candidates with no job prospects but a desire to work, as in Marseilles.

In France, most mayors thus play an essential role in the depoliticization of crime and violence; they have incentives for minimizing social and ethnic conflicts among their constituents and for pushing constructive policies. Urban disorders clearly have an adverse impact on the city's image and on the local business climate. In the USA, at the local level, for the same reasons, the incentives also work against politicization and in favor of public safety. Local authorities' daily immersion in the struggles of the criminal process makes it difficult to ignore the structural correlates of street crime (Scheingold 1991: 186). At election time, mayors and local professionals must answer for their failure to control crime, yet they are sympathetic to the dreadful personal backgrounds of some of the young offenders and call for moderation. This is not to say that some mayors will not be tempted by the politicization of crime control at election time.

What Are the Police Doing?

Riots and disorders in France result in the questioning of political authorities, and the politicization of a fear of violence calls into question the job done by the police. It is obvious that the police do not repress or even slow down delinquency, juvenile delinquency in particular. French citizens feel that the police do not care about larceny, small burglaries, and "crimes against the quality of life." Only 14 percent of all cases of theft in 1993 were clarified. Citizens cannot help but connect their urgent demands for better police protection with the extraordinaryly small number (5 percent) of community police officers who are able to reassure them and the three days a week worked by ordinary police officers (Monjardet, 1996: 150, 224, 236). It is then easy for the far right to play on citizens' discontent and politicize the question of public safety, fed by all kinds of alarming rumors disseminated by the media. The French are somewhat less punitive in the polls than their American counterparts: 48 percent of them (55 percent in 1995) versus 75 percent in the USA support the death penalty (*Le Monde*, 15 September 1999). But their repressive attitudes, even if latent (the structural analysis of the causes of delinquency still prevails among the public and even among the young police officers over the idea that delinquents' parents are totally personally responsible) are fueled by demagogues and the media.

The police claim that it is not in their job to repress delinquency, which is a societal problem. As one French police chief remarked:

> We are not here to put delinquents in prison. They belong to our society. Fifteen years ago, the police were only reactive. If young people are disturbed, we have to make them understand that they have to incorporate themselves in the mainstream . . . It is out of the question to support a zero tolerance approach . . . We forbid ourselves to follow such methods. The police hierarchy must calm down their men and, when confronted with 50 youths armed with iron bars, the only civic reaction for the police is to leave and not to treat the problem when it is inflamed: it would only worsen the situation and make it impossible to redress later on. The police refuse to contemplate a Pyrrhic victory.[5]

This attitude, shared by the police elites, does not necessarily trickle down to the young police officers, who are tired of being humiliated, sneered at, and physically attacked by disenfranchised youths.

But, unlike the Kerner Report in the USA and the Scarman and MacPherson Reports in the UK,[6] no inquiry commission on urban violence in France has directly emphasized the role of the police and its racism as a catalyst for disorders. In previous research, I have argued, however, that the quality of the relationship between young people and the police in poor neighborhoods was the best predictor of social peace or of disorders (Body-Gendrot, 1993a; 1996). Unlike the British police, whose community police officers send useful information from the field,[7] French police are not visible enough in poor neighborhoods. The Brigades anti-criminalité, or BAC, claim that they have to proceed with "a reconquest of the territory." But the very appearance of police in a delinquent neighborhood is an aggravating factor. Obsessed by the fear of committing abuse, many young, untested policemen prefer not to intervene in cases of disorder. The action of repression is then left to specialized anti-riot police forces, the unpopular CRS.

The major problem comes from the fact that the police have been trained to confront organized crime, but not small, urban, large-scale delinquency,[8] which is a delinquency of another type; it is not the start of a continuum that links stealing pigeons as a kid to becoming Al Capone in Chicago. The current delinquency of adjustment or of exclusion is produced by a shattered social fabric, by a non-transmission of norms by the family, the church, the party, the school, or the labor market. Social control in the neighborhood is jeopardized by the flux of populations moving from dormitory communities to jobs or to leisure parks and by the crisis of unemployment or of migration which annihilates the family's authority. Law is not transgressed, it simply does not function anymore. Whereas traditional crime, as studied by Shaw and McKay (1942), marked a departure from law-abiding citizenship, this delinquency expresses a normative dissolution. In the first case, two professionals, the cop and the gangster, confront each other; in the second, the very social fabric is vanishing, and the police share with others responsibility for bringing the ghettoized populations back to the normative mainstream. This is what the majority of the French police and their unions refuse to do. Confronting organized crime gives average police officers the status of Inspector Columbo. Specializing in petty delinquency, they think, will make them a petty cop, a second-rate one.

Any reform in favor of community policing will require massive support from the very top (monthly meetings with the Minister, for instance, massive communication work, merit awards) to change the French police culture and weaken the intermediate hierarchies' power. The current training of 17,000 police officers to replace those retiring, in addition to 16,000 new adjuncts to be sent to designated urban zones in twenty-six *départements* at the end of 1999, may be helpful.

The situation of helplessness and the crisis of authority that France is experiencing is not a new story for the USA. There used to be a time when, as in France now, 80 percent of cases were closed and when the public in major cities was exasperated. After the "nothing works" phase, a series of alternative policing measures were tried, the most famous being community policing, designed to relegitimize the police in urban areas. One major idea was to delegate decision-making authority to beat patrols in order to increase their efficiency and to concentrate their energy on problem solving (POP – problem-oriented policing). Instead of reacting to external stimuli, the police have to establish links between incidents in order to initiate proactive strategies to solve the problem. Another option, already demonstrated in New York and Chicago, consists of reacting to petty delinquency and "quality of life" crimes in response to citizens' demands. Finally, the co-production of safety implies partnerships under the leadership of the police. Safety and care rather than coercion characterizes a "soft" line of policing. Ideally, community policing relies on a beat patrol and sometimes on Japanese-type *kobans* (cf. p. 249). It takes the lead of initiatives in the community; for instance, beat meetings in Chicago allow citizens to express their needs to the police and become mobilized to take measures to improve their own safety. The Citizen Oriented Police Enforcement (COPE) program involves having a small team of police identify the sources of the fear of crime in a community after talking to the various residents. This approach is not without problems. Pragmatically, community policing has only been partially adopted. Citizens have shown their will to reconquer their neighborhoods and repel drug-dealers. They have held vigils and participated in neighborhood-watch schemes. Some television programs have helped in the search for culprits. Local ordinances have attempted to protect public space around schools and other institutions (Brodeur, 1996: 305–11).

Even if community policing has been adopted by only a fraction of the local American police, it demonstrates a capacity for innovation, the recognition of a social problem, and the identification of a solution, a capacity which is atrophied in France.[9] As Monjardet remarks, except for a few symbolic gestures,

> Alternative policing within the *politique de la ville* has not been taken into account by the institution . . . What we get are internal and inefficient policies and the refusal of an alternative policy when it is initiated externally . . . The major difference is elsewhere, however: it is in the system of decision-making which combines three levels in the US: the federal with the leverage of giving grants to the local police adopting community policing (enticing); the mayors, forced to take into account the quality of the relationship between the police and the population with the minorities' political weight (level of decision-making); and the police chiefs who need innovations to get positive grades that will help them on the labor market . . . All these elements are missing in the French system. (1996: 255, 258)

However, Brodeur is cautious in his evaluation of alternative policing. While community policing has been successful in reducing the fear of crime, it has been unable to lower crime rates per se. Researchers now dissociate the fear of crime from crime per se and from incarceration trends which pursue their own trajectories. Based on higher police visibility, better contacts with the residents, the prevention of petty delinquency, and the use of "soft" methods of persuasion, community policing reassures local residents. But only an interventionary police with professional expertise can fight serious crime. Brodeur explains that approaches vary widely: community policing reactivates informal social control in a community, targets youths who are disrupting the quality of life, aims at increasing the visibility of beat police officers on the streets and their continuity of action with tenured policemen living on site, talking and living with the residents. Conversely, an interventionist police force maintains order in critical situations, targets hard-core criminals, intervenes from outside the community, using violent force if necessary (Brodeur, 1996: 324–7). Brodeur then formulates what he calls a "double negative:" it is wrong to claim that the state does not act on drug-trafficking and drug-consumption. State officials claim not that they are efficient but that they simply want to enforce prohibition. Even if it cannot be measured, there must be some law-re-enforcing effect in the state's refusal to accept transgression. Consequently, police repression projects the will of the state rather than the efficient production of a social order. Its real goal is to make prohibition visible and "to give the appearance of policing."

The same could be said about social prevention in France. It is not efficient, it is the projection of the state's will to act. Its real goal is symbolic, it tells the community that the state has perceived the problem and is doing something. But this policy of prevention does not result in efficient social regulation, it only appears to do so.

The Demand for a Better Judicial System

In France, as the state becomes more modest, it is often thought that the justicial system should compensate for its deficiencies. Conflict then can be perceived as an opportunity for (re)socialization. "Democratic society is based on a secret renunciation of unity, on a covert legitimation of its members' fights, on a tacit abandonment of the hope for a political unanimousness" (Gauchet, 1985). It is no longer in the body of a monarch or in the unity of the nation that a democratic society recognizes itself, but in the capacity of the political community to build order from accepted divisions (Garapon, 1996: 46). The transition from divisions to a new unity could be found within the institution of justice. As democracy becomes more heterogeneous and diversified, it is not order but disorder which becomes visible on the new scene.

When the institution of justice is closer to citizens, the power of the judge gains their favor. Claims for rights are diversified enough to kill the illusion of a universal solution. Yet, too much law is as perverse as too little. Too many rights destroy the notion of rights. Too many dissenting powers annihilate a common vision (Garapon, 1996: 49). What is very clear from our study is that ethics do matter in this *fin de siècle*, as illustrated by the resistance of some American judges and, even, part of the public to the politicization of law and order (Estrich, 1998: ch. 3). Still, judges' submission to US law-makers is intriguing for Europeans. It shows contrasting conceptions of democracy's legitimate representatives. If law-makers supposedly express the people's thirst for punishment, then judges are simply asked to apply the law. This is the American version. If, as in France, justice's independence is valued and needs to be protected from political infringements, it is because the judicial institution is thought to compensate for the deficiencies of the people's representatives in parliament and government. It is revealing that the ordinance of 1945 prioritizing education over punishment for juvenile delinquents – somewhat obsolete in that respect and in need of reform – remained in the sole jurisdiction of the Justice Department, lest its reform become politicized and distorted for political gains, should legislators rewrite it.

Does the judicial system then fulfill the need for ethics by default? Is it more than just an administration of the law? Is justice the only institution with the symbolic capacity to protect democratic ideals, to take responsibility for abandoned citizens and to reconstruct logic (of life and death also)? Then it has to send collective signals to city residents. Justice becomes an "identificatory institution," according to Charles Taylor (1994: 93), when the symbolic systems of democratic societies are eroding and when commonalities vanish. In the last resort, and after all the disenchantment surrounding the loss of the social compact, justice is called to cement cohesion. On the other hand, the symbolic function of judicial authority has to be shared with citizens. Take jurors: As Judge Henry Leclerc, the President of the French League of Human Rights remarks:

> Those who, the day before at the bar next door railed against the rise of delinquency and demanded the harshest treatment to nail these troublemakers, now attempt to find the real face of those men and women who talk to them; they question what is fair and unfair and what the function of sentencing is. There are few sites where the evidence of democracy is read so openly. Social responsibility makes citizenry. The voter is more of a citizen than the opinion poll respondent, the elected city councillor more than the resident, the mayor more than the city councillor. Citizen conscience is more important for judging than professionalism. (Leclerc, 1995: 45)

Incorporating citizens in justice functions is already done through citizens' juries in the UK and in Germany, in community courts in the USA, and more and more in Houses of Law and Justice in France.

The "Community" as a Panacea for the Co-Production of Safety

Can civil society and direct democracy be a remedy when residents no longer trust their institutions? We know from Tocqueville that democracy is not only a political regime but also a form of society. The very existence of formal institutions helps us forget that other answers to problems are possible.

> Institutions are based upon convention . . . For most of the 20th century, [they] have been surrounded by a sense of their own appropriateness and transparency . . . but institutions and their regimes are not unshakable nor beyond challenge, particularly when they fail to serve needs, contain conflicts, or answer troublesome questions in a way that is perceived as satisfactory. (Garland, 1990: 4)

In the spheres of welfare and punishment, a growing sense of doubt, dissatisfaction, and frustration is expressed by the feeling that "nothing works," what Stone calls "the overwhelming evidence of . . . social dysfunction" in matters of violence and delinquency reduction (1987: 10; see also Cohen, 1985: 284).

In contrast to France, ever since Jefferson, Americans have always been suspicious of the state and a defining part of the American tradition is to hold individuals as well as communities responsible for their own fates. Contrasting the liberal and the republican traditions, Michael Sandel says the former starts by asking how government should seek principles of justice that treat citizens fairly as they pursue their interests and ends, while the latter begins by asking how citizens can be capable of self-government and seeks the political forms and social conditions that promote its meaningful exercise (1996: 27). This tension between the tyranny of the majority and its institutions and empowered citizens, agents of self-rule, permeates the sociological debates which started with the Founding Fathers and continue today with critics of mass society, social order, and solidarity. What Sandel calls "civic republicanism" could be perceived by others as "communitarianism" and the debate deals with constitutive belonging, a citizenship which elevates without repressing, and the rigors of constructing life together in modern society. What is searched is "a process of social and political empowerment whose long-term objective is to rebalance the structure of power in society in the management of its own affairs, and making corporate business more socially responsible" (Friedmann, 1982). Empowering residents, neighborhoods, and territories to compensate for dysfunctional public services is the goal of a participatory democracy. But for what purpose? The stability of the democratic system or its transformation?

The re-invention of the notion of "civil society," the adoption of post-modern and post-industrial perspectives, and the renewed denunciations of modern individualism place the idea of trust at the center around which the contradictions of modernity are revealed. Neighborhoods' social systems as small bubbles of security could be locations of trust. The construction of trust seems urgent and difficult in our post-industrial societies, because of the feelings of anonymity and insecurity that they have generated. From gated communities employing high technology to neighborhood watches and video cameras, devices tend to reassure people, compensating for the loss of trust or for what Baumgartner (1988) calls "moral minimalism," that is the replacement of neighbors or friends by acquaintances or strangers. However, security devices may also heighten feelings of insecurity by suggesting that there are risks that need to be controled (Beck, 1992).

It is probably erroneous to trust the community – or one's sense of it as an idealized democracy – to solve problems of public safety.[10] The term "community" itself is considerably obfuscated to begin with. By 1971, two authors had already identified more than 90 definitions (Bell and Newby, 1971: 15). Our interest here is to link the notion with the co-production of safety, its meaning and its implications.

An imagined and elusive community?

A brief clarification seems necessary here. In the USA, but less so in France, the notions of neighborhood, of community, of social capital, of social efficacy are loaded with strong ideological and symbolical connotations. For sociologists from the School of Chicago, such as Park, Burgess, and others, the neighborhood is a subpart of a larger entity, that is, an aggregate of people and of institutions in a given space, influenced by ecological, cultural and political forces (Park, 1916; Coulon, 1992). In its ideal form, a neighborhood is built on emotions, traditions, and a specific history. It is the basis of an informal social and political organization and remains the major site for the enforcement of safety in the public space. Social interaction and social networks are at the roots of its dynamics (Tienda, 1991). Whether it is inserted in a wider entity on which it depends or not, it evokes an imbricated structure with moving boundaries, due to networks extending it indefinitely and to forces impacting on it (Suttles, 1972: 59).

Robert Putnam (1996), a professor at Harvard, has traced "the strange disappearance of social capital and civic life" over the course of American history since 1965. Social capital, following Coleman's definition, refers to "networks, norms and trust that facilitate coordination and cooperation for mutual benefits" (1998: 36). Putnam became well known after studying decentralization in Italy in 1970. In his book *Making Democracy Work*, revisiting Tocqueville's insights, he explained that the success of the northern Italian modes of operation lay in *virtù civile*, that is, in the trust and cooperation of citizenry that breeds social capital. By "social capital," he refers to a civic involvement on local or neighborhood issues, allowing northern Italian residents to act together efficiently in the pursuit of collective goals. Whether these goals were noble or not is not at issue, since nobody knows if they were in the idealized past. What is at stake is the evolution of civic commitment in activities revealing a social bond. What began as a bus boycott in Montgomery, Alabama, in 1955–6, for example, became a challenge to local segregation, then a national campaign for equal citizenship and the right to vote. It was a movement of self-government, an instance of empowerment (Sandel, 1996: 314). Simply, drinking coffee with the people next door and discussing neighborhood issues is not a political statement, but it suggests that a social capital exists which can be mobilized whatever the age, the race, the education, and the gender of the citizens. Collective trust is the motor of mobilization. In his essay "Bowling Alone" (1995), focussing on the USA, Putnam suggested a decline in organizational membership and activism (voter turnout, church attendance, union membership, and membership in voluntary associations) betraying a diminution of "social capital." In his view, lower membership in bowling leagues – at the same time that more individuals are bowling on their own – was a metaphor for a society that was increasingly segmented and lacking in community trust.

According to Putnam, the time Americans devote to community activities has diminished by 50 percent. Among the possible reasons for this, he cites the growing number of women in the laborforce, poverty, suburbanization, time pressures, family disruption, progress towards racial equality, the development of the welfare state, and residential mobility. Putnam bases his explanation on the disappearance of a specifically civic generation, born before World War II, that has not been replaced. These individuals were more involved in social issues, they voted more, read more, and did not change when they aged. Most important, they did not start watching television before they were 30. According to this view, there is a direct negative correlation between the time spent in front of the TV and involvement in one or several organizations.

Putnam's thesis was immediately contested on several grounds. When it comes to addressing neighborhood problems and working with one's neighbors, Verba et al. (1995) and Ladd (1996) have demonstrated that participation has slightly increased. Concerned groups do mobilize around the defense of specific interests: they flood their political representatives with letters and phone calls. Success functions as a form of therapy. It fosters pluralism, conflicts, mediation, and compromises which respond better to social micro-mutations. Conflict is to be understood as the expression of a pluralist and multicultural society, open to negotiated solutions. Putnam does not seem to have realized that America is actually more of a society of "joiners," and that affiliations are more diverse today.[11] It may be a good thing that many Americans no longer join such early twentieth-century organizations as the Boy and Girl Scouts, the Elks, etc., which were mostly middle-class and exclusive.

Putnam is not the only one, however, to lament the decline of civic participation. For years in the USA, people have nostalgically deplored the loss of "community" in the sense of a primary society founded on ascriptive links, as defined by Ferdinand Tönnies, Emile Durkheim, Georg Simmel, and Max Weber, and also including the idea of collective work on and collective solutions to problems.[12] But this lament is probably ill-founded. In the 1960s and 1970s, high-quality ethnographic studies revealed the resilience of dynamic neighborhoods with dense social networks and a strong identity (Jacobs, 1961; Gans, 1962; Fainstein and Fainstein, 1974; Stack, 1975). The theory of social disorganization developed by the School of Chicago was challenged and Wellman (1979) concluded that the meaning of community had persisted, or at least had been transformed, in order to survive (Sampson and Groves, 1989, 774–802). The community was less visible in the public space than in former days; the private sphere had gained more importance and neighbors were less necessary to perform essential tasks. Yet the neighborhood and its mythical image remained essential to the accomplishment of common goals relating to a local general interest. This is precisely our point.

Robert Sampson (1997) remarks that there is one factor that Putnam has missed. Beginning around 1965 – when Putnam sees civic life starting to drop off – crime rates began skyrocketing in American cities. Sampson argues that, above all else, crime and its consequences have had important reciprocal effects on community structure and, ultimately, social capital. Crime has symbolized the erosion of the social cohesion that communities are said to sustain; it has been associated with a crisis of social regulation. But I argue that it has also generated strong, yet sometimes ambiguous, reactions.

In my view, self-help is a typically American response. The USA is a horizontal society that works on its own problems because the state is less present in social affairs than in Europe. Thirty million Americans belong to a neighborhood-watch program, for instance, showing substantial diversity in their modes of operation. In contrast, only 3 percent of the French polled in November 1998 (cf. Chapter 3) said they would resort to self-help methods – they said they needed more police and gendarmes for their protection. What is the reason for this contrast? Beyond historical differences (which explains why, after the experience of the collaborationist Vichy regime, it is still difficult for a segment of the French population to "cooperate" with police forces and that self-help may be interpreted as vigilantism), what is at question here is, in the USA, the urgency of problems and the "virtue" of civil participation, compared to trust in state intervention in France. Rather than increasing taxes to support dubious programs or inefficient bureaucracies, American legislators prefer multiplying the number and diversity of committed actors (giving them incentives through tax breaks), encouraging citizens to involve themselves. French mayors rightly argue that tolerance of grassroots innovation means also accepting vigilantes, defensive communalism, the NIMBY syndrome, etc., and that they prefer to avoid such risks by keeping an upper hand.

Community and the fear of crime

It is at this point that one must distinguish between the positive or negative use which is made of social capital, between the content of evil associations and virtuous ones in civil society. Social capital is not value-free. Distinguishing form and content, Abu-Lughod aptly presents social actors in two pairs: the first pair sees the state as evil or fascist and organizes to defend a local space within which an ideology and a praxis of dissidence can flourish (1998). The second pair sees the state more as insufficient than as evil or threatening. The organization operates in the "local space" between the state and the market to achieve collective goals. But both the form and the content of these organizations matter. In her first pair, Michigan militias side with the residents of East Village, New York. What are their similarities? Both lead uncivil advocacy actions and search for their people's empowerment. The comparison can be stretched to skinheads in Germany and to the disruptive group, Right to Housing, in France (Body-Gendrot, 1999). The latter's goal, like that of the East Villagers, is redistributive (the organization occupies state-owned vacant buildings to provide housing to evicted immigrant families in the Paris region), while the former vacillate between fascism and anarchy. In the second pair, the gated communities' residents (close to the political leagues of hunters and fishermen in France in terms of defensive communalism) are next to grassroots community organizations. Again, the content makes the difference. The latter are inclusive and conflict-mediating, the former rejecting and exclusive, "undermining the equity value of the wider society" (Abu-Lughod, 1998).

On the issue of governance and crime, negative examples in the use of social capital are easy to find. For Crawford (1997), the *fin de siècle* crisis of modernity has found its antidote in the concept of community. Crime, as a compelling symbol of lost community, has a particularly salient place in communitarianism. Why community? Whose community? he asks. What constitutes disorder? What definition of order has priority? (1997: 148, 164, 194). It is commonly admitted in the USA that restricted entities where people from the same community think, live, and own in osmosis are better prepared to promote democracy than vast, heterogeneous units. But the twisting of public interests for privatized goals and the NIMBY syndrome, betraying a collective resistance to the imposition of collective public services, present a quite limited interpretation of what democracy is. Participation is encouraged as long as it is centered on the idea of "us." Possessive and defensive communalism suggests that private interests are best nurtured and guarded by their setting in a socially restrictive enclave of kindred interests (Plotkin, 1991: 13).

The tension over the definition of what a community and community mobilization are is recurrent. In the USA, anti-crime programs, "implant projects," constitute one of only a few options local activists have to attract federal or state funding to inner-city neighborhoods. Based on community crime watches, foot patrols, and neighborhood–police liaison programs, they employ surveillance techniques involving residents from local block clubs and community organizations taking down the license-plate numbers of suspicious drivers, contacting residents whose property falls below community standards, and identifying and reporting suspicious behavior to the police and health inspectors. In a "guerilla war that involves all of us together," a war that would be fought, "house to house, neighborhood by neighborhood, community by community" (Gordon, 1994: 203), community mobilizations have their downside. The dominant communal emphasis on maintaining order overshadows minorities' civil liberties. Community is used as a justification for greater punitiveness and arbitrary policing of those "othered" who are seen as deviant. C. Cohen mentions the homeless, and prostitutes, arrested at the demand of residents in rich as well as poor communities (1999).

In his study, Skogan (1990) found that the programs did not mobilize proportionate numbers of renters, people of color, and lower-income residents. The "respectable" elements mobilize against the "bad" and "suspect" elements. Even African-American home-owners vocally support drug raids by local police and express little outrage over the fact that African-Americans are 14 times more likely to be targeted by such a program in the Twin Cities, for example. The redefinition of community development politics and the construction of the American garrison local state is taking place in the context of the polarization of the subcommunities (Goetz, 1996). The community is rarely homogeneous, it is internally polarized. There is little consensus about incivilities associated with "other" elements. Whereas conflict may be the healthy expression of different interests and a sign of pluralism, it is left unaddressed, and power relations are left unmitigated.

Crawford wishes that community participation in the production of safety should be based on bilateral relations of trust. Some disadvantaged people have no freedom of mobility and no alternative to offer to dominant voices campaigning in the name of "moral order." Intra-community relations need to acknowledge dissent and conflict (1997: 200). This echoes Iris Marion Young's postmodern statement:

> An alternative to the ideal of community . . . [is] an ideal of a city life as a vision of social relations affirming group difference . . . without exclusion . . . If city politics is to be democratic and not dominated by the point of view of one group, it must be a politics that takes account of and provides voice for the different groups that dwell together in the city without forming a community. (1990: 227)

We are close here to the "veil of ignorance" and to the "reasonable pluralism" suggested by political philosopher John Rawls to allow questions of what is "reasonable" and "fair" to be negotiated (Rawls, 1971).

Co-producing safety It is the linkage of mutual trust and the shared willingness to intervene for common goals which defines the neighborhood context of what Sampson et al. (1997) call social efficacy. They establish a distinction between social capital – the resources and potential inherent in social networks – and collective efficacy as a task-specific construct referring to shared expectations and mutual engagement by residents in social control.

In a study carried out on 8,782 residents of 343 Chicago neighborhoods in 1995, residents were asked about the likelihood of being able to count their neighbors to take action if children were skipping school and hanging out on a street corner, if they were spray-painting graffiti on a local building, showing disrespect to an adult, if a fight broke out in front of their house, or if the fire station closest to home was threatened with budget cuts. The social cohesion concept was also tested with statements such as "people around here are willing to help their neighbors; this is a close-knit neighborhood; people in this neighborhood can be trusted or conversely, they don't get along with each other; they do not share the same values." Sampson and his team found via these measures that collective efficacy had a strong negative connection with the rate of violence in the neighborhood, that is, that neighborhoods high in collective efficacy had significantly lower rates of violence than places with low shared expectations and mistrust, weak organizational bases, low participation in local voluntary associations, etc. The association of disadvantage and population stability with rates of violence was significantly reduced when collective efficacy was controlled.

The Professionalization of Community Organizations in the USA

Low-income residents and associations in France explain that their efforts are frustrated by experts and professionals when they try themselves to make up for inadequate services. Resistance by professionals who control the services and bureaucratic agencies is present both in the USA and in France. But in France, governments never really follow through on trying to "professionalize" residents who are attempting to solve their problems on their own, for fear of bureaucratic reactions. Street-level civil servants feel threatened by citizens' participation which could show that their work is redundant or inefficient. Some organizations in the USA, such as the Pratt Institute Center for Community and Environmental Development in New York, CANDO in Chicago, and the Center for Community Change in Washington, offer professional training to low-income residents. The evolution of some community development corporations (CDCs), such as Banana Kelly in the Bronx, exemplifies what can be done.[13] But as a rule, CDCs are not in the business of social change, defined as changing the political and economic power of poor communities (Gittell and Newman, 1998).

Some foundations, such as the Washington-based Milton Eisenhower Foundation, contribute to spreading projects that work in low-income neighborhoods from city to city. One spectacular experiment in providing after-school safe havens introduced the Japanese concept of *koban* (see note 8) to US cities. What works best are multiple solutions – community organizing, advocacy, trips by police and civilians to schools, remedial education, job training and placement, youth media enterprises and sports as a way for mentors to guide young people. But such endeavors are fragile and do not meet needs, which are colossal. And when they do, a lot of community initiatives – such as Project Redirection in Chicago, the Responsive Fathers Program, the Dorchester Youth Collaborative in Boston, the Delancey Street Program in San Francisco – reveal the vitality of American civil society with its more than six million community organizations and the flexibility and pragmatism of institutions. In the latter case, the Delancey Street's success at reducing recidivism among adult offenders prompted the state of California and the mayor of San Francisco to offer the program a $5 million grant to reform the juvenile justice system.

What we see at work is an arduous way of stimulating efforts by society to solve society's problems. Federal, state, and city funds, as well as private moneys, are more and more difficult to get for community-based organizations. Banks like South Shore in Chicago, modeled on the Grameen Bank in Bangladesh, support poor residents' initiatives, but large banks are frequently reluctant to do so.

Financing Community Work

The comparative study of rhetoric is of interest here. In France, the official rhetoric continues to focus on the state as a motor of progress and on the goal of social justice, even though, as we have shown, public budgets are shrinking, more public work is being subcontracted to private entrepreneurs, and more corruption unveiled. But, national elites appeal to society's commitment. When eight homeless people died in the winter of 1998, the President said that "every French person should feel personally concerned." The Minister of Employment and Solidarity added: "All does not depend on the state or the mayor. Much can also depend on [yourself]. There's no point in shedding crocodile tears. Everyone can do something to make things better" (Trueheart, 1998).

Centuries of dependency and governmental assistance programs, however, which were meant to service poor communities, have destroyed simultaneously their social capital. Instead, the French are used to turning to the state for their needs and are less involved in community tasks than other Europeans and North-Americans. The French give approximately

0.15 percent of their income per capita to non-profit organizations, compared with 1.2 percent for Americans. The tax system offers some explanation: tax deductions for charitable contributions in France are limited to 6 percent of annual taxable income, whereas the figure is 50 percent in the USA. (Tax deductions are limited to $400 per contribution in France.) Moreover, only one household out of two pays direct income tax because the others are below the income level or cannot show benefits in France: those who are not taxed have no incentive other than altruism to donate, while taxpayers are more overburdened with taxes than any other European (or American). Another explanation is that the French trust the government more to distribute the money evenly where it is needed rather than put their faith in privately run community – or charitable – ad hoc organizations, many of which have proved to be inefficient, politicized, or corrupt.

The discourse that the state could do better has been largely diffused by street-level civil servants every time. For instance, youths from inner-city areas suggested that they could do the work themselves (Body-Gendrot, 1999). Unless there is a very high level of distress, mayors in general have also been suspicious of grassroots initiatives that they are unable to control. There are exceptions, as the case, described below of Pierrefitte, in the Paris region, demonstrates.

French Community Participation: the Cité des Poètes at Pierrefitte

I have chosen to complete this research with this story because it contradicts all the clichés one has about French civil society – which counts 700,000 non-profit associations – and because it may anticipate many changes currently taking place in a society torn apart by many forces, including globalization, Europeanization, immigration, Americanization, distrust of government, and the will of the younger generation to act differently. It shows that the act of participation – as part of a group or in a community – by its associational aspect, is more likely to result in an enhanced sense of efficacy, which in turn encourages further participation (Gittell, 1998a).

This rather unusual story is that of the reconquest of a territory, a story of residents who became actors in their collective fate. It is a universal story, which in more ways than one also reflects other ignored, modest, innovating democratic forms of resistance. Every day, such scripts link themselves to one another around the planet and ensure that hope does not disappear from problematic areas. It is a story that has been singular enough to cause a sensation in France.

In the beginning of the 1990s, the Cité des Poètes, a housing project in the town of Pierrefitte, Seine St-Denis, would have been perfect to illustrate the broken window theory (see Chapter 4): "About fifty youths exerted their domination on Brassens Square, at the core of the neighborhood." Although they did not directly target institutions, "they drank, they had dogs, they fooled around . . . we were scared to death for our children and also for our wives," a resident complained (Dollé, 1998). Too many assaults, the absence of a police presence, the inability of adults to respond (except for petitions sent to local authorities), educational and administrative laxity, and the deficiencies of public-housing managers all contributed to a downward spiral that made collective life unbearable. Mistrust or fear – what Baumgartner (1988) calls "moral minimalism" – prevented residents from intervening in their collective space.

This neighborhood of Pierrefitte was not much different from many other pauperized areas in the largest cities of the world. It suffered from isolation – the residents lived at the margins of the larger city – and from a concentration of socio-economic handicaps, as in Chicago or Harlem (W. J. Wilson, 1996). Immigrant families were overconcentrated in public-housing units, having settled there as a result of the pressures of bureaucracy or their inability to find anything else affordable or acceptable on the rental housing market. The area suffered from both public and private disinvestment that resulted in a lack of services, most acutely, police services. Stigmatization linked to space also took its toll through unemployment and ill-paid jobs.

Residents' involvement in the co-production of safety

As in other places in other countries, the residents' mobilization followed six steps:

(1) For mobilization to get going, a triggering event is needed (Alinsky, 1976). A French television station's coverage of the situation at Pierrefitte – which presented a very negative image of the neighborhood – played that role in 1993. When they saw themselves on a prime-time program, the residents were shocked. Those who had spoken to the journalists felt guilty. Such coverage reveals the ambiguity of television: it demonizes the residents of problematic areas, while at the same time staging their grievances, dramatizing their pain, and letting it be known to the rest of the country (Body-Gendrot, 1993a: ch. 5).

(2) For the residents to mobilize, a social organizer has to find common sores and rub them. Alinsky emphasized the need for well-trained organizers to build confidence and hope among the people. Here, the

head of the social center played that role. And concerns about street violence committed by juvenile delinquents united all the residents. Everyone was concerned because these adolescents were often their own children or those of their neighbors. I would like to underline that, to my knowledge, mobilization around the issue of safety is not common in poor areas. Who wants to call the police when they are perceived as enemies? Who is not afraid of retaliation? There are ambiguous attitudes in the community, here as elsewhere.

(3) A community mobilization is not a calm, uphill operation. It works in leaps and bounds, it backtracks. The decision by residents to provide a social center run by youths themselves proved disastrous. The same failures have been observed in different places: either the youths in charge are quickly overwhelmed by a more cunning, fiercer, or stronger gang, or the leader acts as boss and imposes his own decisions, or the place becomes a shooting gallery, a center for all kinds of deals. Here, the residents responded very positively to the problems at the social center. They asked city hall to close it for six months, before reopening it on another basis with stricter rules.

(4) Transforming young negative heroes into positive characters, involved in community solutions, became the central project. This approach was used after the Los Angeles riots of 1992, when the Bloods and the Crips gangs developed the best plans for rebuilding the community of South Central. More recently, demonstrations by youths in the Paris region, on a "Stop Violence" campaign, expressed the same desire of some inner-city youths to transform their image and let the whole world know. At some point, adults have to trust these young people.

(5) Profound distress is frequently the first catalyst for the residents' involvement when institutions become dysfunctional. In Melrose Commons, in the South Bronx, people were their own firefighters, their own police, the collectors of the garbage that people from other neighborhoods came to throw in their neighborhood. They had no choice. "In distressful places like these," Yolanda Garcia, the organizer of Nos Quedamos, a grassroots organization, would say, "if we believe in a cause, our projects will take shape . . . We had people sign petitions, we held surveys during snow storms, we met five nights a week, sometimes very late. In the winter time, the place is unsafe and night falls early. Yet hundreds of residents came to our meetings, because each individual felt concerned for specific problems and wanted to speak his/her mind. We have a community vibrancy here, because we are all survivors" (Body-Gendrot, 1994). Similarly, the co-production of safety, or self-help, at Cité des Poètes is part of a long fabric of mobilizations woven through the collective memories

of the neighborhoods. Residents seized the right to put an end to violence in the collective space and oversee the children's socialization according to rules that they respect. The social center was the catalyst for the *cité*'s efforts to improve itself. Residents met, created two organizations, and decided to reconquer the public space. Fathers and older brothers would check the youths in the evening, mothers would work with the younger ones and the artists in the project, partnerships would be formed between the social center and the schools nearby, and efforts would be pursued to improve the relationship of the residents and the local police. The teamwork there is similar to that of the Youth Guidance programs in schools in tough areas in Chicago. Its multidimensional approach suggests that no one actor can succeed alone, while working in solidarity provides actors with the energy necessary to confront crucial problems. Immediately praising any kind of success redresses the permanent feeling of failure. Dynamics prevail. The residents kept repeating that they were not the problem but part of the solution.

(6) That the moblization of residents might falter after yielding so many positive results is part of the normal cycle. Residents get discouraged and stop applying themselves to every action that seems framed by professionals or committees. This reminds me of a small experiment that I observed in Diadema, near São Paulo. For years, mojabitos built their homes themselves, quite anarchically by French standards. Some time ago, the left-wing municipality had social-housing projects built on nearby vacant lots. The city offered the potential tenants the opportunity to organize and decide for themselves what color doors they wanted or how they wanted their playgrounds shaped, for instance. The tenants adamantly refused the offer. When life is hard and the first priority is survival, meeting after work for collective decisions is a luxury. Time, energy, and the capacity to listen to others' points of view are not equally distributed in the city. The defense of noble causes crumbles under sordid calculations and rivalries. This is the darker version of community participation.

The positive value of multiculturalism

The experience of the Cité des Poètes demonstrates at its best the importance of shared social capital and of social efficiency. One could argue with Sampson (1997) that there is no socio-economic determinism, implying that low-income residents expressing their solidarity cannot develop initiatives to take their streets and public spaces back from drug-dealers and gangs and take action to protect their young. Poor and racially segregated communities can express "social efficacy" as well as others. Free time is a necessary ingredient for the consolidation of such attitudes that are constantly shaken by external and internal forces (like unemployment or jealousy). Yet it cannot be denied that space effects may have a positive influence.

At the Cité des Poètes, residents come from many diverse countries and 56 percent of the population is of foreign origin, with a large contingent from Africa. Multiculturalism favors the sharing of ideas, of know-how, and of how to know. We are told that in that city, "things have been tried, new modes of operation invented, the term citizen redefined, participation redesigned, collective life improved" (Dollé, 1998). I claim that this "invention of daily life" in the sense of millennial cunning (*ruses millénaires*), as defined by de Certeau (1980), is facilitated by multiculturalism. Residents submitted to constraints reappropriate the physical and symbolical spaces where they live to mark their identity according to strategies that vary with generation, gender, class, and the social environment. They adjust, reinterpret, and retranslate their environment and daily experiences as a function of their imported memories. Each generation of residents produces its own sense, constructs its specific history and identity. Spatial concentration and the succession of immigrant populations in these neighborhoods favors mobilization. Integration in the receiving society varies according to the diversity of cultural origins. Adaptation to daily life in a given space, forms of resistance expressed against the articulations linking economy, social occupation, and national origins cannot be dissociated from the space in which they emerge. While networks move beyond geographical boundaries and persist across generations, migrants are supported by more or less efficient survival strategies.

The African fathers and mothers of the Cité des Poètes are similar to those from other neighborhoods in the world, sharing the conviction that a whole village is needed to bring up a child. When they left their homeland, they lost the traditional notion of social control. In Africa, it is exerted, acknowledged, and validated by the community. The African parents of Pierrefitte organized night watches and sent children back to their parents' home. These watches, a form of curfew for the youngest, look like the initiatives sponsored by organizations such as (Mad or) Responsible Fathers from diverse backgrounds who patrol the streets at night in the US to prevent adolescents from falling prey to the street culture. Once they reconquer their collective space, according to Jane Jacob's (1961) expression, residents become "the eyes of the street" and collective trust takes over.

The fact that Tabib, an Iranian political refugee and a former lawyer, could become the head of the social center in Pierrefitte is not without importance for the mixing of ideas, the risks taken in the approach, the innovation. The import of ideas from other contexts contributes, indeed, to advancing the search for solutions and the imagination that it can be done. Among exiles, micro-solidarities, adjustments, and arrangements emerge. At the same time, learning the cultural modes in French society is a slow, often painful process. For instance, it is difficult for African fathers to learn how to moderately punish their adolescent sons. As other immigrant fathers before them, Algerian fathers here, Mexican fathers there and others in the world, they are asked by street-level bureaucrats to respect the rule of law of the receiving country.

At the Cité des Poètes, I met the sister-warriors of immigrant women in charge of community organizations whom I had come across in a previous study (Body-Gendrot, 1993b). In the low-income neighborhoods of the Paris region, they developed survival strategies and reappropriated the spaces in which they lived. They defined themselves as between two cultures or two generations. They said they were universal, fast, energetic. "We are all the communities at the same time and all the generations as well," they would add. I thought that they were "speeding mutants, hyphenated women." Ms. Sabia, who started a soccer team, could be the sister of Ms. Berebout, an older Algerian woman, involved in the training of police officers and educators in Marseilles, or the sister of V., who created a hairdressing/music parlor for the young in Nanterre because, she says, "brushing and playing the violin" represent the same cause. "If a people loses its cultural memory, its books and its music, this people is extinct, it has no expression in daily life. It is dangerous. Whereas what we mean is, 'We want to stay here with what we are,'" V. said. Such women invent every day a more congenial, serene, secure life out of a chaotic environment. Yet their vigilance must never cease. Older artists, the residents of the community, also provide intergenerational and gender linkages.

A social laboratory

Along with the place in which it is embedded, the Pierrefitte story is interesting because of its timing. It has taken place in an era marked by doubt, questions, and transition for French society. Somehow, the micro-local arena anticipates evolutions which will take place throughout the country in the years to come.

A demand for sanctions One factor that emerged from that Cité des Poètes episode is that the residents in this housing project expressed their fatigue toward juvenile delinquents. "Men and women living at Cité des Poètes, like many others, can no longer hear discourses on the 'tolerance' of phenomena of violence, discourses excusing everything, while discouraging accountability. This attitude is completely passé. As generous as it is arrogant, it is useless" (Dollé, 1998). This statement is firm and violates a taboo still widely supported by leftwing intellectuals not living in those areas and who are thus easily able to be generous and compassionate. It is still irrelevant for them to express sympathy for the victims' victims – residents, storekeepers, or public employees who are scared – rather than for the young delinquents who "when they knock, knock on a door," as a psychoanalyst put it, adding "and I want to be that passage since they knock to be certain that the world exists or that they exist". This is an admirable conviction coming from someone who unwillingly justifies the denunciations of the troublemakers forcefully formulated by the far right. A majority at the Cité des Poètes are in favor of sanctions that should reflect a third way between the criminalization of juveniles and institutional laxness.

Residents and professionals A second factor that emerged relates to the difficult relationship of the residents and the social caretakers. According to pyramidal schemes in France, broad designs are conceived at the summit. The conceivers want to further the well-being of the people without consulting them. "On the one side, there would be the poor, with all kinds of meaning given to the term, and on the other, maybe, those who know, the professionals with tools for doing what is necessary" (Dollé, 1998). This attitude, which was widespread hitherto, is beginning to change; police, social workers, and even sometimes national education institutions, are learning how to delegate and work with people. No one would say that the police can be replaced by self-help. In a state of law, safety is like freedom. It is a right. People are not supposed to extinguish their fires or do their own policing, unless they experience "decivilization".[14] But for residents to get involved in the co-production of safety, social services need to help them solve crucial problems and become more confident in their own abilities. A protected environment, support to parents, health clinics, family budget advisory committees, educational aid, and convenient public transportation may contribute to those goals. In that perspective, the binary or colonial logic of "us" and "them" expressed by some professionals cannot be tolerated. Work in problematic neighborhoods is another kind of job requiring adequate resources, an understanding of the situation and an empathy for the residents. Public-housing managers, for instance, can improve or ruin the tenants' lives when it comes to safety, based on the view of community participation that they have.[15]

Finally, some education is needed to "unfreeze" mayors and councilors and eliminate their fears of politicization every time residents take action in the public space. Elected officials interpret such initiatives as the formation of an opposition which will be heard during the next election campaign. Few attempts have been made to recognize the fact that residents have skills as actors in the city. Regarding residents-actors (including the youth) as critical interlocutors and supporting their political representation reveals a healthy, conflicting practice of democracy. In the USA, administrations at first were reluctant to grant "maximum feasible participation" to low-income residents at the time of the Great Society of President Johnson. Later, at the local level, police forces would despise the Guardian Angels who, they said, were amateurs.[16] US local officials also felt destabilized by grassroots initiatives. But the acuteness and breadth of the problems made them welcome the commitment of a diversity of actors. Former dissident militants became involved in numerous service-delivery tasks. Some of them were even elected mayors. In the name of civic culture, convictions, pragmatism, or survival, everyone found some benefit in seeing problems reduced.

It seems that in Pierrefitte, the mayor also understood the nature of the residents' mobilization and their strong collective will. Neighborhood boards were formed, anticipating a more intensive evolution of civil society's participation in France.

The Virtue of Civil Society or Democratic Resistance

In a third scenario (see Chapters 1 and 2 for my two earlier scenarios), I describe how times of crisis call for the mobilization of local democracy and for the development of a political culture of solidarity that embraces new collective practices enabling residents to express their citizenship. Resistance is the primary way global forces are mediated at lower spatial scales (Beauregard, 1995: 235). Ruby, a French philosopher, offers the image of the archipelago to refer to solidaristic dynamics (by gender, age, race, neighborhood, workplace) and to a new way of thinking and living in the city. The notion of the archipelago makes explicit the authority of an infinite movement coordinating common actions which are constantly rebuilt, a "doing together" actualized, not a "being together." It resists a laissez-faire attitude (Ruby, 1999). Yet in France, provoking the involvement of a society molded by the state and its administration for so many centuries, is a slow process. Sometimes, the concept of participation is perceived as compromising with a logic of struggle and of claims, as an injunction formulated by the "haves" to the "have-nots" or as an instrumentalization of the latter's energy. Tocqueville praised mobilization by citizens who are organized, acting as quasi-professionals, and given legitimacy. These mobilizations go against the notion of representative democracy: the public interest is pursued by the people's representatives. Yet, once elected, they rarely regard themselves as accountable. With the crisis of institutions and the end of a model *à la française*, however, convergences in interactive practices of local governance are taking place on both sides of the Atlantic.

Conclusion

In this comparative examination of the correlation between globalization and urban disorders, European cities emerge as robust and better protected from deadly competition for several reasons. First, as splendidly demonstrated by Max Weber (1958), the European city, which from the start formed a local and political society, has continued to function as an arena where interest groups are amalgamated and represented "together in difference," and where the term "citizenship" is not devoid of meaning, despite all the trade-off necessary to sustain it in the urban imagination (cf. St-Denis and Lyons). With the difficulties experienced by nation-states, even if the city today is no longer an enclosed space as it was in the Middle Ages, this imagination is returning.

Cities founded during the Middle Ages are still the backbone of the current urban fabric in Europe. There, industrial urbanization has had a lesser impact in terms of segregation, for instance, and the cycles of urbanization have been more homogeneous. The most blue-collar neighborhoods in Paris today were already working class a century ago; the same is true of Lyons and Marseilles: there is demographic stability (Body-Gendrot and Beauregard, 1999). In contrast – and Jean Baudrillard (1986) brilliantly developed this theme – in the American city the "urban imagination" is stimulated by a hyper-reality of simulations and simulacra into the real, creating imagined spaces. The city lives in the present and projects itself into the future.

This is, of course, a rhetorical generalization. New York, Boston, Chicago, and San Francicsco are cities apart. There, the tracks of prior cities are still visible through the collective weight of the past, a resistance of the agglomerated form, and a resilience of the core, perceived as the valued culture of American cities. They are close relatives of European cities in this respect and, as Paul Berman once remarked, "It is the strange tenacity of its ancient neighborhoods [that] governs New York's creativity . . . The determinism is geographical. Ghosts run the city" (1988). Well-off residents have not deserted Manhattan or the Loop. Greenwich Village retains a bohemian culture, much like St-Germain des Près.

From the perspective of architectural shapes and identity, cities differentiate themselves because they are in cultural competition. In that respect, Paris is more Paris than ever and New York is more and more New York, as Castells remarks (1998b). The process is specific and local. It is the relation between flux and space that makes the new city. This is whv the identity and modes of operation of New York are different from those of Chicago and from those of Los Angeles. A prosecutor or a judge does not behave in the same way in Dallas, Texas, or in Brooklyn, in Paris, or in Nice, in London or in Liverpool. Expectations, pressures, and collective cultures have a lingering effect on individual behaviors. They reveal the power of place. People invest places with social and cultural meaning (Hayden, 1996: 78). "Material, representational and symbolic activities . . . find their hallmark in the way in which individuals invest in places and thereby empower themselves collectively by virtue of that investment" (Harvey, 1991: 39). Apart from non-places (in reference to which Gertrude Stein has been oft-quoted as saying, "There is no there, there"), cities (yet fewer and fewer) have an identifiable and instantaneous identity.

Different historical arrangements affect the current differentiation and social control that cities choose to exert. The civic cultures of the cities, the traditions of community activism or philanthropy, and the cities' repertoires of surveillance mold the tools that they use to deal with the problems of fractures and of social deviance generated by global forces. As a consequence, one observes an increasing differentiation of cities. One should add to local politics, institutional and demographic arrangements, the role of local elites and of some charismatic characters to explain why "only in San Francisco," "only in Marseilles" could such and such innovation take place. This is what emerges from city effects.

Second, spatial polarization does not seem to be as severe in French cities as in American cities, due to state authorities' master plans, incentives, and national policies mitigating centrifugal forces, and to the unwillingness of the middle classes to exile themselves to the peripheries of cities. The growth of inequalities does not produce a dichotomy between the yuppies and the homeless, as it does in the USA. And it has been shown that state transfers in France in fact have kept pauperization under control since the early 1980s. At the same time, the destruction of industrial jobs and the lack of corporate dynamism in France have imposed enormous costs on cities in terms of unemployment, social destructuring, anomia, violence, etc. Segregation and polarization processes do take place, and sometimes at the initiative of state policies (cf. St-Denis; Lyons), but for the last eighteen years various mechanisms orchestrated within the local governance initiated by the *politique de la ville* have allowed cities to slow down destructuring trends.

Third, a great majority of the French support the welfare state and state intervention as proof of social cohesiveness, an obsessive topic for them. Symbolically, the rhetoric of government officials and parliament members places solidarity and solicitude with the disadvantaged urban areas at the core of national efforts, and national elites try to draw their European partners into the consolidation of a social Europe. The European Community has become involved in assembling greater resources for distressed urban areas; specific programs backed by financial transfers contribute to their regeneration (cf. Marseilles) and, most of all, regional partnerships allow the sharing and coordination of resources, ideas, and modes of action. Cities also emancipate themselves from confining national parameters and use opportunities to redefine their role. "With cities, it is as with dreams: everything imaginable can be dreamed but even the most unexpected dream is a rebus that conceals a desire or, its reverse, a fear," Calvino once wrote (1972: 44).

At a time of increased world exchanges and financial flux, cities appear, then, more and more as sites for the modernization of institutions and for innovative practices in social control. But far from being palimpsests, cities are continuously working upon themselves, each in its own way. History has told us that to each generation of city corresponds a new vision of the world. Not all the choices are yet made. It is at the core of the city of the new millennium that the future of democracy and of cohesive values will be played. Future generations will have to dream of the Third Age city before it actually takes shape.

Notes

1 This remark refers to a brilliant essay by Delmas on the modernity of public action, *Le maître des horloges*, 1991.
2 It is difficult to make a comparison with the USA, which has 50 different kinds of local government. But, according to the 1990 census, the 19,296 autonomous municipalities (1% of which represent 40% of the urban population) added to 16,666 townships would mean a total of about 40,000 "mayors" in the USA.
3 The situation is similar in the UK where there are no black or Asian chief constables in the police force and minorities make up 2% of the police force (*The Economist*, 24 October 1998: 46).
4 Interview with the author, fall 1997.
5 Interviews with police chiefs, spring 1998.
6 The Kerner Report was issued to President L. B. Johnson after a series of riots erupted in American cities during the 1960s (Kerner Report, 1968). The Scarman Report was written in the UK after the Brixton disorders of 1981. The 1999 MacPherson Report is a UK government-appointed inquiry on the handling by the Metropolitan police of a racist murder. Stephen Lawrence, a bright and popular black student, was stabbed by five white jobless youths in south London in April 1993. After five years the police had still failed to reach a conclusion. In all three reports, the racism of a largely white police force was stigmatized and blamed as a cause for urban disorders in minority communities.
7 In the fall of 1998, Home Secretary Jack Straw announced recruitment targets for black and Asian police officers in the UK. After the publication of the MacPherson Report, he aimed at compelling police chiefs to tackle the

problem of institutional racism in the constabularies (interview with the London Metropolitan force Superintendant, Mr French).

8 These considerations are expressed by Monjardet (1996: 245–8).

9 The example of *kobans* illustrates the innovation process. *Koban* police officers live in the low-income housing projects with their families. The *kobans* operate 24 hours a day. Educators, social workers, and volunteers are present to reassure the population. Adults act as mentors for the young. These community police officers are often the only male adults who dare be present at night in the community and coordinate recreation such as basketball and other athletic activities. The concept of "community equity policing" implies that community organization and police work together on an equal basis, sharing resources and control, and providing safe places for the young. The police serve as counselors from 1 to 10 p.m. and the programs are run by community or non-profit youth organizations. As a result of the *kobans*, in tough sections of Baltimore, Boston, Chicago, Philadelphia, San Juan, and Puerto Rico, serious crime dropped by 22% to 27% (Curtis, 1997). As a result of the initial success, police chiefs traveled to Japan and HUD is currently financing safe havens in other low-income public housing around the country. A *koban* project in Rotterdam has met with similar success.

10 I follow Plotkin (1991) on this critique of the concept of community.

11 As for the general involvement of Americans, according to Gallup's Survey of Giving and Volunteering, about 54% of Americans, 18 years old and over, had done some volunteer work during the 12 months prior to the survey (Hayghe, 1991: 19); almost one elderly person out of two said he/she does volunteer work (Broder, 1997); 10 million people and 100 million volunteers work for non-profit organizations in America (*The Economist*, 14 November 1998: 85); 89% of Americans admit that they want to spend time and make efforts to improve the life of their neighborhood, while 52% add that they already do it twice a week (Stoskopf and Stern Strom, 1990). Many programs following the "adopt a child" and "adopt a building" models provide links between social categories and sectors that would never otherwise meet.

12 I follow Sampson (1997) in this development.

13 Banana Kelly restored a deprived neighborhood and created jobs for the residents (Body-Gendrot, 1994; Harris, 1995).

14 I allude here to the thesis Elias develops in *The Civilizing Process* (1939).

15 Public-housing managers can allow juvenile delinquents to make up for their offenses through community work in their project, or refuse to consider that option; they can work with the mayor or not, send in trained janitors, or reduce their numbers in the buildings, etc. Too often, 20–30 different managers are in charge of public housing in the same locality, and incoherence prevails.

16 The Guardian Angels – like the French Older Brothers – consisted of young minority males providing a sense of security in subways and streets. Police unions saw them as a threat to their jobs and after a few scandals they disappeared and were replaced by community police officers.

References

Abramsky, S. 1999: When They Get Out, *Atlantic Monthly*, June: 30–6.

Abu-Lughod, J. 1995: Comparing Chicago, New York, and Los Angeles: Testing Some World Cities Hypotheses. In P. Knox and P. Taylor (eds), *World Cities in a World System*. Cambridge, Cambridge University Press.

Abu-Lughod, J. 1998: Civil/Uncivil Society: Confusing Form with Content. In M. Douglas and J. Friedmann (eds), *Citizens in Cities*. Chichester, UK, J. Wiley.

Akers, R. 1997: *Criminological Theories: Introducing an Evaluation*. Los Angeles, Roxbury Publishing Company.

Alinsky, S. 1976: *Manuel de l'animateur social*. Paris, Le Seuil-Politiques.

Amin, A. (ed.) 1994: *Post-Fordism: A Reader*. Oxford, Blackwell.

Anderson, D. 1995: *Crime and the Politics of Hysteria: How the Willie Horton Story Changed American Justice*. New York, Random House.

Anderson, E. 1990: *Streetwise: Race, Class and Change in an Urban Community*. Chicago, University of Chicago Press.

Assael, S. and Lobia, J. 1992: New York's Ten Worst Judges, *Village Voice*, 15 September: 38–41.

Auletta, K. 1982: *The Underclass*. New York, Random House.

Austin, J. and Krisberg, B. 1981: Wider, Stronger and Different Nets: The Dialectics of Criminal Justice Reform, *Journal of Research in Crime and Delinquency* 18, 1 January: 165–96.

Bachmann, C. and Le Guennec, N. 1996: *Violences urbaines*. Paris, Albin Michel.

Bagnasco A. and Le Galès, P. (eds) 1997: *Villes en Europe*. Paris, La Découverte.

Barrett, W. 1995: Rudy's Shrink Rap, *Village Voice*, 9 May.

Bartkowiak, I. 1998: *Discours présidentiels et répression aux Etats-Unis*. Paris, Sorbonne-Paris IV, MA mimeo.

Baudrillard, J. 1986: *Amérique*. Paris, Grasset.

Baumgartner, M. 1988: *The Moral Order of a Suburb*. New York, Oxford University Press.

Beauregard, R. 1995: Theorizing the Global–Local Connection. In P. Knox and P. Taylor (eds), *World Cities in a World System*. Cambridge, Cambridge University Press.

Beauregard, R. and Body-Gendrot, S. (eds) 1999: *The Urban Moment*. London, Sage.

Beck, U. 1992: *Risk Society: Towards a New Modernity*. London, Sage.

Beckouche, P. 1994: Comprendre l'espace parisien. Fausses questions et vrais enjeux, *Le Débat* 80, May–August.

Beiser, V. 1995: Why the Big Apple Feels Safer, *Maclean's*, 11 September: 39–40.

Begag, A. and Delorme, C. 1994: *Quartiers sensibles*. Paris, Le Seuil.

Bell, C. and Newby, H. 1971: *Community Studies: An Introduction to the Sociology of Local Communities*. London, George Allen and Unwin.

Belluck, P. 1998: Razing the Slums to Rescue the Residents, *New York Times*, 6 September.

Belorgey, J. M. 1991: *La police au rapport: études sur la police de la commission d'études des réformes de la police*. Nancy, Presses Universitaires de Nancy.

Bennett, W., Dilulio, J., and Walters, J. 1996: *Body Count*. New York, Simon & Schuster.

Berman M. 1997: Justice/Just Us: Rap and Social Justice in America. In A. Merrifield and E. Swyngedouw (eds), *The Urbanization of Injustice*. New York, NY University Press.

Berman, P. 1988: Mysteries and Majesties of New York, *Village Voice*, 15 March.

Beuve-Méry, A. 1997: Le tiers des revenus des Français proviennent des prestations sociales, *Le Monde*, 13 August: 5.

Bittner, E. 1990: *Aspects of Police Work*. Boston, Northeastern University Press.

Block, R. 1997: Risky Places in Chicago and The Bronx. Robbery in the Environs of Rapid Transit Stations, working paper.

Block, C. and Block, R. 1984: Crime Definition, Crime Measurement, and Victim Surveys, *Journal of Social Issues* 40(1): 137–60.

Blumstein, A. 1996: *Report on Violence by Young People*, Washington, DC, National Institute of Justice.

Body-Gendrot, S. 1992: Migration and the Racialization of the Post-modern City in France. In M. Cross and M. Keith (eds), *Racism, the City and the State*. London, Routledge.

Body-Gendrot, S. 1993a: *Ville et violence. L'irruption de nouveaux acteurs*. Paris, Presses Universitaires de France.

Body-Gendrot, S. 1993b: Pioneering Muslim Women in France. In R. Fisher and J. Kling (eds), *Mobilizing the Community*. Newbury Park, Sage.

Body-Gendrot, S. 1994: *Ensemble, cela fait une différence. Etude dans les quartiers américains de villes en difficulté*. Paris, Report for the French American Foundation.

Body-Gendrot, S. 1995: Urban Violence, a Quest for Meaning, *New Community*, October: 525–36.

Body-Gendrot, S. 1996: *Redynamiser les quartiers en crise: l'expérience américaine*. Paris, DIV et Documentation française.

Body-Gendrot, S. 1998a: *Les villes américaines*. Paris, Hachette.

Body-Gendrot, S. 1998b: *Les villes face à l'insécurité*. Paris, Bayard Editions.

Body-Gendrot, S. 1999: Marginalization and Political Responses in the French context. In P. Hamel (ed.), *Urban Fields – Global Spaces*, forthcoming.

Body-Gendrot, S. and Beauregard, R. 1999: Imagined Cities, Engaged Citizens. In R. Beauregard and S. Body-Gendrot (eds), *The Urban Moment*. London, Sage.

Body-Gendrot, S. and Le Galès, P. 1997: Introduction. In J. Carré and S. Body-Gendrot (eds), Gouvernance locale, pauvreté et exclusion dans les villes anglo-saxonnes, *Frontières* 9.

Body-Gendrot, S., Le Guennec, N., and Herrou, M. 1998: *Mission sur les violences urbaines*. Paris, La Documentation française/IHESI.

Body-Gendrot, S., Maslow-Armand, L., and Stewart, D. 1984: *Les Noirs américains aujourd'hui*. Paris, Armand Colin.

Bok, D. 1996: *The State of the Nation*. Cambridge, MA, Harvard University Press.

Bok, S. 1998: *Mayhem*. New York, Addison Wesley.

Bonafé-Schmitt, J.-P. 1998: quoted by C. Simon, Jours de violence ordinaire au collège, *Le Monde*, 11 June 1998.

Bonafé-Schmitt, J.-P. 1999: Mediation: From Disputes Resolution to Social Integration. In M. Martiniello and S. Body-Gendrot (eds), *Migrants and Minorities in European Cities*. Oxford, Macmillan.

Bonnemaison, G. 1987: *La Sécurité en libertés*. Paris, Syros.

Bonneville, M. 1997: *Lyon, métropole régionale ou euro-cité?* Paris, Anthropos.

Bordreuil, J. S. 1997: Les gens des cités n'ont rien d'exceptionnel. In *Ces quartiers dont on parle*, collective work, La Tour d'Aigle, Editions de l'Aube: 231–52.

Bourdieu, P. 1984: *Questions de sociologie*. Paris, Editions de Minuit.

Bourgois, P. 1992: Une nuit dans une 'shooting gallery': enquête sur le commerce du crack à East Harlem, *Actes de la recherche en sciences sociales* 94, September: 59–78.

Bousquet, R. (ed.) 1998: *Nouveaux risques, nouveaux enjeux*. Paris, L'Harmattan.

Bragg, R. 1995: New Orleans's hopes rise as crime rate decreases, *New York Times*, 25 December.

Brantingham, P. and Brantingham, P. 1980: Crime, Occupation and Economic Specialization. In D. Georges-Abeyie and K. Harries (eds), *Crime: A Spatial Perspective*. New York, Columbia University Press.

Braudel, F. 1986: *L'Identité de la France*. Paris, PUF, Espaces et Histoire.

Braun, S. 1998: During the gang war, school goes on, *Courrier International* 381, 19–25 February: 25.

Breen, P. 1994: *Advocacy Efforts on Behalf of Children of Incarcerated Parents*. San Francisco, Center Force.

Briggs, X. de Souza and Mueller, E. 1996: *From Neighborhood to Community: Evidence on the Social Effect of Community Development*. New York, Community Development Research Center, New School for Social Research.

Bright, J. 1991: Crime Prevention: The British Experience. In K. Stenson and D. Cowell (eds), *The Politics of Crime Control*. London, Sage.

Broder, D. 1997: America's "Social Capital": Good News and Bad News, *IHT*, 17 December.

Brodeur, J. P. 1996: Policer l'apparence, *Canadian Journal of Criminology*: 285–332.

Brodeur, J. P. 1990: Police et sécurité en Amérique du nord, bilan des recherches récentes, January: 203–41.

Brun, J. and Rhein, C. (eds) 1994: *La Ségrégation dans la ville*. Paris, L'Harmattan.

Brunet, R. 1989: *Les villes "européennes"*. Paris, La Documentation française.

Bui-Trong, L. 1993: L'insécurité dans les quartiers sensibles, *Cahiers de la sécurité intérieure* 14, August–October: 235–6.

Bui-Trong, L. 1998: Quartiers difficiles, état des lieux, Working Paper, April.

Burnham, D. 1996: *Above the Law*. New York, Scribner's.

Butterfield, F. 1995a: Major Crimes fell in 95, Early Date by FBI Indicate, *New York Times*, 6 May.
Butterfield, F. 1995b: Crime Continues to Decline but Experts Warn of Coming "Storm" of Juvenile Violence, *New York Times*, 29 November.
Butterfield, F. 1998: As Crime Falls, Pressure Rises on Police to Alter Data, *New York Times*, 3 August.
Calvino, I. 1972: *Invisible Cities*. New York, Harcourt Brace Janovich.
Canada, G. 1996: *Fist Stick Knife Gun*. New York, Beacon Press.
Caraley, D. 1977: *City Governments and Urban Problems*. Englewood Cliffs, NJ, Prentice-Hall.
Cardo, P. 1991: *Mouvements collectifs et violence*. Paris, Conseil national des villes et du développement urbain.
Carnoy, M. 1994: *Faded Dreams. The Politics and Economics of Race in America*. New York, Cambridge University Press.
Carrez, J. F. 1991: *Le Développement des fonctions tertiaires supérieures internationales à Paris et dans les métropoles régionales*. Paris, La Documentation française.
Castaing, M. 1994: Les squats de Paris, *Le Monde*, 28 July.
Castel, R. 1996: *Les Métamorphoses de la question sociale*. Paris, Fayard.
Castellan, M., Goldberger, M. F., and Marpsat, M. F. 1992: Les quartiers prioritaires de la politique de la ville, *INSEE Première*: 234.
Castells, M. 1972: *La Question urbaine*. Paris, Maspéro.
Castells, M. 1998a: *The Information Age: Economy, Society and Culture*. Vol. II: *The Power of Identity*. Oxford, Blackwell.
Castells, M. 1998b: L'Invité, *Urbanisme*, September: 6–13.
Césari, J. 1997: *Faut-il avoir peur de l'islam?* Paris, Sciences Po Presse.
Chaskin, R. 1995: *Defining the Neighborhood: History, Theory and Practice*. Chicago, Chapin Hall Center for Children at the University of Chicago.
Chevalier, L. 1984: *Classes laborieuses et classes dangereuses*. Paris, Hachette.
Chevigny, P. 1995: *Edge of the Knife. Police Violence in the Americas*. New York, The New Press.
Choay, F. 1988: Le Règne de l'urbain et la mort de la ville. In *La ville*. Paris, Editions Centre Georges Pompidou.
Choldin, H. 1984: Subcommunities: Neighborhoods and Suburbs: An Ecological Perspective. In M. Micklin and H. Choldin (eds), *Sociological Human Ecology*. Boulder, CO: Westview.
Christie, N. 1994: *Crime Control As Industry*. London, Routledge.
Clarke, R. V. 1995: Situational Crime Prevention. In M. Tonry and D. Farrington (eds), *Building a Safer Society: Strategic Approaches to Crime Prevention – Crime and Justice, A Review of Research*, vol. 19. Chicago, Chicago University Press.
Cloward, R. and Ohlin, L. 1960: *Delinquency and Opportunity: A Theory of Delinquent Gangs*. New York, Free Press of Glencoe.
Cohen, C. 1999: Social Capital, Intervening Institutions and Political Power, paper presented at the Ford Foundation Conference, Social Capital in Poor Communities, New York.
Cohen, S. 1985: *Visions of Social Control*. Cambridge, Polity Press.
Coing, H. 1966: *Rénovation urbaine et changement social*. Paris, Editions ouvrières.

Coleman, J. 1988: Social Capital in the Creation of Human Capital, *American Journal of Sociology* 94: 595–620.

Commission des maires sur la sécurité, 1982: *Face à la délinquance: prévention, répression, solidarité.* Paris, La Documentation française.

Committee of Ways and Means, 1997: Greenbook, Washington, DC, Government Printing Office.

Costanza, C., Halperin W., and Gale, N. 1986: Criminal Mobility and the Directional Component in Journeys to Crime. In R. Figlio, S. Hakim, and G. Rengert (eds), *Metropolitan Crime Patterns.* Monsey, NY: Criminal Justice Press.

Coulon, A. 1992: *L'École de Chicago.* Paris, PUF.

Crawford, A. 1997: *The Local Governance of Crime.* Oxford, Pergamon.

Crozier, M. 1963: *Le Phénomène bureaucratique.* Paris, Le Seuil.

Crutchfield, R. D. 1995: Ethnicity, Labor Markets, and Crime. In D. Hawkins and F. Darnell (eds), *Ethnicity, Race, and Crime.* Albany, SUNY Press.

Currie, E. 1985: *Confronting Crime: An American Challenge.* New York, Pantheon.

Currie, E. 1998: *Crime and Punishment in America.* New York, Owl Books.

Curtis, L. (ed.) 1995: *The State of Families.* Washington, DC, Milton Eisenhower Foundation.

Curtis, L. (ed.) 1997: *Youth Investment and Police Mentoring.* Washington, DC, Milton Eisenhower Foundation.

Dahrendorf, R. 1985: *Law and Order.* London, Stevens and Sons.

Davenas, L. 1998: *Lettre de l'Himalaya.* Paris, Le Seuil.

Davis, M. 1992: *City of Quartz.* Berkeley, University of California Press.

Debarbieux, E. 1998: Violence et ethnicité dans l'école française, *Revue européenne des migrations internationales*, July.

De Certeau, M. 1980: *L'Invention du quotidien*, vol. 1, *Arts de Faire*, Paris, 10/18.

De Lataulade, B. 1995: Ville, image et dynamiques sociales, *Les Annales de la recherche urbaine* 68–9: 107–13.

Delmas, P. 1991: *Le Maître des horloges.* Paris, Odile Jacob.

De Rudder, V., Poiret, C., and Vourc'h, F. 1997: La prévention de la discrimination raciale, de la xénophobie et la promotion de l'égalité de traitement dans l'entreprise. Une étude de cas en France. Paris, Urmis, Universités Paris VII and VIII, mimeo.

Deutsch, J., Hakim, S., and Weinblatt, J. 1998: A Micro-Model of the Criminal's Location Choice, *Journal of Urban Economics* 22: 198–208.

Devine, J. 1996: *Maximum Security. The Culture in Violence in Inner City Schools.* Chicago, Chicago University Press.

Dilulio, J. 1995: Why Violent Crime Rates have Dropped, *Wall Street Journal*, 6 September.

Doble, J., Immerwaht, S., and Richardson, A. 1991: *Punishing Criminals: The People of Delaware Consider the Options.* New York, The Edna McConnell Clark Foundation.

Doble, J. and Klein, J. 1989: *Punishing Criminals. The Public's View – An Alabama Survey.* New York, The Edna McConnell Clark Foundation.

Dollé, N. 1998: *La Cité des poètes à Pierrefitte.* Paris, Le Temps des cerises.

Donzelot, J. and Estèbe, P. 1994: *L'Etat animateur. Essai sur la politique de la ville.* Paris, Esprit.

Donzinger, S. (ed.) 1994: *The Real War on Crime*. New York, Harper.
Dourlens, C., Vidal-Naquet, P. 1994: *L'Autorité comme prestation*. Paris, CERPE.
Drozdiak, W. 1996: French Won't Give Up "Womb-to-Tomb" Perks Quietly, *IHT*, 25 June.
Dryfoos, J. 1994: *Full-Service Schools*. New York, Jossey-Bass.
Dryfoos, J. 1998: *Safe Passages: Making It Through Adolescence in a Risky Society*. Oxford, Oxford University Press.
Dubedout, H. 1983: *Ensemble, refaire la ville*. Paris, La Documentation française.
Dubet, F. 1994: Les Mutations du système scolaire et les violences à l'école, *Cahiers de la sécurité intérieure* 15: 11–26.
DuBois, W. E. B. 1961: *The Souls of Black Folk*. New York, Fawcett World Library.
Duneier, M. 1992: *Slim's Table. Race, Respectability and Masculinity*. Chicago, Chicago University Press.
Durkheim, E. 1933: *The Division of Labor in Society*. New York, Free Press.
Eck, J. and Weisburg, D. 1995: Crime Places in Crime Theory. In J. Eck and D. Weisburg (eds), *Crime and Place*. Monsey, NY, Criminal Justice Press.
Edelman, P. 1999: Confronting Clinton With Facts on His Poverty Tour, *New York Times*, 7 July.
Edsall, T. and Edsall, M. 1991: *Chain Reaction*. New York, Norton.
Elias, N. 1933: *La Dynamique de l'Occident*. Paris, Calmann-Levy, reprint 1990.
Elias, N. and Scotson, J. L. 1965: *The Established and the Outsiders*. Newbury, CA, Sage.
Ellsworth, P. and Gross, S. 1994: Hardening of the Attitudes: American Views on the Death Penalty, *Journal of Social Issues* 30(2): 19–52.
Ellwood, D. and Bane, M. J. 1985: The Impact of AFDC on Family Structure and Living Arrangements. In R. Ehrenberg (ed.), *Research in Labor Economics*. JAI Press.
Ericson, R. 1994: The Division of Expert Knowledge in Policing and Security, *British Journal of Sociology* 45(2): 149–75.
Ericson, R. and Haggerty, K. 1997: *Policing the Risk Society*. Toronto, University of Toronto Press.
Estèbe, P. (ed.) 1990: *Diagnostic local de sécurité*. Paris, Report to DIV.
Estèbe, P. 1994: Police, justice et politiques locales: de l'antagonisme au contrat, *Les cahiers de la sécurité intérieure* 16: 25–35.
Estrich, S. 1998: *Getting Away With Murder*. Cambridge, Harvard University Press.
Ewing, C. 1990: *Children Who Kill Children*. Lexington, Lexington University Press.
Fagan, J. 1996: Crime in Public Housing: Two-Way Diffusion Effects in Surrounding Neighborhoods, working paper.
Fagan, J. 1997: The Comparative Advantage of Juvenile Versus Criminal Court Sanctions on Recidivism among Adolescent Felony Offenders, *Law and Policy* 18(1&2), January–April 1996: 77–114.
Fagan, J. and Davis, G. 1998: The Social Context and Functions of Adolescent Violence. In D. Elliott and B. Hamburg (eds), *Violence in American Schools*. Cambridge, Cambridge University Press.
Fagan, J. and Wilkinson, D. 1998: Situational Contexts of Adolescent Violence in New York City, *Revue européenne des migrations internationales*, 14(1): 63–76.
Faget, J. 1992: *Justice et travail social – Le rhizome pénal*. Volle-Trajets, Editions Eres.

Fainstein, S. 1997: Justice, Politics and the Creation of Urban Space. In A. Merrifield and E. Swyngedouw (eds), *The Urbanization of Injustice*. New York, New York University Press.

Fainstein, S. 1999: Can We Make the Cities We Want? In R. Beauregard and S. Body-Gendrot (eds), *The Urban Moment*. London, Sage.

Fainstein, S. and Fainstein, N. 1974: *Urban Political Movements*. Englewood Cliffs, NJ, Prentice-Hall.

Fainstein, S., Gordon, I., and Harloe, M. 1992: *Divided Cities: New York and London in the Contemporary World*. Oxford, Blackwell.

Feeley, M. and Kamin, S. 1997: The Effect of "Three Strikes and You're Out" on the Courts. In D. Shichor and D. Sechrest (eds), *Three Strikes and You're Out. Vengeance as Public Policy*. London, Sage.

Feeley, M. and Sarat, A. 1980: *The Policy Dilemma: Federal Crime Policy and the LEAA, 1968–1978*. Minneapolis, University of Minnesota Press.

Felson, M. 1987: Routine Activities and Crime Prevention in the Developing Metropolis, *Criminology* 25(4): 911–31.

Felson, M. and Cohen, L. 1980: Human Ecology and Crime: A Routine Activity Approach, *Human Ecology* 8(4): 389–406.

Fenoglio, J. and Herzberg, N. 1998: La Seine St Denis peine à émerger, *Le Monde*, 25 April: 8.

Fine, R. 1997: Civil Society Theory, Enlightenment and Critique, *Democratization* 4(1): 7–28.

Flamm, M. 1996: Law and Order: The Conservative Critique of Street Crime and Civil Rights, working paper.

Flanagan, T. and Longmire, D. 1996: *Americans View Crime and Justice*. Thousand Oaks, CA, Sage.

Foucault, M. 1977: *Discipline and Punish: The Birth of the Prison*. New York, Pantheon.

Fraser, N. 1995: From Redistribution to Recognition? Dilemmas of Justice in a "post-socialist" Age, *New Left Review* 212: 68–93.

Freeman, R. 1995: The Labor Market. In J. Q. Wilson and J. Petersilia (eds), *Crime*. San Francisco, ICS Press.

Friedman, L. 1993: *Crime and Punishment in American History*. New York, Basic Books.

Friedmann, J. 1995: Where We Stand: A Decade of World City Research. In P. Knox and P. Taylor (eds), *World Cities in a World System*. Cambridge, Cambridge University Press.

Friedmann, J. and Wolff, G. 1982: World City Formation. An Agenda for Research and Action, *International Journal of Urban and Regional Research* 6(3): 309–44.

Froment, E. and Karlin, M. 1993: L' Implication du domaine bancaire et financier. In A. Sallez (ed.), *Les Villes, lieux d'Europe*. La Tour d'Aigle, Datar/Editions de l'Aube.

Frost, M. 1998: Migrants, Civil Society and Sovereign States: Investigating an Ethical Hierarchy, *Political Studies* XLVI: 871–85.

Fustenberg, F. 1971: Public Reaction to Crime in the Streets, *American Scholar* 40: 601–10.

Gangi, R. and Murphy, J. 1990: *Imprisoned Generation*, a report by the Correctional Association of New York and New York State Coalition for Criminal Justice, New York, September.

Gans, H. 1962: *The Urban Villagers*. New York, Free Press.

Garapon, A. 1996: *Le Gardien des promesses*. Paris, Odile Jacob.

Garland, D. 1990: *Punishment and Modern Society*. Chicago, University of Chicago Press.

Gaubatz, K. T. 1995: *Crime in the Public Mind*. Ann Arbor, Michigan University Press.

Gauchet, M. 1976: L'Expérience totalitaire et la pensée du politique, *Esprit*: 7–8.

Gauchet, M. 1985: *Le Désenchantement du monde*. Paris, Gallimard, NRF.

Gaudin, J. P. 1998: Modern Governance, Yesterday and Today: Some Clarifications to be Gained from French Government Policies, *International Social Science Journal*, March: 47–56.

Gerbner, G. 1994: Television Violence: The Art of Asking the Wrong Questions, *Currents in Modern Thought*: 385–97.

Gibbs, J. T. (ed.) 1988: *Young, Black, and Male in America: An Endangered Species*. Dover, MA, Auburn House Publishing.

Giddens, A. 1984: *The Constitution of Society*. Cambridge, Polity Press.

Giddens, A. 1985: *The Nation State and Violence*. Cambridge, Polity Press.

Giddens, A. 1990: *The Consequences of Modernity*. Cambridge, Polity Press.

Girard, R. 1982: *Le Bouc émissaire*. Paris, Grasset.

Gittell, M. 1998a: Expanding Civic Opportunity. Urban Empowerment Zones, *Urban Affairs Review* 33(4), March: 530–58.

Gittell, M. 1998b: *Participation, Social Capital and Community Change*, paper presented at the Aspen Institute Round Table on Comprehensive Community Initiative.

Gittell, M. and Newman, K. 1998: Empowerment Zone Implementation: Community Participation and Community Capacity, mimeo.

Goetz, E. G. 1996: The US War on Drugs as Urban Policy, *International Journal of Urban and Regional Research* 20(3), September: 539–49.

Goldstein, 1973: *New York Times*, 9 May, sec.1: 1.

Gordon, D. 1991: *The Justice Juggernaut: Fighting Street Crime, Controlling Citizens*. New Brunswick, NJ, Rutgers University Press.

Gordon, D. 1994: *The Return of Dangerous Classes*. New York, W. W. Norton.

Gordon, D. 1999: Crime Control Unlimited: The American Approach to Governing, *Le Monde des débats*, June.

Green, N. 1998: *Du sentier à la 7è Avenue. La confection et les immigrés – Paris–New York, 1880–1980*. Paris, Le Seuil.

Greenberg, S. and Rhoe, W. 1984: Neighborhood Design and Crime: A Test of Two Perspectives, *Journal of the American Planning Association* 50: 48–61.

Gregory, S. 1998: *Black Corona*. Princeton, Princeton University Press.

Gottfredson, D., Gottfredson, G., and McNeil, R. 1991: Social Area Influences on Delinquency: A Multilevel Analysis, *Journal of Research in Crime and Delinquency* 28(2): 197–226.

Gurr, T. 1989: Historical Trends in Violent Crime: Europe and the United States. In T. Gurr (ed.), *Violence in America: The History of Crime*. Newbury Park, CA, Sage.

Gurrey, B. 1998: La Seine St Denis, test pour le gouvernement, *Le Monde*, 23 April: 1.

Habermas, J. 1990: La Crise de l'Etat Providence ou l'épuisement des énergies utopiques, *Ecrits politiques*. Paris, Editions du Cerf.

Hagan, J. and Peterson, R. D. 1995: Criminal Inequality in America: Patterns and Consequences. In J. Hagan and R. D. Peterson (eds), *Crime and Inequality*. Stanford, Stanford University Press.

Hamnett, C. 1994: Social Polarization in Global Cities: Theory and Evidence, *Urban Studies* 31(3): 401–24.

Hardy, J. P. 1999: *Guide de l'action sociale contre les exclusions*. Paris, Dunod.

Harries, K. 1996: Cities and Crime, *Criminology* 14(3): 369–86.

Harris, L. 1995: Banana Kelly's Toughest Fight, *The New Yorker*, 24 July.

Harvey, D. 1973: *Social Justice and The City*. Baltimore, Johns Hopkins Press.

Harvey, D. 1991: From Space to Place and Back Again: Reflections on the Condition of Postmodernity, text for UCLA GSAUP Colloquium, 13 May.

Haut Conseil à l'Intégration, 1991: *Pour un modèle français d'intégration*. Paris, La Documentation française.

Hawkins, D. (ed.) 1995: *Ethnicity, Race and Crime*. Albany, SUNY.

Hayden, D. 1996: *The Power of Place*. Cambridge, MA, MIT Press.

Hayghe, H. 1991: Volunteers in America: Who Donates the Time? *Monthly Labor Review*, February: 17–23.

Hénu, E. 1998: La Résorption du bidonville de La Lorette. Histoire de vies, histoire de villes, *Hommes et migrations* 1213, May–June: 89–98.

Herbert, B. 1998: Harlem's People Keep Their Cool Despite the Makings of a Riot, *International Herald Tribune*, 8 September.

Héritier, F. 1999: *De la violence*, vol. 2. Paris, Editions Odile Jacob.

Hernandez, R. 1996: Who Gets the Credit? Views Differ, *New York Times*, 4 January.

Heydebrand, W. 1979: The Technocratic Administration of Justice, *Research in Law and Sociology* 2.

Hillery, G. 1984: Definitions of Community: Areas of Agreement, *Rural Sociology* 20: 11–123.

Hirsch, A. von. 1976: *Doing Justice. The Choice of Punishments*. New York, Basic Books.

Hirsch, A. 1983: *Making the Second Ghetto. Race and Housing in Chicago 1940–1960*. New York, Cambridge University Press.

Hirschman, A. O. 1970: *Exit, Voice, Loyalty*. Cambridge, MA, Harvard University Press.

Hobsbawn, E. 1959: *Primitive Rebels*. Manchester, Manchester University Press.

Hoffmann, A. von. 1997: Good News! From Boston to San Francisco, the Community-Based Housing Movement is Transforming Bad Neighborhoods, *The Atlantic Monthly*, January.

Home Office 1990: *Partnership in Crime Prevention*. London, Home Office.

Human Rights Watch 1998: *Shielded from Justice, Police Brutality and Accountability in the United States*. New York, Human Rights Watch.

INSEE 1994a: *Les Étrangers en France – Portrait social*. Paris, INSEE.

INSEE 1994b: *Atlas du Grand Lyon*. Lyon, INSEE.

Irving Jackson, P. 1997: Minorities, Crime and Criminal Justice in France. In I. Marshall (ed.), *Minorities, Migrants, and Crime*. Thousand Oaks, CA, Sage.

Jacob, H. 1984: *The Frustrations of Policy: Responses to Crime by American Cities*. Boston, Little, Brown.

Jacobs, J. 1961: *The Life and Death of American Cities*. Harmondsworth, Penguin.

Jargowsky, P. 1996: *Poverty and Place: Ghettoes, Barrios, and the American City*. New York, Russell Sage Foundation.

Jazouli, A. 1992: *Les Années banlieue*. Paris, Le Seuil.

Jean, J. P. 1995: L'Inflation carcérale, *Esprit*, October.

Jencks, C. 1992: *Rethinking Social Policy. Race, Poverty and the Underclass*. Cambridge, MA, Harvard University Press.

Jessop, J. 1995: The Regulation Approach, Governance and Post-Fordism: Alternative Perspectives on Economic and Political Change, *Economy and Society* 24(3): 307–33.

Johnson, D. 1997: Chicago School System Gets Tough on Students, *International Herald Tribune*, June 7–8.

Johnston, D. 1993: *Report N° 13: Effects of Parental Incarceration*. Pasadena, CA, The Center for Children of Incarcerated Parents.

Johnstone, J. 1978: Social Class, Social Areas and Delinquency, *Sociology and Social Research* 63: 49–72.

Jones, M. 1953: *The Therapeutic Community: A New Treatment Method in Psychiatry*. New York, Basic Books.

Judd, D. and Swanstrom, T. 1998: *City Politics*. New York, Longman, 2nd print.

Judge, D., Stoker, G., and Wolman, H. (eds) 1995: *Theories of Urban Politics*. Thousand Oaks, CA, Sage.

Kane, J. 1989: Shock Incarceration: A Dose of Discipline for First Time Offenders, *Time*, 16 October: 17.

Kasarda, J. 1993: Inner-City Poverty and Economic Access. In J. Sommer and D. Hicks (eds), *Rediscovering Urban America; Perspectives on the 1980s*. Washington, DC, Dept of Housing and Urban Development, Office of Housing Policy Research.

Katz, M. 1989: *The Undeserving Poor: From the War on Poverty to the War on Welfare*. New York, Pantheon.

Katzman, M. 1981: The Supply of Criminals: A Geo-Economic Examination. In S. Hakim and G. Rengert (eds), *Crime Spillover*, Beverley Hills, Sage Publication.

Kelling, G. and Coles, C. 1998: *Fixing Broken Windows*. New York, Simon & Schuster.

Kennedy, R. 1997: *Race, Crime and the Law*. New York, Pantheon.

Kensey, A. and Tournier, P. 1998: French Prison Numbers Stable, *Overcrowded Times* 9(4), August.

Kepel, G. 1987: *Les Banlieues de l'islam*. Paris, Le Seuil.

Kerner Report 1968: Advisory Commission on Civil Disorders. New York, Bantham.

King, A. 1990: *Global Cities*. London, Routledge.

King, D. 1995: *Separate and Unequal*. New York, Oxford University Press.

Kirschenman, J. and Neckerman, K. 1991: We'd Love to Hire Them, But . . . : The Meaning of Race for Employers. In C. Jencks and P. Peterson (eds), *The Urban Underclass*. Washington, DC, Brookings Institute.

Krauss, C. 1995: The Commissioner vs. the Criminologists, *New York Times*, 19 November.

Ladd, E. 1996: The Data Just Don't Show Erosion of America's Social Capital, *The Public Perspective*, June/July, 1: 5–6.

Lane, R. 1997: *Murder in America. A History*. Columbus, Ohio State University.

Lardner, J. 1997: The Drop in Crime, *New York Review of Books*, August.

Lasswell, H. 1950: *Power and Society*. New Haven, Yale University Press.

Lazerges, C. and Balduyck, J. P. 1998: *Réponses à la délinquance des mineurs*. Paris, La Documentation française.

Léauté, J. 1990: *Les Prisons*. Paris, PUF.

Leclerc, H. 1995: Faut-il en finir avec le jury populaire? *Esprit*, March.

Lefèvre, C. 1995: La Recomposition territoriale en question: position d'acteurs, *Revue de géographie de Lyon* 70(2).

Lefèvre, H. 1974: *La Production de l'espace*. Paris, Anthropos.

Le Galès, P. and Mawson, A. 1994: *Management Innovations in Urban Policy. Lessons from France*. London, Report to the Local Government Management Board.

Lévy, R. and Ocqueteau, F. 1987: Police Performance and Fear of Crime: The Experience of the Left in France between 1981 and 1986, *International Journal of the Sociology of Law* 15: 259–80.

Lévy, R. 1996: La Crise du système policier aujourd'hui: de l'insertion aux enjeux locaux européens, working paper.

Lii, J. 1997: When the Saviors Are Seen As Sinners, *New York Times*, 18 May.

Lin, A. C. 1998: The Troubled Success of Crime Policy. In M. Weir (ed.), *The Social Divide. Political Parties and the Future of Activist Government*. New York, Russell Sage Foundation.

Linhart, V. 1992: Des Minguettes à Vaux-en-Velin, les réponses des pouvoirs publics aux violences urbaines, *Cultures et conflits* 6, June.

Lloyd, C. 1999: Antiracist Responses to Fortress Europe. In R. Koopmans and P. Statham (eds), *Challenging and Defending the Fortress: Political Mobilization and Ethnic Difference in Comparative and Transnational Perspective*. Oxford, Oxford University Press.

Lorrain, D. 1995: La Grande Entreprise urbaine et l'action publique, *Revue de sociologie du travail* 2.

Madison, J. 1961: Federalist Papers, nos 10 and 46. In A. Hamilton, J. Madison, and J. Jay, *The Federalist Papers*. New York, Mentor.

Maguire, K. and Pastore, A. L. (eds) 1997: *Sourcebook of Criminal Justice Statistics 1996*. Washington, DC, Government Printing Office.

Maillard, J. de, 1997: *L'Avenir du crime*. Paris, Flammarion.

Mann, C. R. 1993: *Unequal Justice: A Question of Color*. Bloomington, Indiana University Press.

Mansuy, M. and Marpsat, M. 1994: La Division sociale de l'espace dans les grandes villes françaises hors Ile de France. In J. Brun and C. Rhein (eds), *La Ségrégation dans la ville*. Paris, L'Harmattan.

Marchand, B. 1993: *Paris, Histoire d'une ville, XIXè–XXè siècle*. Paris, Editions du Seuil.

Marsh, D. and Rhodes, R. (eds) 1992: *Policy Networks in British Government*. Oxford, Clarendon Press.

Marshall, I. (ed.) 1997: Introduction. In *Minorities, Migrants, and Crime*. Thousand Oaks, CA, Sage.

Martison, R. 1974: What Works? Questions and Answers About Prison Reforms, *Public Interest* 35: 25–54.

Marx, G. 1988: *Undercover: Police Surveillance in America.* Berkeley, University of California Press.

Marx, K. 1968: *Oeuvres choisies.* Paris, Gallimard.

Massey, D. and Denton, N. 1992: *American Apartheid.* Cambridge, MA, Harvard University Press.

Massey, D. 1995: Suburbs and Cities: Changing Patterns in Metropolitan Living, *Report*, Washington, DC, Aspen Institute.

Massey, D. 1997: Space/Power, Identity/Difference: Tensions in the City. In A. Merrifield and E. Swyngedouw (eds), *The Urbanization of Injustice.* New York, NY University Press.

Mauer, M. 1995: *America Behind Bars, The International Use of Incarceration.* Washington, DC, Sentencing Project.

Mayer, M. 1999: Urban Movements and Urban Theory: Assessing Social Movements in the Late 20th Century City. In Beauregard and Body-Gendrot 1999.

Mayer, M. (with S. Clarke) 1986: Responding to Grassroots Discontent: Germany and the United States, *International Jounal of Urban and Regional Research* 10(3): 401–17.

Mazetier, S. and Portelli, S. 1998: Une exception française: la détention d'innocence, *Libération*, 17 September.

McConnell, G. 1966: *Private Power and American Democracy.* New York, Random House.

McKenzie, E. 1993: *Privatopia.* New Haven, Yale University Press.

Meares, T. and Kahan, D. 1999: Law and (Norms of) Order in the Inner City. *Law and Society Review* (in press).

Merryman, J. H. 1969: *The Civil Law Tradition: An Introduction to the Legal Systems of Western Europe and Latin America.* Stanford, Stanford University Press.

Merton, R. 1938: Social Structure and Anomie, *American Sociological Review* 3: 672–82.

Miller, J. 1997: *Search and Destroy; The Plight of African-American Males in the Criminal Justice System.* Cambridge, Cambridge University Press.

Mitchell, D. 1991: *Income Transfers in Ten Welfare States*, quoted in D. Bok, 1996.

Mollen Report. 1994: Commission to Investigate Allegations of Police Corruption and the Anti-Corruption Procedures of the Police Department 7 July.

Mollenkopf, J. 1997: *Hollow in the Middle. The Rise and Fall of New York City's Middle Class.* Report to the New York City Council, December.

Mollenkopf, J. and Castells, M. 1991: *Dual City, Restructuring New York.* Russell Sage Foundation.

Monjardet, D. 1996: *Ce que fait la police.* Paris, La Découverte.

Monjardet, D. 1988: Moderniser, pour quoi faire? *Esprit* 1: 5–18.

Monnoken, E. 1995: Racial Factors in New York City Homicides, 1800–1874. In D. Hawkins (ed.), *Ethnicity, Race and Crime.* Albany, SUNY.

Moore, B. 1978: *Injustice. The Social Bases of Obedience and Revolt.* White Plains, NY, M. E. Sharpe.

Morales, E. 1997: New CCRB appointee named in Mollen Report. Benefit of the Doubt? *Village Voice*, 23 April.

Moran, M. 1995: More Police, Less Crime? Wrong, *New York Times*, 27 February.

Muir, W. K. 1977: *Police: Streetcorner Politicians*. Chicago, Chicago University Press.

Murray, C. 1984: *Losing Ground, American Social Policy, 1950–1980*. New York, Basic Books.

National Advisory Commission on Civil Disorders (Kerner Commission), 1968: *Report*. New York, Bantham.

National Commission on the Causes and Prevention of Violence, 1969: *Report*.

Neveu, C. (ed.) 1999: *Espace public et engagement politique*. Paris, L'Harmattan.

Newman, O. 1972: *Defensible Space, Crime Prevention Through Urban Design*. New York, Macmillan.

Nichols, W. 1980: Mental Maps, Social Characteristics and Criminal Mobility. In D. Georges-Abeyie and K. Harries (eds), *Crime: A Spatial Perspective*. New York, Columbia University Press.

Normandeau, A. 1995: *Un panorama des politiques et pratiques pénales de la nouvelle pénologie* made in America: *1980–2005*, working paper.

Odubekun, L. 1993: *The Vera Institute Atlas of Crime and Justice in New York City*. New York, Vera Institute of Justice.

O'Malley, P. 1996: Indigenous Governance, *Economy and Society* 25(3): 310–26.

Page, J. 1993: Urban Criminal Justice: No Fairer than the Larger Society, *Fordham Urban Law Journal* XX(3): 609–19.

Park, R. 1916: The City: Suggestions for the Investigations of Human Behavior in the Urban Environment, *American Journal of Sociology* 20: 517–612.

Park, R., Burgess, E., and McKenzie, R. 1925: *The City*. Chicago, Chicago University Press, reprint. 1967.

Paugam, S. 1993: *La Société française et ses pauvres*. Paris, PUF.

Peraldi, M. 1996: *Migrants in European Cities – City Summary of Marseilles*, working paper.

Peraldi, M. 1999: Migrants' Careers and Commercial Skills in Marseilles: An Urban Alternative? In M. Martiniello and S. Body-Gendrot (eds), *Migrants and Minorities in European Cities*. Oxford, Macmillan.

Peralva, A. 1997: Des collégiens et de la violence. In B. Charlot and J. C. Emin (eds), *Les Violences à l'école. L'état des savoirs*. Paris, A. Colin.

Perkins, D., Wandersman, A., Rich, R., and Taylor, R. 1993: The Physical Environment of Street Crime: Defensible Space, Territoriality and Incivilities, *Journal of Environmental Psychology* 13: 29–49.

Perrineau, P. 1999: L'Abstention du 3 juin démontre l'ampleur du malaise démocratique, *Le Monde*, 1 July: 14.

Peyrefitte, A. (ed.) 1977: *Réponses à la violence*. Paris, Presses-Pocket.

Phillips, P. 1980: Characteristics and Typology of the Journey to Crime. In D. Georges-Abeyie and K. Harries (eds), *Crime: A Spatial Perspective*. New York, Columbia University Press.

Pickvance, C., Préteceille, E. (eds) 1991: *State Restructuring and Local Power: A Comparative Perspective*. New York, Pinter.

Pinheiro, P. 1998: Safety in Latin American Cities: Uncivil Societies under Democratic Rule, *Revue européenne des migrations internationales*, 14(1): 47–61.

Pitts, J. 1963: Continuité et changement au sein de la France bourgeoise. In S. Hoffmann (ed.), *In Search of France*. Cambridge, MA, Harvard University Press.

Plotkin, S. 1991: Community and Alienation: Enclave Consciousness and Urban Movements, *Comparative Urban and Community Research* 3: 5–25.

Poiret, C. 1999: Au travail, le bonneteau de l'ethnicité, *Mouvements* 4, May–July: 17–23.

Popkin, S. 1996: Criminological Research on Public Housing: Towards A Better Understanding of People, Places and Spaces, *Crime and Delinquency* 42(3), July: 361–78.

Préteceille, E. 1999: *Divisions sociales et services urbains*, vol. 1. *Inégalités et contrastes sociaux en Ile-de-France*. Paris, CSU.

Procacci, G. 1993: *Gouverner la misère*. Paris, PUF.

Prothrow-Stith, D. 1991: *Deadly Consequences*. New York, Harper Perennial.

Putnam, R. 1993: *Making Democracy Work*. Princeton, Princeton University Press.

Putnam, R. 1995: Bowling Alone: America's Declining Social Capital, *Current* 373, June: 3–9.

Putnam, R. 1996: The Strange Disappearance of Civic America, *The American Prospect*, winter: 34–48.

Pyle, G. 1976: Spatial and Temporal Aspects of Crime in Cleveland, Ohio, *American Behavioral Scientist* 20(2): 175–98.

Racawich, L. 1987: Lock'em Up, *Progressive*, August: 16–19.

Raine, J. W. and Wilson, M. 1993: *Managing Criminal Justice*. London, Harvester Wheatsheaf.

Rand, A. 1986: Mobility Triangles. In R. Figlio, S. Hakim, and G. Rengert (eds), *Metropolitan Crime Patterns*. Monsey, NY, Criminal Justice Press.

Rawls, J. 1971: *A Theory of Justice*. Cambridge, MA, Harvard University Press.

Reed, D., Gill, J., and Permultter, W. 1997: *Juvenile Trend from the Statistics of the Juvenile Court of Cook County, Ill*, Children and Family Justice Center, March.

Reider, J. 1985: *Canarsie*. Cambridge, MA, Harvard University Press.

Reinarman, C. and Levine, H. 1989: Crack in Context: Politics and Media in the Making of a Drug Scare, *Contemporary Drug Problems*, winter.

Remnick, D. 1997: The Crime Buster, *The New Yorker*, 24 February and 3 March: 94–109.

Rhein, C. 1998: Globalization, Social Change and Minorities in Metropolitan Paris: The Emergence of New Class Patterns, *Urban Studies* 35: 429–47.

Rhodes, R. A. W. 1981: *Control and Power in Central–Local Relations*. Farnborough, Gower.

Rhodes, R. A. W. 1991: Policy Networks and Sub-Central Government. In G. Thompson (ed.), *Markets, Hierarchies and Networks*. London, Sage.

Robert, P. and Pottier, M. L. 1997: Sur l'insécurité et la délinquance, *Revue Française de Science Politique* 47(5), October: 630–44.

Roncek, D., Bell, R., and Francik, J. 1981: Housing Projects and Crime: Testing a Proximity Hypothesis, *Social Problems* 29(2): 151–66.

Rorty, R. 1997: *Achieving our Country. Leftist Thought in Twentieth Century America*. New York, William E. Massey.

Ross, B. and Levine, M. 1996: *Urban Politics*. Itasca, IL, E. E. Peacock.

Rothman, D. 1980: *Conscience and Convenience: The Asylum and its Alternatives in Progressive America*. Boston, Little, Brown.

Rubin, L. 1996: *Quiet Rage. Bernie Goetz in a Time of Madness*. Berkeley, California Press.
Ruby, C. 1999: (Promised) Scenes of Urbanity. In Beauregard and Body-Gendrot 1999.
Rusk, D. 1993: *Cities Without Suburbs*. Washington, DC, Woodrow Wilson Press.
Saegert, S. 1996: What We Have to Work with: the Lessons of the Task Force Surveys. In Michelle Cotton with Susan Saegert and David Reiss, *No More "Housing of Last Resort"*. New York, The Task Force on City-Owned Property. Distributed by the Parodneck Foundation.
Salas, D. 1997: Mineurs: une justice à refonder, *Informations sociales* 62: 84–92.
Sampson, R. 1983: Structural Density and Criminal Victimization, *Criminology* 21(2): 276–93.
Sampson, R. 1985: Neighborhood and Crime: The Structural Determinants of Personal Victímization, *Journal of Research in Crime and Deliquency* 22(1): 7–40.
Sampson, R. 1986: The Effects of Urbanization and Neighborhood Characteristics on Criminal Victimization. In R. Figlio, S. Hakim, and G. Rengert (eds), *Metropolitan Crime Patterns*. Monsey, NJ, Criminal Justice Press.
Sampson, R. 1987: Does an Intact Family Reduce Burglary Risk for Its Neighbors? *Sociology and Social Research* 71: 204–7.
Sampson, R. 1997: What "Community" Supplies, working paper.
Sampson, R. and Groves, W. 1989: Community Structure and Crime: Testing Social-Disorganization Theory, *American Journal of Sociology* 94(4): 774–802.
Sampson, R. J. and Lauristsen, J. 1997: Racial and Ethnic Disparities in Crime and Criminal Justice in the US. In M. Tonry (ed.), *Ethnicity, Crime and Immigration: Comparative and Cross-national Perspectives*. Chicago, Chicago University Press.
Sampson, R. and Wilson, J. W. 1995: Towards a Theory of Race, Crime, and Urban Inequality. In J. Hagan and R. Peterson (eds), *Crime and Inequality*. Stanford, Stanford University Press.
Sampson, R. and Laub, J. 1993: Structural Variations in Juvenile Court Processing: Inequality, the Underclass, and Social Control, *Law and Society Review* 27: 285–313.
Sampson, R. J., Raudenbush, S. W., and Felton, E. 1997: Neighborhoods and Violent Crime: A Multilevel Study of Collective Efficacy, *American Association for the Advancement of Science* 277, August: 918–24.
Sandel, M. 1996: *Democracy's Discontent*. Cambridge, MA, Belknap Press.
Sanjek, R. 1998: *The Future is Us*. New York, Random House.
Sante, L. 1992: *Low Life*. New York, Vintage.
Sassen, S. 1991: *The Global City*. Princeton, Princeton University Press.
Sassen, S. 1995: Identity in the Global City: Economic and Cultural Encasements. In P. Yaeger, *The Geography of Identity*. Ann Arbor, University of Michigan Press.
Sassen, S. 1994: La Ville globale. Éléments pour une lecture de Paris, *Le Débat* 80, May–August, Paris, Gallimard.
Sassen, S. 1996: *Losing Control? Sovereignty in an Age of Globalization*. New York, Columbia University Press.
Sassen, S. 1999: Whose City Is It? In Beauregard and Body-Gendrot 1999.
Saunders, P. 1986: *Social Theory and the Urban Question*. London, Hutchinson.

Scarman Report 1981: *The Brixton Disorders*. London, HMSO.
Scheingold, S. 1984: *The Politics of Law and Order*. New York, Longman.
Scheingold, S. 1991: *The Politics of Street Crime*. Philadelphia, Temple University Press.
Schmitter, P. 1974: Still the Century of Corporatism? *Review of Politics* 36: 85–131.
Schmitter, P. 1985: Neo-corporatism and the State. In W. P. Grant (ed.), *The Political Economy of Corporatism*. Basingstoke, Macmillan.
Schnapper, D. 1996: *La Communauté des citoyens*. Paris, Gallimard.
Schwartz, A. 1992: Corporate Service Linkages in Large Metropolitan Areas: A Study of New York, Los Angeles and Chicago, *Urban Affairs Quaterly* 28(2): 276–96.
Schwartz, B. 1982: *L'Insertion professionnelle des jeunes*. Paris, La Documentation française.
Sellin, T. 1938: *Culture Conflict and Crime*. New York, SSRC.
Shapiro, B. 1995: One Violent Crime, *The Nation* 3(4).
Shaw, C. and McKay, H. 1942: *Juvenile Delinquency and Urban Areas*. Chicago: University of Chicago Press.
Shaw, C., Zorbaugh, H., McKay, H., and Cottrell, L. 1929: *Delinquency Areas: A Study of Geographical Distribution of Schools Truants, Juvenile Delinquents and Adult Offenders in Chicago*. Chicago, University of Chicago Press.
Sherman, L., Gartin, P., and Buerger, M. 1989: Hot Spots of Predatory Crime: Routine Activities and the Criminology of Place, *Criminology* 27(1): 27–55.
Silberman, C. E. 1978: *Criminal Violence, Criminal Justice*. New York, Random House.
Silver (Judge) 1996: On the Johnson Case, *New York Law Journal* 216(9), December.
Silver, H. 1993: National Conceptions of the New Urban Poverty: Social Structural Change in Britain, France and in the United States, *International Journal of Urban and Regional Research* 17(3): 185–203.
Simcha-Fagan, O. and Schwartz, J. 1986: Neighborhood and Delinquency: An Assessment of Contextual Effects, *Criminology* 24(4): 667–703.
Simmons, A. 1996: *Announcement*. Chicago, J. T. and C. D. MacArthur Foundation.
Simon, J. 1993: *Poor Discipline*. Chicago, University of Chicago Press.
Simon, J. 1996: The Rhetoric of Vengeance, *Criminal Justice Matters* 25, autumn: 8–9.
Simon, J. and Feeley, M. 1994: True Crime: The New Penology and Public Discourse on Crime. In T. Blomberg and S. Cohen (eds), *Punishment and Social Control. Essays in Honor of S. Messinger*. New York, Aldine de Gruyter.
Skogan, W. 1990: *Disorder and Decline: Crime and the Spiral of Decay in American Neighborhoods*. New York, Free Press.
Skogan, W. (ed.) 1996: *Community Policing in Chicago – Year Three*. Illinois Criminal Justice Information Authority.
Skolnick, J. 1998: The Color of the Law, *The American Prospect* 39, July.
Smith, N. 1997: Social Justice and the New American Urbanism: The Revanchist City. In A. Merrifield and E. Swyngedouw (eds), *The Urbanization of Injustice*. New York, NY University Press.
Soulignac, F. 1993: *La Banlieue parisienne*. Paris, La Documentation française.
Spierenburg, P. 1991: *The Prison Experience: Disciplinary Institutions and their Inmates in Early Modern Europe*. New Brunswick, NJ, Rutgers University Press.

Stack, C. 1975: *All our Kin. Strategies for Survival in a Black Community*. New York, Harper and Row.
Stern, V. 1998: *A Sin Against the Future. Imprisonment in the World.* Boston, Northeastern University Press.
Stoker, G. 1997: The Economic and Social Research Council Local Governance Programme: An Overview. In J. Carré and S. Body-Gendrot (eds), Gouvernance locale, pauvreté et exclusion dans les villes anglo-saxonnes, *Frontières* 9.
Stone, C. 1989: *Regime Politics*. Lawrence, KA, University of Kansas Press.
Stone, L. 1987: *The Past and the Present Revisited*. London.
Stoskopf, A. and Stern Strom, M. 1990: *Choosing to Participate*. Brookline, MA, FHAO.
Sueur, J. P. 1998: *Demain la ville*, Report to the French Minister of Social Affairs, Paris, La Documentation française.
Sullivan, M. 1989: *"Getting Paid": Youth Crime and Work in the Inner City*. Ithaca, NY, Cornell University Press.
Sullivan, M. 1991: Crime and the Social Fabric. In J. Mollenkopf and M. Castells (eds), *Dual City: Restructuring New York*. New York, Russell Sage Foundation.
Sunquist, J. 1968: *Politics and Policy, The Eisenhower, Kennedy and Johnson Years*. Washington, DC, Brookings.
Sutherland, E. 1934: *Principles of Criminology*. Chicago, Lippincorn.
Suttles, G. 1972: *The Social Construction of Communities*. Chicago, Chicago University Press.
Sutton, J. 1997: Punishment, Labor Markets, and Welfare States: Imprisonment Trends In Five Common-Law Countries, 1955–1985, working paper.
Tabard, N. 1993: Des quartiers pauvres aux banlieues aisées: une représentation sociale de territoire, *Economie et statistique* 270: 5–22.
Tarrius, A. 1992: *Les Fourmis d'Europe, migrants riches, migrants pauvres et nouvelles villes internationales*. Paris, L'Harmattan.
Tarrius, A. 1995: Le Quartier maghrébin de Belsunce à Marseille: ville-monde ou nouveau comptoir colonial méditerranéen? In R. Gallissot and B. Moulin (eds), *Les Quartiers de ségrégation*. Paris, Khartala.
Taylor, C. 1994: Les Institutions dans la vie nationale, *Esprit*, March.
Taylor, R., Gottfredson, S., and Brower S. 1984: Block Crime and Fear: Defensible Space, Local Social Ties, and Territorial Functioning, *Journal of Research in Crime and Delinquency* 21(4): 303–31.
Tienda, M. 1991: Poor People and Poor Places: Deciphering Neighborhood Effects on Poverty Outcomes. In J. Huber (ed.), *Macro–Micro Linkages in Sociology*. Newbury, CA, Sage.
Tilly, C. 1973: Do Communities Act? *Sociological Inquiry* 43: 209–40.
Tocqueville, A. de, 1945: *Democracy in America*, vol. 1. New York, Vintage Books.
Tocqueville, A. de, 1967: *L'Ancien Régime et la Révolution*. Paris, Gallimard.
Tonry, M. 1995: *Malign Neglect. Race, Class and Punishment in America*, Oxford, Oxford University Press.
Touraine, A. 1991: Face à l'exclusion, *Esprit* 169, February: 7–13.
Tournier, P. and Robert, P. 1991: *Etrangers et délinquants*. Paris, L'Harmattan.

Tribalat, M. 1995: *Faire France.* Paris, La Découverte.

Trueheart, C. 1998: French Charity: A Weak Culture of Giving in a Land of Fraternity, *International Herald Tribune*, 27 November.

Tyson, A. 1995: What Comes Down Must Go Up – Some Cities Face Sharp Crime Hikes, *Law Enforcement News*, June.

US Department of Justice, 1994, 1995, 1996: *Sourcebook of Criminal Justice Statistics.* Washington, DC, Bureau of Justice Statistics.

US Department of Justice, 1995: *Criminal Victimization in the U.S.* Washington, DC, Bureau of Justice Statistics.

Veblen, T. 1915: *Imperial Germany and the Industrial Revolution.* New York, Viking Press.

Venkatesh, S. 1997: An Invisible Community: Inside Chicago's Public Housing, *The American Prospect* 34, September–October: 35–40.

Verba, S., Lehman, K., and Brady, H. 1995: *Voice and Equality: Civic Volunteerism in American Politics.* Cambridge, MA, Harvard University Press.

Vertet, H. 1995: Exclusion dans le judiciaire et le pénitentiaire, presentation at the European conference of Strasbourg (unpublished).

Vichery, G. 1997: Le Grand Projet urbain de Marseille, in Bordreuil 1997.

Vital-Durand, B. 1999: Les prisons sous surveillance, *Libération*, 9 July.

Vogelstein, F. 1997: Cons, Don't Leave Without It, *Us News and World Report*, 21 July.

Vourc'h, C. and Marcus, M. 1994: *Sécurité et démocratie.* Paris, Forum européen de la sécurité urbaine.

Waldinger, R. 1996: *Still the Promised City? African-Americans and New Immigrants in Post-industrial New York.* Cambridge, MA, Harvard University Press.

Walzer, M. 1983: *Spheres of Justice: A Defence of Pluralism and Equality.* Oxford, Blackwell.

Weber, M. 1958: *The City* (trans). New York, Free Press.

Weisburd, D. and Green, L. 1995: Measuring Immediate Spatial Displacement: Methodological Issues and Problems. In J. Eck and D. Weisburd (eds), *Crime and Place.* Monsey, NY, Criminal Justice Press.

Weisburd, D., Maher, L., and Sherman, L. 1992: Contrasting Crime General Crime Specific Theory: The Case of Hot Spots of Crime. In F. Adler and W. Laufer (eds), *Advances in Criminological Theory*, vol. 4. New Brunswick, NJ, Transaction Publishers.

Wellman, B. 1979: The Community Question: The Intimate Networks of East Yorkers, *American Journal of Sociology* 84: 1201–31.

White, G. 1990: Neighborhood Permeability and Burglary Rates, *Justice Quarterly* 7(1): 57–67.

White, R. 1932: A Relation of Felonies to Environmental Factors in Indianapolis, *Social Forces* 10: 498–509.

Whyte, W. 1943: *Street-Corner Society: The Social Structure of an Italian Slum.* Chicago, Chicago University Press.

Wieviorka, M. 1999: *Violences en France.* Paris, Le Seuil.

Wilkerson, I. 1991: In Chicago, Drug Turf Wars Raise Murder Rate, *International Herald Tribune*, 25 August.

Wills, G. 1997: The New Adam, *New York Review of Books*, June.

Wilson, W. J. 1987: *The Truly Disadvantaged.* Chicago, Chicago University Press.

Wilson, W. J. 1996: *When Work Disappears: The World of the New Urban Poor.* New York, Alfred A. Knopf.

Wilson, J. Q. 1977: *Thinking About Crime.* New York, Vintage.

Wilson, J. Q. and Herstein, R. 1985: *Crime and Human Nature.* New York, Simon & Schuster.

Wilson, J. Q. and Kelling, G. 1982: The Police and Neighborhood Safety, *Atlantic Monthly*, March: 29–38.

Wilson, J. and Howell, J. 1993: *A Comprehensive Strategy for Serious, Violent and Chronic Juvenile Offenders.* Washington, DC, Office of Juvenile Justice and Delinquency Prevention.

Wolman, H. and Goldsmith, M. 1992: *Urban Politics and Policy. A Comparative Approach.* Oxford, Blackwell Publishers.

Young, I. M. 1990: *Justice and the Politics of Difference.* Princeton, Princeton University Press.

Zauberman, R. and Robert, P. 1995: *Du côté des victimes, un autre regard sur la délinquance.* Paris, L'Harmattan.

Zimring, F. and Hawkins, G. 1988: The New Mathematics of Imprisonment, *Crime and Delinquency* 34(4): 425–36.

Zimring, F. and Hawkins, G. 1997: *Crime is Not the Problem. Lethal Violence in America.* New York, Oxford University Press.

Zukin, S. 1997: Cultural Strategies of Economic Development and the Hegemony of Vision. In A. Merrifield and E. Swyngedouw (eds), *The Urbanization of Injustice.* New York, NY University Press.

Zukin, S. and Zwerman, G. 1988: Housing Ethnic and Racial Minorities in New York City: Jews and Blacks in Brownsville, *New Community* XIV(3), Spring: 347–55.

Index